Ecological Economic

AN INTRODUCTION

GARETH EDWARDS-JONES
School of Agriculture & Forest Sciences
University of Wales, Bangor

BEN DAVIES
Department of Land Economy
University of Cambridge

SALMAN HUSSAIN
Scottish Agricultural College
Edinburgh

Blackwell
Science

Editorial Offices:
Osney Mead, Oxford OX2 0EL
25 John Street, London WC1N 2BL
23 Ainslie Place, Edinburgh EH3 6AJ
350 Main Street, Malden
MA 02148 5018, USA
54 University Street, Carlton
Victoria 3053, Australia
10, rue Casimir Delavigne
75006 Paris, France

Other Editorial Offices:
Blackwell Wissenschafts-Verlag GmbH
Kurfürstendamm 57
10707 Berlin, Germany

Blackwell Science KK
MG Kodenmacho Building
7–10 Kodenmacho Nihombashi
Chuo-ku, Tokyo 104, Japan

First published 2000

Set by Graphicraft Limited, Hong Kong
Printed and bound in Great Britain at
MPG Books Ltd, Bodmin, Cornwall.

A catalogue record for this title
is available from the British Library

ISBN 0–865–42796–8

Library of Congress
Cataloging-in-publication Data

Edwards-Jones, Gareth.
Ecological economics : an
introduction / Gareth Edwards-Jones,
Salman Hussian, Ben Davies.
p. cm.
ISBN 0–8654–2796–8
1. Environmental economics.
2. Ecology Economic aspects.
I. Hussain, Salman. II. Davies,
Ben. III. Title.
HC79.E5 E328 2000
333.7—dc21 99–089561

DISTRIBUTORS

Marston Book Services Ltd
PO Box 269
Abingdon, Oxon OX14 4YN
(*Orders*: Tel: 01235 465500
Fax: 01235 465555)

USA
Blackwell Science, Inc.
Commerce Place
350 Main Street
Malden, MA 02148 5018
(*Orders*: Tel: 800 759 6102
781 388 8250
Fax: 781 388 8255)

Canada
Login Brothers Book Company
324 Saulteaux Crescent
Winnipeg, Manitoba R3J 3T2
(*Orders*: Tel: 204 837 2987)

Australia
Blackwell Science Pty Ltd
54 University Street
Carlton, Victoria 3053
(*Orders*: Tel: 3 9347 0300
Fax: 3 9347 5001)

For further information on
Blackwell Science, visit our website:
www.blackwell-science.com

Contents

Preface

This book is an introduction to the field of ecological economics, particularly for natural scientists. It introduces readers to some of the key themes in the field from first principles, and focuses particularly on the areas of economics, ethics, and environmental management. At the time of writing, we feel there is no single text which introduces these areas in a way that covers not only core principles, but equally importantly something of their history and context, their interrelationships, their analytical limitations, and their current roles in decision-making. This book is a first bridge into this field for readers including students, environmental practitioners, policy makers, and others involved or interested in environmental issues. We also hope that some economists will find ideas and methodologies within it that are stimulating, and which perhaps should be included in conventional economics training but are sadly usually omitted.

Ecological economics itself is a new and dynamic field, and also a pluralistic one. This presents a challenge for anyone writing an introduction to the subject. In our approach, we have been guided by what we consider a central ethos of ecological economics: exploring the value of multiple perspectives on environmental problems, and an awareness of the limitations of using particular techniques and disciplines in isolation. The integration of these different approaches and perspectives still remains a major challenge, but the recent expansion of research in interdisciplinary approaches to sustainable development augurs well for the future. Ecological economics itself is fast becoming recognised at the forefront of this work on applied sustainable development.

Given the pluralism inherent in the field, our own choice of material and emphasis warrants some explanation in two main areas. Firstly, space constraints have forced us to exclude introductory material on environmental science and ecology, much as we wanted to include it. Whilst non-scientists may regret this limitation in coverage, we are still confident that those without a science training will have little trouble in following the text. Secondly, there is also a fair portion of the book which can be classified as 'conventional' environmental economics. Whilst we do not support conventional economic approaches in all contexts, nor indeed a number of the principles on which they rest, we none the less believe it is vital to explore these fundamental principles and related methodologies: only then is it possible to separate what is genuinely valuable from what is questionable practice. Our experience of teaching students from a variety of backgrounds is that they often find these insights into alternative approaches the most rewarding, not least so they can engage with them critically where appropriate.

A special word is also perhaps due on the emphasis we have given to ethics throughout the book. There is often a suspicion amongst environmentalists that economists are allergic to moral questions; and an equal suspicion amongst economists that environmental campaigners are blind to the costs of environmental protection. Such views can quickly polarise environmental debates into slogans concerning 'morality' versus 'money', rather than stimulating a genuine exploration of the alternative beliefs and ethical values held on both sides. Ethical assumptions are inherent in all aspects of environmental issues, but they are often submerged under the details of complex planning and assessment methodologies. We are keen to emphasise the importance of engaging with underlying principles whenever they have an influence on decision-making. We feel such an emphasis is another distinguishing feature of ecological economic analysis.

This book arose during several years when we were involved as both staff and students with the MSc in Ecological Economics at the University of Edinburgh, Scotland. We have our first degrees respectively in biology, philosophy and economics, and there are few departments in the UK and indeed in academia at large where collaboration between those disciplines is likely, let alone actively encouraged. We were therefore fortunate to have this opportunity. Whilst initial training suggests an obvious division of labour amongst some of the chapters, we should also say that it is not always a reliable guide. We argued over many areas and contributed actively to each other's work throughout the text. The whole text is thus a product of much discussion, with the final authorship being assigned almost arbitrarily.

As always, thanks are due to a large number of past and present colleagues and friends, and in particular Herman Meyer, Barry Dent, Murray McGregor, Fraser Quin, Alice van Harten, Wendy Kenyon, Julie Whittaker, John McInerney, EJ Milner-Gulland, David Oglethorpe, and staff and past students of the Centre for Human Ecology and Edinburgh University Ecological Economics Masters programme. We would like to thank Ian Sherman at Blackwell Science for initially suggesting the book and for his cheerful perseverance in seeing it through to fruition; and Dave Frost for his work on the manuscript. We are particularly grateful to Vito Cistulli for permission to use training materials he had prepared for FAO as a basis for much of Chapter 6.

On a more personal note Gareth would like to thank his wife, Emma for her love and unerring support during the production of this book, and he would also like to dedicate his part of the book to his mother and father, who were always so certain that education and learning were good things—diolch am bob peth. Ben would like to thank those friends who offered encouragement and criticism with equal enthusiasm from start to finish, and he dedicates his work to his parents. While Salman would like to thank his partner Christele for her laughter and love, and hopes that his parents can now better understand his pursuit of academia—shukrya.

Part 1
Foundations for Ecological Economics

1 Introduction

'There are no passengers on spaceship Earth. We are all crew.' [Marshall McLuhan]

1.1 What is ecological economics?

Ecological economics is a transdisciplinary field of study which examines the interactions between economic and ecological systems from a number of related viewpoints. It is focused on areas where economic activity is increasingly seen to be in conflict with both the well-being of the ecological system, and the well-being of the human social system. The first of these systems ultimately supports all activities, while the second is the system to which the benefits of economic activity should ultimately be directed.

Ecological economics, therefore, is concerned with three analytical strands: the economic system, the ecological system and the social system. It emphasizes the relationships between these systems at a number of hierarchies and scales from the local to the global. It searches for ways in which analyses of these different systems complement and support each other and aid a wider understanding of complex environmental problems.

Both academics and practitioners working in the field stress that ecological economics is not a new discipline but rather a synthesis of many separate disciplines, all of which bear in some way on the central problems of unsustainable human impacts on the natural environment. Within this field, there is a commitment to deliberately seek multiple perspectives on problems rather than relying on any particular viewpoint or methodology. To this end, different disciplines are mutually supportive of each other rather than seeking intellectual supremacy. An understanding of the fundamentals of different viewpoints taken together highlights the limitations of any one particular discipline.

Central to the ability to develop solutions to environmental problems is an informed understanding of the fundamentals of different methodologies. This text has been written with that end in mind, and particularly to explain from first principles key ideas in economics and ethics, placing them within the context of their historical and intellectual development, and exploring their contribution to the analysis of environmental problems in combination with ecological analysis.

1.1.1 The birth of ecological economics

As different chapters in this book will show, ideas now considered central to ecological economics transcend both disciplinary and faculty boundaries. Insights come from diverse sources, stretching from philosophy through branches of the social sciences to natural sciences, including ecology and thermodynamics. Ideas related to social justice go back to Ancient Greek philosophy, for example, while economic concerns with resource limits are apparent in the work of many early economists writing in the eighteenth century. The history of some of these ideas is traced in Chapters 2, 4 and 5.

Despite these early precursors, it was not until well into the second half of the twentieth century that these separate strands became increasingly woven together. It is therefore from the late 1960s onwards that what are now viewed as seminal works in ecological economics appeared. These works, including ideas such as Kenneth Boulding's 'spaceship Earth' and Herman Daly's 'steady state economy', among others, stressed the closely intertwined relationship between economic growth and the biophysical capacity of the environmental system that supported it. Perhaps their key contribution was that they challenged the conventional wisdom that Western society could continue indefinitely along its high-growth trajectory of economic development.

In the views of these writers, there were both physical and social limits to the success of high-growth economies. An economy that focused on producing more and more goods, and measured its success according to how many more goods it could produce, had limits of two kinds. It had physical limits because it could outgrow the environment that supported it; and it had social limits because as the economy grew it squeezed out sources of welfare not related to material goods. The early ecological economists thereby raised the question of limits to growth and the purpose of economic activity. Some of these concerns came to be recognized in the core concept of 'sustainable development'.

1.1.2 The sustainability paradigm

Questioning the purpose and progress of economic activity is central to the concept of sustainable development, famously defined by the landmark 1987 UN World Commission on Environment and Development (WCED 1987) as 'development which meets the needs of the present without compromising the ability of future generations to meet their own needs'. This conference elevated the concept of sustainable development from an academic interest of earlier 'antigrowth' economists to a key principle of political, economic and social reform. Although, as later chapters show, it is still highly contested how to translate this broad policy goal into practice, the WCED 'sustainability' paradigm recognized explicitly that the ability to sustain worldwide economic activity into the distant future was dependent on the form of that activity and how it related to wider social and environmental concerns.

The WCED report discusses many issues relevant to achieving sustainable development, but of particular note are several critical objectives. These are:

- reviving growth;
- changing the quality of growth;
- meeting essential needs for jobs, food, energy, water and sanitation;
- ensuring a sustainable level of population;
- conserving and enhancing the resource base;
- reorienting technology and managing risk; and
- merging environment and economics in decision-making.

While it is principally the last of these issues that concerns us in this book, it is almost impossible to merge environment and economics without considering the other six factors listed alongside this objective. Furthermore, the main purpose of merging environment and economics in decision-making is primarily to assist in meeting all other objectives, in ways that ensure the stability of social and environmental systems over the long term.

Ecological economics as a field of study is focused clearly on helping to define what 'sustainable development' or 'sustainability' might mean for different societies. Key papers from its first international conference in 1990 are published as a collection titled 'Ecological Economics: the science and management of sustainability' (Costanza 1991). In a more recent publication of practical applications of ecological economics the authors suggest:

> 'Probably the most challenging task facing humanity today is the creation of a shared vision of a sustainable and desirable society, one that can provide permanent prosperity within the biophysical constraints of the real world in a way that is fair and equitable to all of humanity, to other species, and to future generations.'
> (Costanza *et al.* 1996)

It is to this end that ecological economics is directed.

1.1.3 The key issues

To put the very broad policy concerns listed above in a more direct form, the issue of defining sustainable development as it relates to environmental concerns can be summarized in four key questions (adapted from Jacobs 1996).

1 What is the correct level of environmental protection?
2 How should this be decided?
3 How should this level be achieved?
4 Who should bear the cost?

It is clear that no single discipline can produce definitive answers to these questions. The active ingredients required in an answer encompass questions of moral values, issues of economic feasibility, biological limits, technological resources and many others. The definition of a 'correct' level of environmental quality will vary according to the emphasis placed on human as opposed to animal welfare, on the rights allocated to individual members of society, and on the importance attached to future consumption over the present. It will also depend on the capacity of the environment to assimilate wastes, its productive capacity, the certainty with which that capacity can be estimated, and the reversibility of any degradation that occurs. It will depend too on the technical possibilities open to provide environmental management, on the willingness of individuals to pay for that management, on society's ability to invest in research to reduce the costs of management, and so on. The web of interconnections involved in assessing sustainability options is sufficiently vast to leave any one discipline heavily overstretched.

1.2 The scope of ecological economics covered in this book

This book stresses the contribution of three central but different disciplines within ecological economics—economics, ecology and ethics—and emphasizes their assumptions and where and how they may interrelate in analysing environmental problems. In this sense they are each taken as providing distinct visions of the purpose and process of analysis, prior to suggesting policy or action on the basis of that analysis. From one point of view, this can be understood as understanding the 'preanalytic vision' of different disciplines.

1.2.1 Disciplines as filters: the preanalytic vision

In simple terms, the theories and techniques that make up any particular discipline can be thought of as a filter through which the analyst sees the structure of a particular problem. The economic theorist Joseph Schumpeter referred to the basic elements of this filter as a 'preanalytic vision'. The 'vision' refers to the concepts that the researcher has in his or her head before beginning any analysis. The preanalytic vision in chemistry would thus involve concepts such as atoms and molecular bonds. Explanation and theory within chemistry explains all problems in terms of these preanalytic concepts. It does not analyse these concepts themselves, except perhaps on very rare occasions, because these concepts actually define the extent of the discipline itself. They are a black box in the core beliefs of the science.

Skills and knowledge within particular academic disciplines are therefore skills and knowledge related to particular visions of the way the world operates. Each discipline has its own problem-structuring approach. As a result, each discipline also presents its own vision of possible solutions to problems based on that approach. Where economists may see a problem stemming from inefficient market signals or price distortions, ecologists may see a problem stemming from biological uncertainty or irreversible pollution damage. It is important to recognize that your discipline filters your vision of the world. You need to know how it filters it and what preconceptions you therefore hold about it, and how far other visions create alternative problem-framing and problem-solving approaches. In some cases the preanalytic vision of one discipline may not simply construe problems differently, but fail to recognize that they exist at all.

The approach to ecological economics adopted in this text is centred on this issue of multiple perspectives. The objectives of ecological economic study are a fuller understanding of the complex interrelationships between the economic, the social and the environmental systems directed to identifying and solving real world problems where conflicts between those systems are apparent. The book outline goes through these issues in more detail below but we put them in context here by introducing in thumbnail the three 'filters' which many practitioners believe need to be combined to be able to create a more effective problem-structuring framework for issues in sustainable development.

ECONOMICS

Economics is concerned with studying the efficient use of scarce resources: 'economizing on effort' in the common sense view. Chapter 3 explains from first principles what this means within neoclassical economics (which is the prevalent form), and here we make just one observation. That is, the key characteristic of conventional economic theory is the use of 'market price' as an organizing principle. Prices signal when there are shortages, when there are gluts, what to produce, what is worth buying and so on. Conventional economic theory can be thought of as rotating around different aspects of how prices, and consequently people, behave under a vast number of different scenarios. Only a small fraction of this work has focused directly on environmental issues, and this is generally in two fields: resource and environmental economics.

Resource economics is concerned with the use of natural resources as inputs to economic processes; it deals with issues such as the extraction rate of minerals from the earth, the harvesting of fisheries and forestry and the management of other resources, such as water and renewable energy sources. It analyses *how price regulates the quantities produced of desired environmental goods.*

Environmental economics was traditionally concerned with the use of the environment as a 'sink' for the waste products of economic activity, and traditionally researchers in this field were concerned with pollution and its control. These were situations where prices did not take account of damaging effects. This work was concerned with *how prices failed to regulate the use of the environment, and how they can be corrected.* Over time the scope of work undertaken by environmental economists gradually widened and they began to consider other environmentally related issues, such as the conservation and management of biodiversity, environmental decision-making and development issues, such as agriculture, forestry and soil conservation.

Economics is of central concern within ecological economics for at least three reasons. It can be a foundation for public decision-making, through cost–benefit analysis; it can both explain and predict behaviour in relation to natural resources (for example, in harvesting rates) in certain circumstances; and it can develop possible improvements to decision-making in both these contexts through specific tools of analysis, which we develop at length in later chapters.

ETHICS

Ethics can be considered as the study of just actions or, as the Ancient Greek father of philosophy, Plato, put it, 'no small matter but the way we ought to live'. Fundamentally, we seek ethical reasoning as a means to justify behaviour, at least when that behaviour impacts to any extent on other people. Whatever the result of such reasoning may be, there is a sense in which the search for moral justification is itself a universal concern of humankind. Indeed, some would consider an awareness of moral issues or dilemmas to be a defining characteristic of humanity. While moral codes and creeds may be in flux or crisis, varying across cultures and changing through history, the quest for some kind of moral insight itself seems, by contrast, universal and enduring.

Ethics is of essence both a private and a public matter. It is private insofar as all individuals are 'keepers of their own consciences', and in that sense all individuals have no choice but to take responsibility upon themselves for their behaviour in response to moral issues that the world throws up. Even to deny such responsibility is itself, by default, a moral act. Moral issues are also public, however, in two important respects. First, governments have to act on behalf of citizens (at least when they are democracies), and all such actions therefore implicitly contain a moral viewpoint. Secondly, moral discussion is public in the sense that it is engaged whenever relations *between people* are involved, which is to say any activity that is not wholly personal in its effects. Indeed, for many people it is important even beyond this (for example, in the case of animal welfare and species loss).

An ethical awareness is important in ecological economics for a number of reasons. It is a useful guide in decision-making. It is focused on clarifying and determining ultimate or end objectives, without which solution-driven methodologies are useless. It also governs many aspects of social behaviour, and in this sense it is highly relevant to know how these inform and regulate activity. The interpretation given to ethics within this text therefore encompasses some understanding of the social situation, in addition to exploring in a direct sense the meaning of concepts of 'right and wrong' in environmental management.

ECOLOGY

Traditionally ecology has been defined as 'the scientific study of the interactions that determine the distribution and abundance of organisms' (Krebs 1972). This study has been conducted at the individual, community and ecosystem level, investigating the complex relationships between individuals and species, and between organisms and their biogeochemical environment. Despite intensive research, ecologists remain humble in their levels of understanding in the face of the enormous complexity governing many of these relationships. The complexity of the natural environment and its dynamic nature tends to confuse and complicate traditional reductionist scientific analyses, making ecology perhaps the least directly reductionist and most holistic of all natural sciences.

The focus of ecological science has tended to take two different orientations. In Europe it has involved a predominantly process-orientated approach to ecological questions, concentrating on themes such as patterns of behaviour, individual species relationships and the dynamics of populations. In American research there has been greater emphasis on systemic behaviour, investigating the interrelationships between elements of particular ecosystem types. While there is a great deal of cross-over between these two aspects, they do emphasize two elements of interests in ecological economics. On one hand there is great concern over the survival situation facing many species, and here research is directed to understanding the particularities of the species under threat. On the other, there is growing interest in the systemic features of ecosystems that provide valuable life-support services, such as crop production, nutrient cycling, climate stabilization and pollution assimilation. Although the science of ecology does not differentiate itself strongly along these lines, these are probably the two elements where the science has greatest input into ecological economics.

Ecological science thus has a pivotal role in ecological economics in several areas. It provides data on important ecological parameters, such as sustainable harvesting rates for biological resources; it identifies and predicts changes occurring in natural systems in response to human-caused stresses, such as pollution; and it thereby provides guidance on the management of human economic activity to sustain environmental quality in the long term. In addition, and by no means least of its contributions, it also generates and communicates a deeper and richer understanding of the immense diversity of life on Planet Earth.

As an introductory text, this book was written with the thought that ecologists and environmental scientists would form the bulk of its readership. A treatment of ecological themes equivalent to that on economics and ethics would have taken up a large number of pages, without adding any significant knowledge for such readers. A thorough coverage of foundational concepts in ecology has therefore been omitted.

1.2.2 Orientation of this book

One principle of sustainable development is that there is no single blueprint for achieving such development which is universally applicable to all places and peoples. Different cultures will develop different approaches, guided only by the fundamental concerns of sustainability—ecosystem integrity, economic efficiency, equity and social justice. With this in mind, we recognize that the issues considered important in one context may not transfer to another; and that the style and focus of analysis will vary accordingly.

In this book we acknowledge three general biases. The first is that, by and large, we have a Western background and experience, and much of our emphasis is therefore on issues and problems related to this experience. The priorities and targets for analysis are often very different in other countries and cultures, although the analytical skills needed to tackle these problems may be very similar. In the course of this book we give examples and base analysis on problems drawn from around the world, but our Western perspective will be self-evident in many cases.

Secondly, we have concentrated on what might be called a mid-range or mid-scale level of analysis, targeted predominantly at the community level and the individual project scale (though by no means only this scale). This is one of a number of scales on which ecological economists operate; among others are very fundamental ones, such as energy flows and modelling; global issues, such as climate change and international trade; and longer-term evolutionary perspectives on economic and cultural change. These scales interrelate in many respects, but to try to encompass all of them in a single introductory volume would be very difficult. For those coming to the subject for the first time, and interested in direct application to environmental concerns, we feel this mid-range scale is the most appropriate and accessible.

Thirdly, and most evidently, we have devoted significant attention to issues concerning species and habitat conservation, which is only one among many interests in ecological economics. Concomitantly, we have not presented issues such as energy or transport, although these and other topics within ecological economics would certainly warrant equivalent treatment to the issues of biodiversity. This particular emphasis arises out of the combination of our particular interests and the space constraints of the current text.

1.3 Outline of the book

This book is separated into four main parts. Each part indicates the general orientation of chapters within it, which are:

Part 1: Foundations for Ecological Economics.
Part 2: Value and Valuation Tools.
Part 3: Frameworks for Decision-Making.
Part 4: Applications: Theory and Practice.

1.3.1 Part 1: Foundations for Ecological Economics

In Part 1 we introduce economic, ethical and environmental theory and place each in a historical context.

Chapter 2 is a historical review of the development of economic theory, setting out the most important insights. The first half deals with classical and neoclassical analysis, while the second concentrates specifically on latter twentieth-century work and particularly the issues of resource limits and sustainability. A short introductory section distinguishes economic 'growth' from 'development'. This sets current thinking about environmental management and the social organization of economies in context. Readers who are unfamiliar with the basic principles of economics may prefer to read Chapter 3 before Chapter 2, as this will clarify concepts only briefly introduced in this historical survey.

Chapter 3 is an introduction to key principles within neoclassical economics, which form the basic analytical approach for economic analysis. The first, shorter part of the chapter explains how and why economies may be organized in different ways, and the key features that economists look for in judging different forms of organization. In the second part, the microeconomics of how a market economy functions are outlined. This involves grasping the concepts of consumer choice on the basis of indifference, and the interaction of supply and demand.

The history of ethics and environmental philosophy is introduced in Chapter 4. In similar fashion to Chapter 2, key ideas in the historical development of ethical theories are reviewed first, including utilitarian and rights-based theories. Section 4.4 then reviews the evolution of cultural attitudes to the environment over time, and Section 4.5 again covers the most recent period of environmental concern in more detail. This establishes the many perspectives on the environmental crisis as seen from philosophical as opposed to economic viewpoints.

1.3.2 Part 2: Value and Valuation Tools

Part 2 examines the key building-block of decision-making: the issue of appropriate methods of valuation. Chapter 5 explores the different meanings given to the concept of 'value'. Section 5.2 explores very fundamental issues in metaethics, which question whether value judgements have any justification at all. Subsequently, three important strands of value are covered in more detail: ethical values within the economic system, focusing on the concept of 'economic justice'; the meaning and components of human well-being or 'welfare'; and moral values and the natural world. All these issues can be thought of as exploring alternative value criteria to the economic idea of value as expressed in market prices for goods.

Chapters 6 and 7 then review practical environmental evaluation tools. Chapter 6 covers the full range of economic valuation techniques currently utilized by economists; each technique is explained and their main limitations noted. Chapter 7 reviews ecological evaluation criteria, primarily related to designation of rare species and habitats. The second section of the chapter introduces the concept of ecosystem services, which provide direct human benefits, such as water filtration and climate stabilization. Both Chapters 6 and 7 therefore review methods for identifying and quantifying 'valuable' aspects of the environment from the perspectives of ecology and economics.

1.3.3 Part 3: Frameworks for Decision-Making

The frameworks considered in Part 3 are all ways in which the evaluative techniques outlined in Part 2 can be applied to assist directly in environmental decision-making. Three of the frameworks—cost–benefit analysis, environmental impact assessment and multicriteria analysis—are generally applied at the project level although they are not limited to this level. The fourth, national income accounting, is a national level framework which essentially 'takes the temperature' of the economy as a whole in terms of its economic and, more recently, its environmental performance.

Cost–benefit analysis (CBA) in Chapter 8 is a very widely employed appraisal method which is also widely criticized by many environmentalists. The important steps of the methodology are explained, with problems noted, and the issue of discounting future costs and benefits is covered in detail. A short example also demonstrates the different levels of complexity that can be undertaken in considering costs and benefits.

In Chapter 9 we review the philosophy and practice of environmental impact assessment (EIA), which the European Union has identified as a key element in achieving sustainable development. Environmental impact assessment is mainly employed by planners concerned with identifying the potential social and environmental impacts of plans or projects. The methodology is considered at three levels: the legal framework; the structure of an EIA; and the tools and techniques employed in evaluating social and environmental impacts.

Multicriteria analysis (MCA) is a very diverse style of project planning and assessment introduced in Chapter 10. Of the three microlevel appraisal frameworks, it is the least well practised although it perhaps holds the greatest potential for combining numerous different kinds of value within a single framework. The chapter outlines the basic principles and components of the MCA approach, and gives simplified examples of two of the more popular techniques. In a short second section, the philosophy and practice of participatory rural appraisal (PRA) is explained. This can be seen as an alternative participation-based approach to planning, which utilizes multicriteria in decision-making but without a formal framework.

Part 3 concludes with national income accounting (NIA) in Chapter 11. The national income accounts are a key indicator of economic performance used by government which have long been criticized for failing to take account of environmental factors. The basic theory of the accounts is covered, and their limitations identified. Three new developments are then examined: national satellite environmental accounts; the index of sustainable economic welfare; and the concept of environmental space. All these approaches seek new ways of judging how well the economy is performing, using environmental and social criteria rather than relying on traditional economic methods.

1.3.4 Part 4: Applications: Theory and Practice

In Part 4 we extend ideas introduced earlier into three areas: natural resource harvesting; nature conservation; and pollution and waste. We concentrate particularly on analytical insights drawn from economics but with an awareness of the limitations of those insights and their interrelationship with social and ecological factors.

Natural resource harvesting begins with a brief section on the concept of property rights. Subsequently, it introduces interesting issues in three areas of harvesting: agriculture, forestry and fisheries. In agriculture, a basic framework for viewing environment-production trade-offs—the production possibility frontier—is presented. In forestry, the basic economic approach to harvesting is first explained; the importance of informal property rights in forest management is then shown in an example drawn from Nepal; and finally the principles of rural development forestry are presented. In the final section on fisheries, the interrelationship between a biological stock and harvesting effort is traced in more detail. The ramifications of various ocean fisheries management options are then reviewed.

Nature conservation (Chapter 13) is divided into two parts: the first considers species and the second, habitats. Both parts consider direct economic arguments for species and habitat protection. In considering species, the concept of the safe minimum standard is also introduced, as well as ways of prioritizing conservation spending. There follows an example of applying marginal economic analysis to the costs of protecting the Spotted Owl, which throws up a number of interesting issues. In considering protected areas, attention is focused on the UK system of sites of special scientific interest (SSSIs) and environmentally sensitive areas (ESAs). The operation of the SSSI scheme is analysed from an economic and ethical perspective, and a way of comparing the cost-effectiveness of the two conservation instruments is then reviewed.

Chapter 14 considers the management of pollution and waste. The first part outlines the theory of market-based instruments for the management of pollution, which include taxes and tradable permit systems. An example of an effective implementation of the theory is then given, the Californian RECLAIM programme. The second part considers the problem of municipal solid waste. The concept of the 'waste hierarchy' which orders the preferred environmental disposal options is first introduced. The methodology of lifecycle analysis (LCA) is then explained, which is a means of assessing the waste and pollution impacts of products and production processes. The chapter concludes with a partial LCA applied to test the waste hierarchy.

1.4 A final comment

Each chapter in this book is written to be as self-contained as possible, although later chapters build upon concepts developed earlier in the book. As far as possible, more emphasis is placed on understanding how the repertoire of analytical frames can be applied and where they interrelate, than on specific issues. Thus there are relatively few policy prescriptions and relatively few broad analyses of environmental problems, such as deforestation, global warming or food security. The emphasis is on ways of thinking about aspects of these problems through economics, ethics and ecology, rather than blueprints for action or principles for policy. Readers should be placed to develop their own principles as a result of an understanding of the themes developed here, and there is in any case no shortage of other texts offering such principles. This text is presented in the spirit of an intellectual toolbox that encourages understanding of the range and limitations of particular methodologies and how they may be applied in complex environmental situations.

2 A Brief History of Ecological Economic Thought

'We should all be concerned about the future because we are going to have to spend the rest of our lives there.' [Anon]

2.1 Introduction

In this chapter we review the historical development of important theoretical insights within ecological economics, but focusing on those drawn from economic theory. The related strand of social and ethical issues is covered in Chapters 4 and 5. This chapter falls essentially into two parts. In the first we review briefly some of the historically interesting developments in economic theory in general. In the second part we concentrate on more recent twentieth-century work identified specifically within the field of ecological economics.

Initially, it is useful to distinguish between the history of economic theory and the history of economic development. The history of economic theory is the study of the evolution of key ideas and concepts within the analytical discipline that is economics. This is the main interest of this chapter, which introduces such concepts as Adam Smith's theory of the 'invisible hand', Marx's theory of capital and Pareto's theory of optimal efficiency. This is the history of how economists have come to see the world and to approach problems in a particularly individual way.

By contrast, economic history seeks to explain the development path of particular national or regional economies through analysis of important economic pressures. Its purpose is the explanation of historical events and periods by reference to economic causes, such as shifts in technology, labour supplies and productivity. We introduce patterns of economic development very briefly below, but concentrate on economic theory in the rest of the chapter.

2.2 A thumbnail sketch of economic development

Theories of economic history seek to account for the direction taken by particular economies in history (for an introduction see Cameron 1993). To go into any detail on such theories is beyond the scope of this book, but it can help to have a thumbnail sketch of the pattern of development as viewed in relation to advanced Western economies. We sketch such a thumbnail below.

A very important starting point is to distinguish between 'growth' and 'development' of national economies, a theme that is developed in more detail throughout this book. Growth is an increase in the size of the total economy, thought of as its total production of goods and services. Development is a broader concept that refers to changes in the type and form of economic production and organization; say from a self-sufficient economy to a trade-based one, or from more agricultural to more industrial outputs. Issues of economic development therefore have wide-ranging social, cultural and political ramifications, whereas issues of growth are essentially more limited to environmental and resource concerns. The economy of the Roman Empire, for example, *grew* substantially through both internal expansion and conquest, but it did not *develop* significantly beyond an agriculture-based system centred on private land ownership amongst smallholders.

As a thumbnail, we can consider four epochs of economic development of Western civilization: the Palaeolithic; the Neolithic; the agrarian; and the industrial. To some extent overlapping with these epochs are four stereotypical forms of economic organization: self-sufficient hunter-gathering; communally based pastoralism; feudally organized agricultural production; and trade and wage–labour-based capitalism. These are stereotypes for the purpose of comparison, and should not be taken as identifying precise historical periods.

The Palaeolithic (literally 'ancient stone') era, stretching from the dawn of humankind until approximately 20 000 BC (any precise date is obviously somewhat arbitrary) is the first 'age' and is characterized by a hunter-gatherer lifestyle. Early humans lived off the sustenance that their natural local environments offered them. They interacted intimately with those environments, but they adapted to them and did not in any significant way adapt them. When food or fuel resources were depleted in one area, they moved on to another as required.

The shift from Palaeolithic to Neolithic ('new stone') was the shift from wild hunting to pastoralism; that is, the rudimentary domestication of animals and establishment of more permanent settlements, perhaps of a few families, in areas that were particularly abundant in resources. The shift to settled habitation corresponded with increasing adaptation of the local environment, perhaps making some deliberate efforts to improve food yields from local land; for example, protecting fruit trees or even planting basic crops. In studying tribes in Polynesia, Sahlins (1972) describes them as following an essentially Stone Age economics, part way between hunter-gathering and animal–agricultural production.

The agricultural phase itself can be designated from some time after the turn of the last ice age, around 10 000 BC, until the eighteenth century. This very long and diverse period is delineated in development terms by the overall reliance on agriculture. The increase of agricultural production through cultivation helped to create for the first time significant surplus wealth which could support a religious or ruling caste; hence the opportunity for social relations to move from communally orientated to more hierarchical or feudal ones. In this sense the early city state empires of the Middle East, such as Ur, had forms of economic organization probably not significantly different from much later European kingdoms. Although various forms of organization evolved, involving slaves, free men and land owners, the centre of wealth in land essentially continued until the Industrial Revolution though other activities, such as mining and trade, grew in importance.

The fourth, industrial, age is a significant turning point in three respects (Cameron 1993). Firstly it introduced powerful new fuel sources, such as coal and oil, thereby harnessing mechanical power far beyond any biological power source. Correspondingly, it increased dramatically the non-natural elements within the environment through increased mining, refining and production. From an ecological perspective this changed the human–environment relationship significantly. The diversity of new activities, and corresponding specialization made possible by the technical and scientific innovation of this period, also strained the fixed social hierarchies possible under simple agricultural production, leading to more capitalist economic organization.

The drawing of these divisions is only to establish broad stereotypes, which serve as an overall context for the development of economic theory. It is clear that historically many varied mixtures of these broad types have existed, and coexisted, even within single nations over the same period. Even identifying whether particular relationships are feudal, or capitalist or essentially self-sufficient is not necessarily an easy task in some circumstances. Perhaps the most important feature to be noted is that in terms of social organization, the economic cannot be sensibly separated from political, social or environmental factors, all of which interrelate through aspects such as the ownership of resources, opportunities for exploitation and social responsibilities recognized under different systems.

We now consider briefly three periods in economic development: the preclassical, the classical and the neoclassical. An excellent full-length discussion of these developments is provided by Kula (1998). In the second half of the chapter we consider this theoretical inheritance in relation to current themes in ecological economics, as well as the new themes that are emerging within the subject.

2.3 A synopsis of economic developments: the preclassical school

The birth of economics as a discipline is traditionally associated with the work of Adam Smith in the late eighteenth century (see below), and it is in these and later works that the real theoretical insights of economics developed. In this section we note briefly the evolution of economic systems prior to Smith.

The advance of agrarian societies in Ancient Egypt, Greece and Rome essentially involved management of a mainly peasant economy supporting a relatively tiny ruling class. The quantities and importance of trade was small, limited to high-value goods because of the expense of transport and lack of large-scale production. The scale of the state was limited by the amount of surplus that agriculture could provide, freeing labour and creating wealth to pursue government projects. Social relations were hierarchical: a ruling elite and warrior class supported by a labouring class.

The medieval economic system of Europe (AD 500–1400) replicated this hierarchical system based around a manorial system, with the lord of the manor ruling over his local area of responsibility and in turn responsible to a more powerful local baron, in turn answerable to a yet more powerful duke and so on, up to the ultimate authority of the Crown. The lords oversaw local justice, agricultural production based on traditional techniques, and provided protection for their communities, either in national or local defence. The system of production and exchange was highly regulated (see Section 5.3.1 for comments on the just price and usury) and change relatively slow.

This situation began to change around the fourteenth century with a rapid increase in overseas exploration and greater international trade, much flowing through the Dutch merchants, an era Cameron calls Europe's 'second logistic' (the first being the Mediterranean civilizations of Greece and Rome). Advances in agricultural production, especially the adoption of the three-field rotation, improved productivity. The growth of trade and productivity led to the emergence, for the first time, of a significant middle class, composed of merchants and professionals, such as lawyers to oversee the increase in trade contracts. This class pushed to replace the local dominance and trade restrictions of the baronial system with a more nationally regulated system within which they could gain greater freedom and more stable trading conditions.

The predominant economic ideology of this expansionist period of development in the sixteenth and seventeenth centuries was that of *mercantilism*. Essentially, this saw national prosperity as synonymous with the national stock of wealth in terms of precious metals. To this end, state policy was generally 'protectionist', which means protecting home markets against imports through trade barriers, such as import taxes and quotas. This discouraged the buying in of imports, because imports were seen as causing a flow of these precious metals out of the country. State-sponsored colonial expeditions from European nations opened up sources of wealth, such as Latin America, which could be mined to enrich the mother nation, and domestic production for export was encouraged.

Although mercantilism became the governing ideology of this expansionist period, an alternative school of thought was that of the physiocrats, championed by the French economists Quesnay (e.g. *Tableau Economique*, first published in 1758) and Turgot (Kula 1998). This school considered land to be the dominant source of wealth, not precious metals. The justification for this was that agricultural output, produced by the workers, was a necessary basis for wealth creation which could then be transformed into manufactured output. Neither the brokerage deals of traders nor the collection of rents by landlords was adding value to the economy, according to the physiocrats. Land and labour were the only source of wealth and the quantity of these two primary factors of production embodied in a commodity defined its intrinsic value. The physiocrats had some influence in France, but mercantilism dominated the political arena.

The ferment of the second logistic can thus be seen as leading into a sustained period of economic development. Regional and national trade expanded, production innovations became more frequent and more quickly diffused, technology thus advanced and the economic wealth and power of European nation states increased. It is against this backdrop that the work of the classical political economists emerged.

2.4 The classical economic paradigm

The classical political economy school has provided a significant academic foundation to current thinking in economics and important insights, particularly for environmental and ecological issues. The most noteworthy contributions were made in the late eighteenth century by Adam Smith, Thomas Malthus, David Ricardo and John Stuart Mill. As *political* economists, they did not attempt to abstract from reality to the same degree as later theoretical schools. A common theme was the role of the market mechanism as an efficient resource allocation mechanism and as a stimulus to growth in consumption. However, there was unanimous pessimism regarding the capacity of the market to continue this growth in perpetuity. The predicted conclusion to economic growth was a subsistence existence termed the 'stationary state'. There are important differences with regards the theories of each of the four writers. Each is considered in turn below.

2.4.1 Smith

Adam Smith (1723–90) is famous for developing the concept of the 'invisible hand' to describe a key characteristic of a market economy. In Smith's view, self-interested behaviour by an individual serves to satisfy that individual's wants *but also happens to coincide with the interests of society at large*. It seems thereby that there is an invisible hand coordinating the operation of a market economy to maximize its output, but this is merely an accidental property of freely functioning markets:

> 'As every individual therefore endeavours as much as he can both to employ his capital in the support of the domestic industry, and so direct that industry that its produce may be of greatest value;

> every individual necessarily labours to render the annual revenue of the society as great as he can. He generally, indeed, neither intends to promote the public interest, nor knows how he is promoting it.' (Smith 1966/1776)

This quotation is found in modern-day texts in business studies and business ethics and is used to justify the operation of the free market. But to treat this quotation in isolation as a complete expression of Smith's social theories is inaccurate.

It is noteworthy that Smith was a moral philosopher and he conditioned the 'invisible hand' with the moral underpinning provided by *The Theory of Moral Sentiments* (Smith 1982/1760), written 16 years before *The Wealth of Nations*. Smith considered that the economic structure of society should be ordered so as to allow its citizens to enhance themselves as free and rational beings pursuing self-interest, without hindrance from the state; this coincided with the economy prospering. He defines his agenda as: 'that system of moral philosophy wherein constituted the happiness and perfection of a man, considered not only as an individual but as a member of a family, of a state, and of the great society of mankind'. (Smith 1966/1776).

This notion of self-enhancement is an important one; an increase in aggregate consumption is far from being the be-all and end-all of Smithian economic theory. Thus the Smithian vision for society arguably has links with aspects of the social and ethical critiques of economics discussed in Chapter 5.

Smith was writing in the mid- to late eighteenth century, a time when the Industrial Revolution in Britain was rapidly changing production processes, both in the agricultural and non-agricultural sectors. In continuing the development of economic thinking with respect to the role of land and labour, he proposed that there was no shortage of agricultural land and that the profits from agriculture would stimulate progress in other sectors. At that time, agriculture in the UK was efficient and machinery-intensive relative to international competitors: Jethro Tull invented a drill for planting seeds and pioneered a method for planting seeds efficiently in rows; Charles Townsend developed new feed crops which meant that the land need no longer remain fallow and unproductive (Kula 1998). The manufacturing sector was also booming: for instance, new inventions in cloth manufacture generated efficiencies in the cotton textile industry (Kula 1998). In summary, the UK economy of Smith's day was driven by entrepreneurship and performing well, stimulating his overall optimism.

Smith's designation of the role of the state was in accordance with an economy in ferment. The state thus should not act as a burden on individuals. The only three legitimate roles of the state are the provision and operation of the judicial system, national defence, and the provision of infrastructure and other public goods for which there is no incentive for private sector provision (a 'nightwatchman' state). This designation of the state's functions has been adopted by monetarist economic policy-makers, and applied by the Thatcherite and Reaganite regimes of the 1980s and by some development institutions. Self-interested behaviour is taken as a cure-all. But the individuals that Smith refers to are principally sole proprietors (owner-operators), not employees of firms. He refers to small-scale enterprise in town markets comprised of people of roughly similar income levels, with universally accessible information about prices and local opportunities, and no significant market dominance for any one group.

Although Smith's theory might be applicable when the owner constitutes the whole enterprise, such as a cobbler or ironmonger, the 'invisible hand' does not operate as successfully in situations where there is a conflict of interests between the owners of the enterprise and their workers. This conflict might be alleviated by providing appropriate incentives structures, such as profit-related pay. Further, with today's larger firms exercising some degree of market control, there is more scope for anticompetitive collusion which is against the pubic interest. Smith considers such collusion but surmises that competition would sort things out (Kula 1998). Today, such collusion is prohibited under antitrust legislation.

Smith was writing at a time in which the availability of natural resources was not, in general, a constraint on the economy. Smith developed John Locke's perceptions on the 'value' of nature as merely *instrumental*; that is, valuable only if it serves human wants.

> 'The earth furnishes the means of wealth; but wealth cannot have any existence, unless through industry and labour which modifies, divides, connects, and combines the various production of the soil, so as to render them fit for human consumption.' (Smith 1966/1776)

This human-centred (anthropocentric) view is common to the classical political economists. However, the existence of externalities,[1] that is the imposition of costs—and sometimes benefits—from production or

[1] Externalities as a market failure are discussed in Section 3.6.3.

consumption on a third party who is not compensated, implies a further role for the state. This is particularly pertinent in the case of environmental externalities which are very much a modern-day occurrence. For instance, global warming, ozone depletion and acid rain simply were not issues in the realm of Smith's reasoning, but none the less are the outcomes of rational and self-interested behaviour. This does not make Smithian theories wrong *per se*, but it does imply that there is more scope for dysfunctional behaviour and antisocial outcomes in markets today than was the case at the time that Smith was writing.

2.4.2 Malthus

Thomas Malthus (1766–1834) published his famous book, *An essay on the principle of population, or, A view of its past and present effects on human happiness: with an inquiry into our prospects respecting the future removal or mitigation of the evils which it occasions* (Malthus 1989/1807) at a time when society was undergoing the radical economic transition of the Industrial Revolution. The Industrial Revolution was initiating structural changes in the pattern of employment in the UK, with a move away from agrarian self-sufficiency to commercial production in urban centres. Coupled to this trend was demographic change, with increased life expectancy resulting from medical advances. Malthus predicts that demographics imply a geometric increase in population over time, whereas the growth in food production only increases arithmetically. This dichotomy is represented in Fig. 2.1.

One reason for Malthus's prediction of an arithmetic increase in agricultural output is the economic law of diminishing returns. To understand this, consider the conventional economic assumption that there are three inputs required to produce commodities: natural resources, otherwise termed land or natural capital (NK); human labour; and plant and machinery, called man-made capital (MK). The coordination of these inputs into the production process is the role of the entrepreneur under capitalism. The law of diminishing returns dictates that if the input of one, and only one, factor of production is increased by $\alpha\%$, then aggregate output will increase but by strictly less than $\alpha\%$. Malthusian reasoning is that, because the supply of land is fixed, then increasing the MK input accrues increases in production but at a decreasing rate. He accepted the possibility of converting previously unexploited land for agricultural production, but predicted

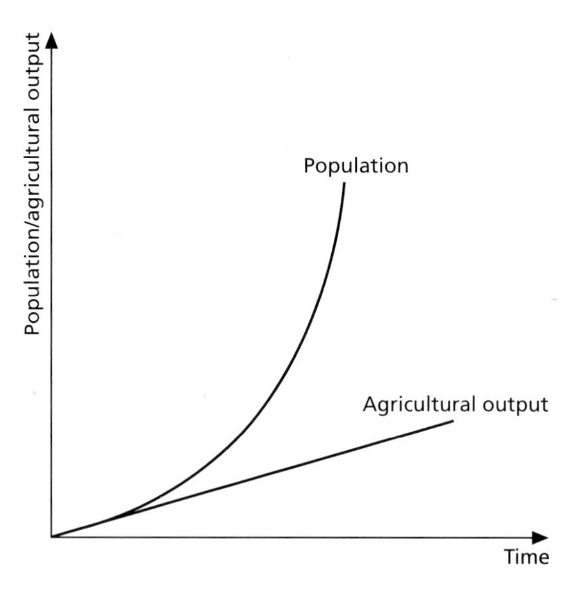

Fig. 2.1 Malthusian model of population growth vs. agricultural output growth. Population grows exponentially whereas agricultural output only increases arithmetically.

low yields from these areas. This seems intuitive in that entrepreneurs are likely to exploit the most productive land first, and thus any unexploited areas are likely to be less productive in comparison.

Malthus was one of the first theorists to express the limitations imposed by the finite nature of the Earth's resources. He considered the critical constraint to be humanity's capacity to grow sufficient food to satisfy the needs of a rapidly burgeoning population. He depicted the interaction between the two variables—food production and population—as a complex one. An increase in population implies a larger endowment of the labour factor of production to the agricultural process, thereby increasing output. However, a countervailing force exists in that an expanded food supply spurs an increase in population and consumption. One reason for this latter force is that the younger generation were, in his time, expected to act as a social security system during their parents' old age. A higher number of offspring implied greater security. In summary then, population is dependent upon food supply and vice versa.

The law of diminishing returns and exponential population growth, treated in conjunction, led to Malthus forecasting that humanity would eventually become trapped in a dismal and chronic state of existence. As food availability became a binding constraint population would have to be checked, either by choice through abstinence, contraception and abortion, or by force through wars, disease and famine.

Malthus's arguments can be challenged. The law of diminishing returns is only applicable under the assumption that the state of technology remains constant. This assumption is certainly not valid in reality. For example, the current technology of high-intensity mechanized farming using artificial fertilizers and irrigation is vastly different to cultivation for self-sufficiency. Malthus's assumption that food availability spurs population growth has also not applied, at least in the developed world, in recent times. As income and education levels have improved, family size has diminished. Further, population growth has not been translated into a proportional increase in the labour factor input to agricultural production because of the processes of urbanization and mechanization. Thus, with hindsight Malthusian theory appears to be flawed.

2.4.3 Ricardo

David Ricardo (1772–1823) was a contemporary of Malthus. He too predicted that an equilibrium steady state would arise but via cyclical fluctuations over time, implying malnutrition and suffering. There are two central assumptions to the Ricardian model: profits alone stimulate growth, and labourers' wages alone determine changes in population. If wages exceed subsistence, then Ricardo assumes that this provides sufficient incentive for procreation. This incentive applies up to the point at which, because of the increase in the labour supply, wages are driven down again to subsistence levels. Ricardo predicts periods of mass starvation because of the systematic lag between the signal to cut back on the rate of procreation and the downward trend in wages.

The other factor in this cycle is profits. Wages and profits are interdependent in that an increase in wages squeezes profits, which in turn implies a decrease in the accumulation of man-made capital as less money is invested. As population growth progresses, wages are cut and profits increase, spurring investment and consequent growth in the economy.

This cycle is depicted in Fig. 2.2. The aggregate output curve increases decreasingly owing to the law of diminishing returns. The law applies because of the successively decreasing quality of land available for agriculture, as opposed to the finite nature of its availability. Further, again paralleling the Malthusian

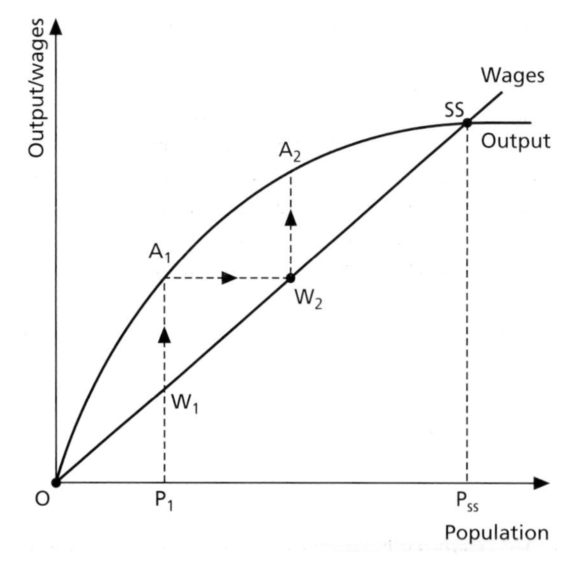

Fig. 2.2 The Ricardian commodity production model. A gap between industry output and the wages bill ($A_1 - W_1$) encourages investment and a (lagged) growth in population. Over time the wages bill increases constantly with increases in population, but investment brings diminishing returns, leading to a final steady state (SS).

argument, there is no technical progress in the model and thus the aggregate output curve remains fixed. Thus, Ricardian theory is subject to the criticisms made of the Malthusian argument above.

Ricardo also provided a critical contribution to the debate on international trade and protectionism. Ricardo's law of comparative advantage demonstrates that there are benefits to specialization and trade if there is a difference in the *opportunity costs* of producing goods between countries. There are many reasons why opportunity costs might vary: costs of labour; natural resource endowments; availability of technology, etc. If countries specialize in the production of commodities that they can produce relatively cheaply, then aggregate world production increases and thus theoretically, after trading, all countries can consume more. There are, however, problems in practice with the comparative advantage logic. For example, with respect to trading arrangements, even though there is a larger overall commodity cake to cut up, developing countries often do not get their fair share.

2.4.4 Mill

John Stuart Mill (1806–73) wrote on a diversity of economic subjects: international trade; natural resources;

welfare issues; and finance. Like his contemporaries, he wrote about the relative merits of mining versus the agricultural sector. He was radical in defining an important distinction between the two: in mining there is a trade-off between current productivity and future productivity. This implies that there is an optimal extraction path for non-renewable resources which requires a different analysis to agricultural production, an analysis that was pioneered by Hotelling in the early twentieth century. Mill also realized that there were two countervailing forces in the mining sector: diminishing returns apply to the extraction of known reserves, as the most cost-effective mining opportunities are exploited first; and through the course of time, new deposits are likely to be discovered (Kula 1998). Overall, Mill was optimistic about the development of the mineral extraction industries.

Although Mill was a contemporary of Malthus and Ricardo, his vision of progress is very different to theirs. Further, his seminal work, *Principles of Political Economy* (Mill 1848), is still a source of reference for conservationists to this day. He is an opponent of the insatiable materialism which his contemporaries consider to be an engine for progress. He admits the theory of a stationary state but adds, 'I sincerely hope, for the sake of posterity, that [people] will be content to be stationary, long before necessity compels them to it' (Mill 1848). The pertinent question is, then, not how many people *can* the planet feasibly support, but what is the most *desirable* population level. The physical carrying capacity of the planet might well exceed the social equivalent if solitude and nature are deemed fundamental human needs. Mill was no Luddite: for him technological innovation is progressive, but with the caveat that once people's basic material needs are satisfied then there should be scope for the development of aesthetics and self-realization. The relevance of this conceptualization can be witnessed in current thinking. Herman Daly expresses a similar view in *Steady-State Economics*:

> 'Is the nature of the Ultimate End such that, beyond some point, further accumulation of physical artefacts is useless or even harmful? . . . could it be that one of our wants is to be free from the tyranny of infinite wants?' (Daly 1992)

The nineteenth-century economic theorists heralded a revolution in perception away from the classical political economy school which came before them. The Marxist paradigm has left an indelible footprint on world history, and is discussed in general in the next section along with Marx's philosophy on resource exploitation and the environment.

2.4.5 Marx

As an economic thinker, Karl Marx (1818–83) has had a truly profound impact on the world politics of the twentieth century. The arms race and hostilities of the cold war between communist and capitalist nations were a major threat to the survival of humanity, and the Stalinist and Maoist regimes in the Soviet Union and China, respectively, accounted for an unprecedented loss of life in the name of Marxism. The question remains as to how far Marx's works themselves were responsible for this state of affairs.

The Marxist interpretation of the outcome of the Industrial Revolution was that the capitalist classes had systematically exploited the working classes to achieve a high level of material consumption and associated political power which served to maintain this *status quo*. The urbanization and industrialization had implied oppressive and dangerous working conditions, the employment of child labour and poor compensation in wages. Adam Smith's vision of capitalism—the goal of self-enhancement and free and rational expression—was not realized for the working classes. Further, Marx argued in the *Communist Manifesto* that one group could only prosper at the *expense* of the other under the capitalist productive structure, and thus depicted capitalism as a new form of feudalism and slavery. Marx claimed that, because the capitalists could dictate conditions of employment, workers were paid only enough to survive. Children were the only form of social security for parents in their old age, and so the supply of labourers increased perpetually, reducing their bargaining position in the workplace. Socioeconomic forces, such as the enclosure movement, wherein peasants were forcibly evicted from the land, propelled further workers involuntarily into the urban market.

Marx predicted a struggle between these two forces, to be resolved post-revolution with the overthrow of the minority capitalist class. One of the mechanisms stimulating this revolution would be the boom-and-bust economic cycle which he proposed as being inherent to capitalism: as output expanded there would be a surplus of commodities on the market as workers' incomes would not be increased to compensate; this would necessitate unemployment and output contraction until this surplus was squeezed out of the system; output would increase post-crisis and stimulate once more an over-supply on the market. The suffering and insecurity of the workers caused by this cycle would set the foundations for revolution. The process of technological innovation would also feed this class conflict, as more capital-intensive production processes robbed the workers of their livelihoods as machinery substituted labour.

Marx's alternative to capitalism was a communist vision of a society free of class distinctions, with an abundant supply of resources which would be distributed according to each citizen's needs. This vision is rooted in classical economics in that he adopted a labour theory of value. Kula (1998) describes this labour theory as having two parts: use value and exchange value. Use value is related to a commodity's material qualities which are realized in consumption. The exchange value is exclusively derived from the collective labour input required per unit output, with labour acting as part of a *social* production arrangement. For Marx, a product, e.g. a shirt, produced under a communist regime embodies the social relationships involved in its collective production, whereas a physically identical shirt produced under capitalism does not.

The labour factor input is the only source of net income in the economy under Marxism. Although the Ricardian and Malthusian theories both focus on labour supply variations as a determinant in achieving the steady state, neither explicitly consider production as a social process. Marx considers production in terms of a class struggle.

Neo-Marxist theory portrays environmental impact as another means of social suppression of the labouring class under capitalism. Capitalists seek short-run profitability, without paying due diligence to long-term impacts. Labour-saving innovations might entail a higher environmental impact in terms of toxicity and waste discharge. This affects a diverse range of habitats and species. For humans, the costs are not evenly distributed across society, as workers are relatively over-exposed, thereby implying class confrontation.

This process of capitalist exploitation can be seen in the way that some transnational corporations (TNCs) operate in developing countries. The phenomenon of *footloose capital*, that is production plants which can be disassembled and transferred to an alternative location at relatively low cost, is commonplace. One example is athletic footwear manufacture. World trade globalization gives TNCs the flexibility and market incentive to set up plants in the most cost-effective locations. Further, because these plants generate local employment and income, various competing host countries might provide incentives for such investment. These incentives can include taxation immunity, the banning of trade unions to avoid potential disruptions to production and exemption from environmental regulation. Even if such regulation is applied, it is likely to be less rigorous than in developed nations, which remain as the eventual destination for the bulk of the production. Ecological degradation is thus exported to these host manufacturing nations.

Marxist theory on natural resources is exploitative, i.e. the natural environment has only instrumental value, with the presumption that science would provide solutions for environmental problems. It is noteworthy that empirically the environmental performance of the former Eastern block is, in general, poor relative to that of the developed capitalist world, at least with regard to pollution control.

2.5 The rise of neoclassicism

Neoclassicism has come to be synonymous with mainstream economic thought. The principles of economics outlined in Chapter 3 are wholly drawn from this school. Neoclassicism challenged the labour theory of value as conceptualized by the classical school up until the 1870s. Writers such as Jevons and Menger argued that value was not based on the labour input embodied in a commodity, but rather in the utility that the commodity yielded to the consumer.

> 'Repeated reflection and inquiry have led me to the somewhat novel opinion, that value depends entirely on utility . . . In this work I have attempted to treat economy as a calculus of pleasure and pain.' (Jevons 1871)

This was a pure demand-side theory of value. Jevons, who studied problems related to the supply of coal in maintaining the UK economy, has been described as 'the forerunner of neoclassical economics' (Stigler 1946). He developed the concept of diminishing marginal utility (see Section 3.7.1). A further critical input to the development of contemporary microeconomic theory was provided by Marshall (1842–1924) whose seminal work *Principles of Economics: an Introductory Volume* was published in 1890. Marshall brought together the supply-side orientated labour theories of value of the classical school and the new demand-side utility-based alternatives in the formulation of a demand and supply interaction model, with price and value being determined by both. Amongst many economic insights, this has developed into one of the most fundamental and widely valued.

Another significant theorist in the neoclassical tradition was Hotelling, who dealt principally with the optimal rate of extraction of non-renewable resources, e.g. oil. His seminal paper (Hotelling 1931) considered the change in price over time (the price path) for a non-renewable commodity. In Hotelling's model, the owner of a non-renewable resource has two options: extract now, and earn interest on the proceeds of sale; or keep the resource in the ground, with the expectation that its price will rise. Hotelling found that the price path that maximized social welfare could be achieved if the extracting industry was perfectly competitive.[2]

This result was particularly appealing to the neoclassical school as it allows social welfare, defined purely in terms of the utility derived from the consumption of goods, in this case non-renewable resources, to be maximized *without state intervention*. A fundamental premise of the neoclassical school is that it is best to leave decision-making to individuals in the market, who are assumed to act rationally in allocating their budgets in accordance with their preferences.

Neoclassicism thus shifted the debate away from the classical school's various theories on cyclical growth patterns and their culmination in a stationary state. If operating efficiently, the price mechanism is deemed sufficient for avoiding such a zero-growth steady state: price provides incentives for innovation, and thereby the market supplies substitutes for commodities which become relatively scarce and relatively expensive. Again, any state intervention other than establishing a framework for competition is considered to be inappropriate.

2.5.1 Welfare economics and Pareto optimality

Once established as the dominant school of thought, neoclassical economic analysis spread itself across a wide range of fields; for example, microeconomics, financial, labour, public and resource. A particularly important area from a government point of view was that of welfare economics, which was devoted to the explicitly value-laden issue of the well-being of society. Thus, welfare economics considered how well the economy was doing the job of raising welfare, rather than simply understanding how parts of the economy worked.

The neoclassical foundations of welfare economics were established in the work of an Italian, Vilfredo Pareto (1848–1923). Pareto defined a particularly important concept known as Pareto efficiency, which was a measure of how efficient the economy was at improving social welfare. A Pareto efficient economy was one in which, given the initial distribution of resources, no person could be made better off without making at least one other person worse off. In other words, in a fully Pareto efficient (also called a Pareto optimal) situation, there was no opportunity for increasing the overall supply of goods to provide universal enrichment; welfare gains could only come about by taking goods away from some people and redistributing them to others.

The Pareto efficiency concept became important from a neoclassical perspective partly because of its perceived value neutrality. The issue of redistributing goods is highly political—there are winners and losers, and this requires difficult judgements to be made about who should win and who should lose, by how much, and so on. By contrast, suggesting Pareto efficient improvements in the economy (that is, ways in which some people could gain and nobody would lose, called a Pareto gain or improvement) seemed genuinely unproblematic. It seems an ethically neutral concept because an *increase* in Pareto efficiency should be acceptable to everyone: everyone wins as a result, or at least, no one loses anything and some people gain. We can note in passing that a Pareto optimal situation may in itself be extremely *inequitable* in terms of actual wealth. It is not value neutral in terms of the 'goodness' of any particular situation; rather, a Pareto optimal situation cannot be made any more efficient at generating welfare without becoming involved explicitly in redistribution. Most economists insist redistribution is itself a political and ethical decision, which lies beyond their remit as economists.

Free trade itself should enable society to make constant Pareto improvements, because a trade of goods requires two individuals to agree mutually to the exchange; if they agree freely, they presumably do so because they each regard their set of commodities after the exchange as being better than that before the exchange. The situation after the exchange is superior for both individuals, simply as a result of reallocating goods between them. It is therefore a Pareto gain.

[2] Perfect competition is an industry structure in which there are many buyers and sellers of the commodity, none of whom individually can exercise any control on the market price; they are said to be 'price-takers' (see Section 3.6.3).

The adoption of Pareto efficiency as the focus of welfare economics explains the economist's interest with free and perfectly functioning markets. Such markets result in the highest possible welfare for society, given the initial distribution of goods. This is summarized as the first theorem of welfare economics: an equilibrium allocation of goods achieved by a set of competitive markets will necessarily be Pareto efficient.[3] As a simplistic summary, this has been interpreted as: establish the conditions for the market to function competitively and social welfare will be maximized.

2.5.2 Interventionist school vs. property rights theorists

Although the doctrine of neoclassicism became synonymous with a *laissez faire* (literally 'let act') approach to welfare, essentially leaving free markets to allocate goods so as to maximize welfare, in the early twentieth century an alternative school was challenging the universal applicability of this doctrine. The interventionist school proposed that, in some instances, the state should intervene to achieve socially desirable outcomes. These interventions were necessary because markets were almost always imperfect in some respects (see Section 3.6.3).

Arthur Pigou (1877–1959) is perhaps the most influential writer in the interventionist tradition. Although he analysed many aspects of appropriate resource allocation, his most famous contribution is that of 'internalizing externalities' associated with environmental damages. If firms pollute rivers in the course of producing goods, but do not have to pay the costs of that pollution, then the price of these goods in the market will not reflect those costs. Pigou advocated imposing a tax on the polluting firm in proportion to its output of pollution, known as a Pigouvian tax. This would force the firm to adapt its output accordingly and should lead to a Pareto optimal outcome under certain conditions. Pigouvian taxes are covered in detail in Section 14.3.2.

An important aspect of the Pigouvian solution to pollution problems was the ability of the polluter to 'get away' with pollution. This is possible because there are no strongly defined property rights to environmental resources, such as the air and rivers. In these cases, no single person is able to sue for damages. In many cases, the effect on each individual may also be small, though added up it may be very significant. The importance of property rights within the analysis of environmental problems was highlighted by Ronald Coase (1960), leading to what is known as the Coase theorem. The theorem basically states that if property rights are clearly defined, polluter and polluted person can come together and bargain over the acceptable level of pollution and come to a Pareto efficient outcome. We analyse the Coase theorem in detail in Section 14.3.1.

Coase's work provided the foundations of the 'property rights' school of thought on environmental problems. This school sees environmental problems as caused by the absence of property rights, whether in air, deforestation, whale depletion, or whatever. In their view, if property rights are assigned in some way, then markets can be established and the welfare logic introduced above—namely that free markets produce Pareto optimal outcomes—applies. Although in certain instances assigning property rights may provide a solution to problems, there are limits to just how far it can be applied; this discussion is taken up in Chapters 12 and 14.

2.6 Institutional economics

Neoclassicism has become the predominant form of economic analysis in the Western economic tradition. Marxist economics has formed one separate branch. One other that is useful to note is institutional economics.

Institutional economics, starting around the turn of the century, focused principally on the relationships between institutions and human behaviour. 'Institutions' means organizational systems for behaviour and involves both formal systems, such as the legal constitution and regulations, and informal institutions, like traditions, habits, moral 'norms' and social codes. Institutional analysis has tended to be much more sociological than neoclassical economics, exploring how behaviour is influenced and, crucially, dependent upon such institutions. Whereas neoclassicism has generally studied rational economic behaviour within the relatively narrow confines of the institution of the market, institutional economics has been interested in how the institutions themselves influence behaviour, and how the institutions have developed, giving the further branch of 'evolutionary economics'. John Commons and Thorsten Veblen are the best known of its early

[3] There are some technical assumptions that apply: agents act rationally and have well-defined utility functions; subsatiation conditions apply; and there are no non-market interdependencies, i.e. one agent's utility is not dependent on another agent's (see Sections 3.7–3.10).

proponents, with Commons focusing on legal institutions and Veblen on more sociocultural variables.

The problem of institutional economics has been that it has been unable to match the rigour and predictive power of neoclassical analysis. According to Blaug (1985), institutionalists offered description without theory and consequently failed to challenge the dominance of neoclassicism. In the later twentieth century John Kenneth Galbraith has been perhaps the most prominent advocate of a more institutionalist-style economics. Galbraith's work has often criticised the *laissez faire* model of affluent US lifestyle as representing a culture of contentment that is focused narrowly on self-satisfaction. In this he is in common with other socially orientated critics who stress the limited criteria that neoclassicism studies (behaviour in response to price signals within narrowly defined markets) and advocate a richer set of concepts for study.

Some aspects of what can be considered institutional-style analysis are introduced in Sections 5.3 and 5.4 concerning the meaning of welfare and the role of social and cultural factors in economic systems. However, in economic analysis we focus more generally on valuable insights from the neoclassical school.

2.7 The emergence of an ecological consciousness in economics

The principal analytical tools of neoclassical economics developed over many decades and were applied to many different areas, including environmental concerns. In these instances, neoclassical economic theories were developed that brought the allocation issues associated with natural resources, such as efficient forest and fisheries management, into the overriding 'supply and demand' framework. Environmental parameters were important only in certain circumstances, for example, where they related to the optimal extraction path of non-renewable resources from a mine, or where they determined the growth rates of renewable resources. Analysts remained focused on maximizing the growth of the economy by improving its efficiency through marginal optimization techniques.

Around the late 1960s and early 1970s, a set of studies appeared which challenged one issue of fundamental importance within prevailing neoclassical theory: namely, that the entire economic system was itself embedded within the wider environmental system. This new perspective on economic activity had potentially enormous implications for the direction of analysis. In the subsequent sections we trace some of the key figures and arguments within this debate.

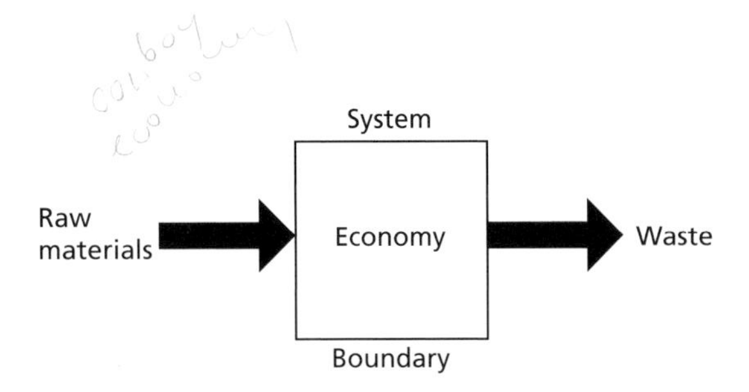

Fig. 2.3 Conventional view of the economic system separate from the environment.

2.7.1 Kenneth Boulding and 'Spaceship Earth'

In 1966, Kenneth Boulding published a short paper entitled 'The economics of the coming spaceship Earth' (Boulding 1966), in which he identified the need, in his view, for a radical revision in the way economists perceived the economic system. In the traditional frame of economic analysis, the wider natural environment was generally ignored; it provided 'inputs' in the form of raw materials, and accepted 'outputs' in the form of waste, but economic analysis was focused entirely on the processes going on between these two poles. This is represented in Fig. 2.3.

The vision behind Fig. 2.3 is appropriate to what Boulding calls the cowboy economy. The isolated cowboy travelling across the open expanse of the prairie can easily ignore worrying about running out of fuel for his next campfire, or the effect of the ashes that he leaves behind the following morning. His path of consumption and production is linear, as Fig. 2.3 suggests, because his impact on the environment is so small compared to the vast size of that wider environment that it is simply negligible. Raw materials are converted to wastes, but the amount he consumes today does not limit the amount he can hope to consume tomorrow because there are always more resources available just over the next ridge.

Boulding suggests that this vision must be replaced by that of the spaceman economy. On board a spaceship, the spaceman has only a limited stock of inputs —whatever he began his journey with—and only a limited capacity to carry wastes. The spaceship is essentially a closed system, and if it is going to continue operating it has to be run on a circular production and consumption system that recycles waste outputs and conserves inputs within limits that enable life on board

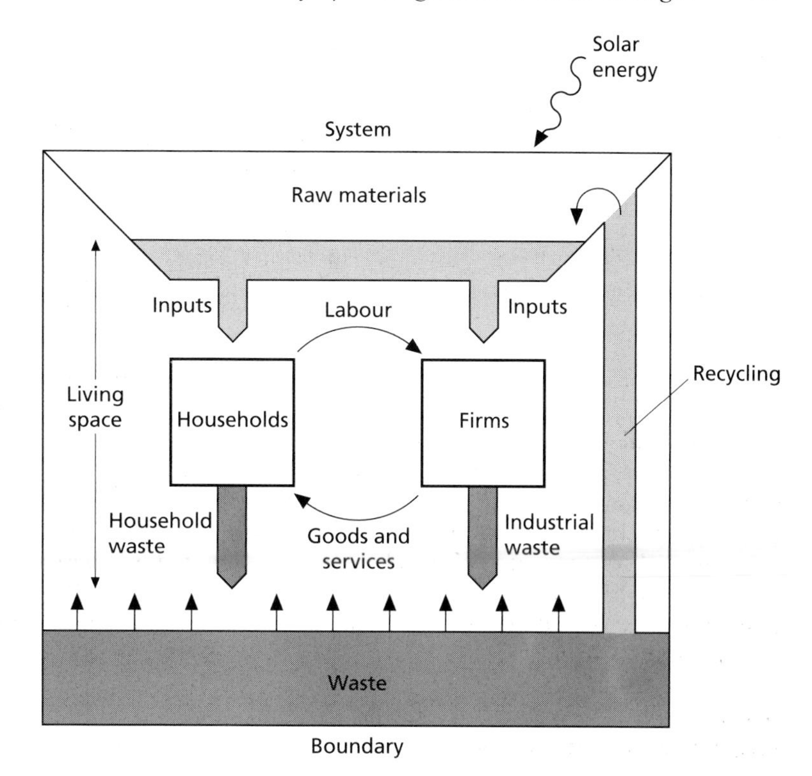

Fig. 2.4 Vision of the economy as dependent upon the wider environment.

to be sustained. The activities that can be undertaken on board the spaceship are therefore circumscribed absolutely by the ability of the on-board environment to cope with the consequences of those actions. This suggests a new vision of the economy, shown in Fig. 2.4.

What difference does this new vision make to the operating principles of those running the economy? Clearly it suggests that the environmental impacts of economic activity must be kept within some limits that enable a tolerable life on the spaceship to be maintained. The problems are likely to arise in determining those limits. Boulding suggests two guides in this regard.

First, if we are uncertain about the total stocks and the total recycling capacity of the spaceship then it is sensible to try to minimize the throughput of material in the economy (that is, the amount of material being used up and recycled in the processes of production and consumption); and to try to ensure that what throughput does go through does so as efficiently as possible (producing as much usable output per unit of raw material as possible). He points out that current measures of economic success, based on the growth of gross national product, are not based on this kind of measure but rather on maximizing the amount of consumption going on, regardless of throughput. These are appropriate guides for cowboys but not for spacemen, an issue considered in detail in Chapter 11.

Secondly, as a general rule, the higher the total capital stock on board, the better. Boulding considers the capital stock on board spaceship Earth to comprise all those things that contribute to human welfare, including man-made artefacts, natural resources and human minds and knowledge. All of these are important factors in maintaining not simply minimum survival conditions on Earth but in maximizing welfare subject to the absolute constraints of the environment (its ultimate carrying capacity). The issue of measuring the on-board capital stock is a complex one that we consider in more detail below (and again in Chapter 11).

2.7.2 Georgescu-Roegen and entropy

The model established by Boulding relies on the characterization of the Earth as a closed system with extremely limited opportunities for exchanging materials and energy with areas outside that system. A similar kind of approach was taken by Georgescu-Roegen (1971) in *The Entropy Law and the Economic Process*, in which he traced out the implications of the laws of thermodynamics for economic activity. Like Boulding, Georgescu-Roegen considered the economic system as embedded within a wider set of environmental conditions than economists generally consider, namely the laws of thermodynamics. His focus was therefore on one aspect of the spaceship's life-support system—one which Boulding also drew attention to—that concerning energy flows.

All activity, including that taking place within the economy, involves material and requires energy. The laws of thermodynamics govern what happens to that material and energy as it is used in the process of production and consumption. The two fundamental laws of thermodynamics state the following:

1 Matter and energy themselves are neither created nor destroyed. In a closed system, the total amount of energy and matter remains constant.

2 In any thermodynamic process (that is, any physical activity of any kind), the entropy of the system either remains the same or increases. Entropy is a measure of the level of organization or structuring of material and energy; high-entropy states are disordered, low-entropy ones are well ordered.

We can now trace some implications of these laws for the economic system, taking it as a subsystem of a closed system which is Planet Earth.

The first law refers to the need for some kind of circular flow of material, because the total mass of material on board remains constant despite raw materials being converted to products and eventually to waste. This insight is associated with 'materials balance' analysis within ecological economics and engineering, which traces the flows of materials through systems. The total mass of materials involved at the start and finish of any process must balance; waste products can be traced according to whether they are recycled, stored or otherwise escape into the larger system. Waste does not just disappear out of the system as conventional economics might suppose; it has to be accommodated somewhere.

This insight is made particularly significant by the second law. The tendency for the entropy of any system to increase as any activity takes place means that the useful, highly structured elements of the system, such as stores of fossil fuel, are used up and the material in them converted to a relatively useless, more highly disorganized form, such as a mix of widely dispersed gases in the atmosphere. Although higher entropy wastes can be converted back into more useful low-entropy forms through recycling, this recycling process itself takes up more energy which in turn increases entropy. Thus, the closed system as a whole slowly increases its overall level of disorder, while we can only manipulate small subsystems of that total system by shuffling the overall limited store of low-entropy energy around. The ultimate end of this process for a completely closed system is a 'heat death' in which the system attains a stationary state at the highest level of entropy or disorder.

How critical are these insights for the Planet Earth system? Two points can be made. First, the Earth is not a completely closed system, as Boulding and Georgescu-Roegen recognize. Most crucially, it receives solar energy, and it radiates (a little) heat to space. The loss of heat is negligible, but the receipt of solar energy is critical. As solar energy is a constant flow of energy received by the Earth, it means that some form of activity can be sustained even if all stocks of fuel have been used. Secondly, extracting raw material from the environment, changing the form of that material and then reintroducing it into the environment at a higher entropy level clearly changes that environment. These changes may be relatively damaging, though they need not be just because they are higher entropically, and the residuals of economic processes must therefore be managed so as not to perturb that environment too far. But this management will itself take more energy, and so on.

The implications are that, as in Boulding's example, we should look to use low-entropy energy stocks as efficiently as possible if they are in short supply. Furthermore, if we are ultimately to run out of stocks of low-entropy materials, we should in the meantime prepare by adapting economic systems to use the fixed flows of solar energy that will remain available.

2.7.3 'The Limits to Growth'

The theme of absolute resource scarcity and the outlook for continual growth in the physical scale of the economy was one addressed by the 'Club of Rome', an informal group of diverse professionals who gathered in Rome in 1968 to discuss a wide range of social themes. The Club tried to address these themes 'holistically': that is, considering them as intimately related parts of a complete system. It subsequently supported the building of a large-scale computer model of the world economic system, which tried to put some quantitative flesh on the conceptual bones of Boulding's spaceship concept. The results of this modelling exercise were subsequently published as *The Limits to Growth* (Meadows *et al.* 1972).

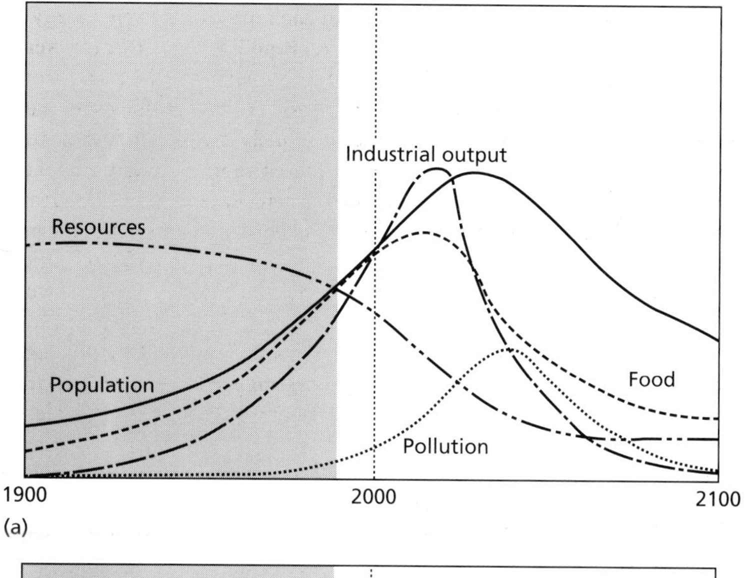

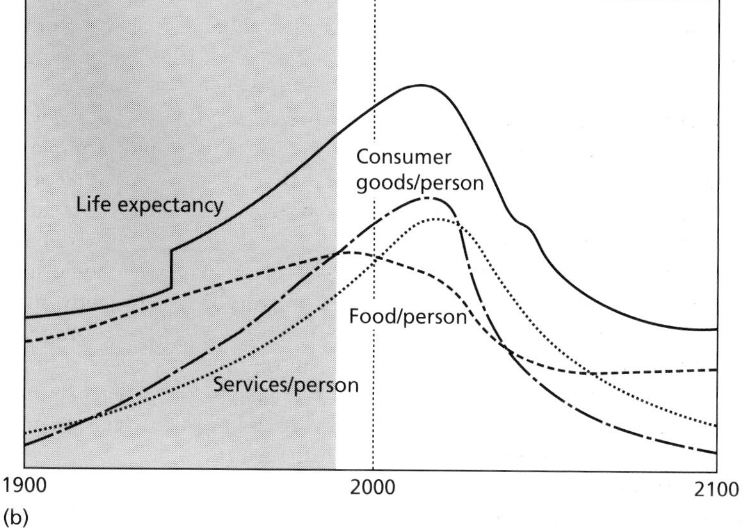

Fig. 2.5 The standard run predictions of *The Limits to Growth* model of the world economy, showing projected changes in the key variables to 2100. (Source: Meadows *et al.* 1992. Reprinted by permission of Sterling Lord Literistic, Inc. Copyright by Dennis Meadows, Donella Meadows, Jorgen Randers.)

The Limits to Growth stimulated a wide-ranging debate, primarily because of its major neo-Malthusian prediction. This was, in the words of the report, that 'under the assumption of no major change in the present system, population and industrial growth will certainly stop within the next century, at the latest' (Meadows *et al.* 1972). The model accounted for some feedback loops between the variables included within it, rather than relying on pure extrapolation of the eight key variables. These variables were population, non-renewable stocks, services, birth rates, death rates, individual output, pollution and food availability.

On the so-called 'standard run' of the model, the failing of the world economic system was caused by non-renewable resource exhaustion. As resources become scarcer, more and more capital is consumed to extract more and more inaccessible reserves, essential as inputs to a system that rotates around these inputs. This starves investment in the future which eventually leads to a collapse in the productive base of the economy and the subsequent end of material growth. The standard run is shown in Fig. 2.5. Simplified models of four possible trajectories for the behaviour of different systems are shown in Fig. 2.6(a–d) (Meadows *et al.* 1992).

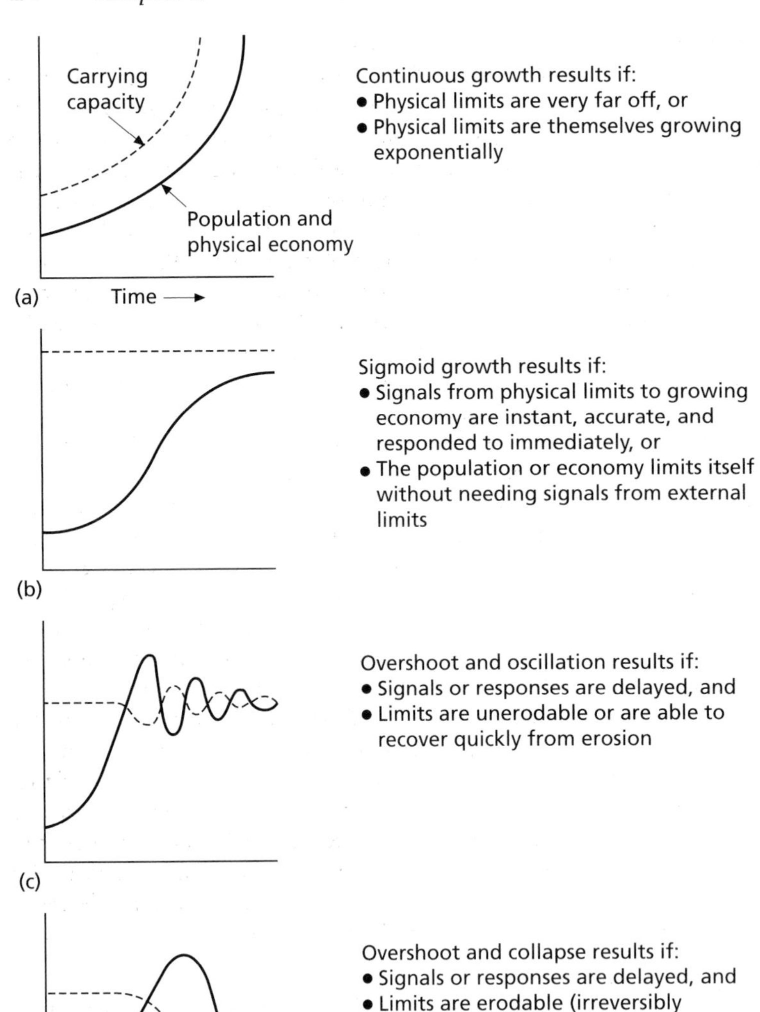

Fig. 2.6(a–d) A range of possible dynamic relationships between carrying capacity and the human population and its economic system. The *Limits to Growth* model predicted a trajectory similar to scenario (d). (Source: Meadows *et al.* 1992. Reprinted by permission of Sterling Lord Literistics, Inc. Copyright by Dennis Meadows, Donella Meadows, Jorgen Randers.)

The Limits to Growth team was careful to point out that their model was run under assumptions of no major changes in the general conduct of governments or business. They note in their final paragraph: 'The final thought we wish to offer is that man must explore himself—his goals and values—as much as the world he seeks to change. The dedication to both tasks must be unending.' If the model's predictions were correct with the given assumptions, they were not in any sense immutable, and social change could very easily prevent the predicted outcome if society could bring itself to make the necessary changes.

2.7.4 Responses and critiques to 'The Limits to Growth'

The Limits to Growth report received a number of significant criticisms, which taken together are really sufficient to undermine any claims to policy relevance for the model. These are well summarized in Kula (1998).

First, model-building as a technique for assessing future developments is always subject to limitations, and as the size and scope of the model increases the limitations also increase. The data requirements for building such a model, trying to capture as it does the entire world economic system and related aspects of the environmental system, are simply vast, and what was available may be considered fundamentally inadequate. Cole and Curnow (1973) claim that if the model is run backwards it fails to trace history successfully, though the Meadows team claim the contrary.

Secondly, the model treats the world as a homogeneous entity, with predictions made for the world economic system. Clearly, different parts of the world exhibit very different patterns of development. Although some pollution problems do now have global reach, it seems quite possible that certain parts of the world may undergo severe problems caused perhaps by resource scarcity, while others isolate themselves from these problems.

Thirdly, Cole and Curnow (1973) also claim that certain assumptions, for instance regarding resource scarcity, are unfairly pessimistic. By altering some of these assumptions a much rosier picture can be achieved. Given the difficulty of knowing the precise relationships underlying many elements of the model, it is difficult to see which set of assumptions, optimistic or pessimistic, is best justified. Additionally, by adopting a 'business as usual' approach to the extrapolation of trends, the model does not fully reflect the potential for technological innovation. A review of the last hundred years indicates that innovation has been the rule rather than the exception, and this changes the parameters of important model relationships.

Finally, economists pointed out that the model did not adequately take account of the role of the price mechanism, which would force changes in behaviour in response to resource scarcity not acknowledged in the model. As non-renewable stocks become scarcer they become more expensive: this stimulates efforts at conservation, more efficient use, and powers research into the development of substitutes. These factors make the image of sudden collapse based on final exhaustion of non-renewable resources highly unrealistic.

These final criticisms summarize the essence of an 'anti-*Limits*' position associated with a number of prominent economists, including Simon and Kahn in *The Resourceful Earth* (1984), Wilfred Beckerman's *In Defence of Economic Growth* (1974) and Kahn *et al.*'s *The Next 200 Years* (1976). The argument of these writers was essentially optimistic regarding the possibility of continual economic progress, based on the ingenuity and creativity of the human intellect. Boulding (1966) himself acknowledges that what he calls the nosphere, the domain of human knowledge, is the most important component of the economic system. Still, continued economic growth remains bound by firm physical and ecological limits. While the technological optimists do not deny that these limits may exist, they do deny that we need to be concerned by them, because in their view we will always invent our way round problems when those problems become too pressing.

The optimists cite a number of powerful examples in support of their case. They point out that, historically, substitutes have been found for dwindling resources and agricultural yields, far from reaching stasis, have shown an ever-upwards trend. Moreover, energy efficiency in production is continually increasing and relatively vast sources of most resources still remain to be tapped. Furthermore, they suggest that development, far from damaging environmental quality, is actually positively correlated with it. Other relationships, for instance, that population tends to stabilize in higher income countries, are presented as further proof that it is the *absence* of a well-advanced and fast-growing economy that damages the environment, rather than its presence. These arguments give rise to a particular growth-orientated interpretation of sustainable development, outlined in more detail below.

2.7.5 Herman Daly and the steady state economy

The steady state economy, championed by the World Bank economist Herman Daly (1973) in his book *Toward a Steady State Economy*, is closely related to Boulding's spaceship theme. Daly stresses, like Boulding, that it is the total scale of the economic system, relative to the wider environment, that is becoming a key issue, rather than allocation issues within the economic system. The problem is that economic theory has no fundamental 'existence' theorem related to resource inputs or waste outputs. An economy can be functioning very efficiently from the point of view of production viewed in isolation, but this may be beyond the capacity of the environment. He draws an analogy with a nautical plimsoll line (Daly 1991), which indicates how well cargo has been arranged on board ship to keep it level. Even if you load the boat so that it keeps perfectly level, it still sinks if you simply load it with too much cargo. Similarly, the economy may be doing a great internal organizational job, but it is ultimately reliant on parameters outside the system itself.

His description for the steady state is given as follows:

> 'A steady-state economy is defined by constant stocks of physical wealth (artefacts) and a constant population, each maintained at some chosen, desirable level by a low rate of throughput—i.e. by low birth rates equal to low death rates and by low physical production rates equal to low physical depreciation rates, so that longevity of people and durability of physical stocks are high. The throughput flow, viewed as the cost of maintaining the stocks, begins with the extraction (depletion) of low entropy resources at the input end, and terminates with an equal quantity of high entropy waste (pollution) at the output end. The throughput is the inevitable cost of maintaining the stocks of people and artefacts and should be minimized subject to the maintenance of a chosen level of stocks.' (Daly 1991)

The steady state economy and the spaceship Earth concept are both important precursors of the concept of sustainable development, which at the current time dominates the debate regarding economic development.

2.8 Sustainable development

Concern over the environmentally damaging nature of certain economic activities can be traced back at least as far as Ancient Greek times. The preceding sections indicate that concern over the ability of the Earth to sustain global economic activity itself goes back several decades. The common theme of all these concerns is the possibility of maintaining economic activity, whether at the local or the global scale, into the future.

In 1983 the United Nations established a Commission on the Environment and Development (UNCED) to address these broad issues, under the chairmanship of Gro-Harlem Brundtland, the future prime minister of Norway. The Commission consulted widely across a spectrum of issues and four years later published its findings in a report, *Our Common Future* (WCED 1987), also known as the Brundtland Report. This report is now acknowledged as a landmark publication in the development of environmental awareness and particularly the need for environmental concerns to be integrated into all aspects of successful development.

The most famous quotation from the Report is that defining the key concept of 'sustainable development', which is described as 'development that meets the needs of the present without compromising the ability of future generations to meet their own needs' (WCED 1987). Within this definition the needs of the poor in all societies were particularly stressed, and the limitations imposed by technological and social factors on the ability of the environment to meet needs across generations.

As befits a broad policy goal, the Brundtland definition was vague in a number of important respects while its overall intention seemed pretty clear. The most obvious problem arose over what constituted a 'need'. Economic analysis has essentially two approaches to needs:

1 from a market perspective, it simply takes market demand as the operating definition of society's needs; and

2 from a government perspective, there is a 'need' to provide welfare-improving public goods, both for the disadvantaged and in view of allocation failures of the market.

The sustainable development debate then asked to what extent should every identifiable market or non-market demand for goods and services be met. The answer depended crucially on the interpretation of what future needs might be, and how in turn they might be met.

2.8.1 The capital issue

The debate concerning limiting future possibilities hinges around the role and definition of 'productive capital', because capital in some form is essential to meeting needs. Capital is the material needed for production of valuable goods and services, and in Section 3.13 is defined as taking basically two forms: natural capital (NK), such as trees, fish, oil, minerals; and man-made capital (MK), such as machinery, houses, roads, railways. Under a very broad definition, these are all things that contribute in some way and in some form towards human welfare. We noted that in building economic models of the economy, analysts frequently use a cover-all term, 'capital', referring to MK, and ignore the natural resource input. This is because they are more interested in what happens when capital is combined with labour, than in the functions of different kinds of capital.

In fact, to be rigorous in definition, labour itself can be considered another form of capital, namely human capital (HK). In an obvious sense, people's labour is needed as an input into economic processes. In addition, not only is human labour required but, in maintaining the economy, human knowledge and understanding are also important. This gives rise to a fourth concept of capital, 'cultural capital', which, in reference to Boulding, is the 'state of human minds and experience' related to the stock of knowledge and skills of society. It might be considered strange to separate human minds from bodies in this way, but it does clarify two different kinds of functions that humans undertake. They both analyse problems (using cultural capital) and act to solve them (using human or labour capital, usually augmented by man-made capital). Below we focus on the central issue of natural capital because this is most clearly implicated in the sustainability debate.

NATURAL CAPITAL

The concept of natural capital, meaning essentially useful elements of the natural environment, has many aspects. An initial division can be made between natural capital as a 'source' of raw materials (timber, fish, minerals), as a 'sink' for waste products (either recycling them or simply storing them), and as providing a range of so-called useful 'services', such as stabilizing erosion-prone soils (forests), preventing flooding and controlling nutrient flows (wetlands), to take just two examples (the services are covered in more detail in Section 7.4). In addition, natural landscapes are often just attractive to look at and they generate welfare for people in this way too.

The 'source' aspect of natural capital can be further subdivided according to whether they are renewable or non-renewable, depletable or non-depletable stocks, and biological or physical. Later chapters deal with aspects of the management of these different forms, concentrating mostly on renewable biological resources.

Clearly, humans derive benefits from being able to harvest these sources of natural capital. In the case of renewables this may be indefinitely, whereas fixed stocks will eventually be used up altogether. They also derive value from the useful services of ecosystems. However, they also derive value from man-made capital, in the form of useful artefacts which either contribute directly to welfare (consumer goods) or assist in manufacturing these goods (machinery and 'productive' capital). These man-made goods ultimately use natural capital as raw materials. More man-made capital therefore seems to mean, on balance, less natural capital.

The central issue of sustainable development (not compromising future generations' ability to meet their needs) can now be addressed in a more concrete form. What kind of total capital stock (which determines how needs can be met) should be left to future generations? We can determine two baseline positions on this question:

1 maintain the total capital stock (NK + MK) between generations; or

2 maintain or increase the natural capital stock (NK) between generations.

Position 1 is known as 'weak sustainability', because it allows diminishing levels of natural capital to be compensated for by more man-made capital. Position 2 is known as 'strong sustainability' because it does not allow man-made capital to substitute for natural capital; the overall stock of natural capital must be maintained.

Although this is a more precise approach to the sustainability definition, it is not without problems. Most importantly, as we noted for natural capital above, capital itself is a many-layered concept. It is the overall blend of specific types of capital, natural and man-made, that overall generates any particular level of welfare. The very broad classifications of natural and man-made only makes some aspects of this division clear. As natural capital provides raw materials, it is clear that if we bequeath future generations zero NK, they will not be able to meet any needs at all. But if some substitution is possible, and fuel efficiency increases, lower stocks of oil in the future might have no particular impact on future levels of welfare.

The issue at stake here is that of complementarity and substitutability. If goods are complements then they are more valuable together than separate (there is a synergy between them); whereas substitutes can replace each other without a loss of value. It seems likely in most cases that natural and man-made capital, and different types of these capitals, are partial substitutes, partial complements. If a certain type of natural capital is very abundant, we can substitute some man-made capital for it, converting that natural capital into desired consumer goods. If it is scarce, we may derive more value from conserving the natural capital and man-made goods will not be effective substitutes.

A similar kind of issue arises in relation to natural capital itself. Such capital is not a homogeneous type of 'stuff': forests and the services they provide are different from wetlands, which differ from estuaries, and from the deep oceans yet again. Is it possible to consider how an increase in forest cover compensates for the reduction of a peat marsh? There is no real agreement on the way in which this question might be answered; we can simply note three possible approaches.

One would be to attempt to make economic valuations of the total various stocks and ensure that the sum total value of stocks did not decline; this is subject to great methodological difficulties. Another approach would be simply to measure the physical quantities of each kind of stock and try to maintain these levels (for example, by replanting forests at the same rate as they are harvested). This approach would mean treating each individual type of natural capital as a separate category, does not enable judgements about substitution between types, and is not really meaningful for non-renewable resources, such as oil and coal, for which any use entails depletion. A third possibility might be to try to measure the flow of services from these stocks and maintain the capacity to maintain these flows. This would also present great methodological challenges. One potential problem with any approach which attempts to measure individual elements of capital is that such measures may not adequately address the need to maintain the overall integrity of the environment which is dependent on aspects of a number of these elements.

A similar concern with system effects is evident in the concept of 'critical natural capital' (Pearce & Turner 1990). This is natural capital which, if destroyed, has profoundly damaging consequences. An example might be the ozone layer. In terms of welfare, it seems impossible that any quantity of man-made capital could realistically compensate for the loss of this critical natural capital. As a rule, therefore, even weak sustainability supporters should advocate safeguarding critical natural capital. The associated problems are twofold. One is designating what is critical natural capital; and the other is determining if it is being eroded or damaged to dangerous levels. These are issues of natural science, but it is important to note that the market economic system has no mechanism for including such critical safeguards within the system.

2.8.2 The sustainability outlook

Having reviewed some of the arguments surrounding capital within the sustainability debate, we can now return briefly to consider the basic definition of sustainable development. Without going through the numerous definitions that have been presented, we simply note perhaps the four most widely recognized possibilities (good reviews of the possible interpretations of sustainable development can be found in Reid 1995). Sustainable development can be taken to mean for society:

1 non-declining welfare over time;
2 non-declining levels of consumption over time;
3 constant or increasing stocks of natural capital; and
4 constant or increasing stocks of all capital (subject to critical natural capital constraints).

Aspects of these positions recur throughout the book and we do not discuss their implications any further here. Definitions 3 and 4 are in some ways more technical, means-orientated positions which stress what conditions may have to be fulfilled in order to achieve the goal-orientated definitions of 1 or 2. The relationship between welfare, consumption and forms of capital is, however, a complex one. In fact, we can note simply that all of these positions are subject to some criticism, both as general policy frameworks and as practical guidelines. In terms of their supporters, the 'strong sustainability' position 3 has generally been endorsed by environmentalists and the 'weak sustainability' position 4 by economists. While both these groups would probably endorse position 1 (though its purely human focus is not acceptable to some environmentalists), many 'antigrowth' proponents would positively reject position 2.

2.8.3 The Earth Summit and Agenda 21

A final chapter in the sustainability debate can be noted in the UNCED Conference on the Environment held in Rio de Janeiro in 1992, also known as the Earth Summit. The Earth Summit was the largest international gathering ever held and produced a number of important publications, although relatively few firm commitments. The scale of the event was, however, itself an indication of the seriousness with which environmental issues were being addressed.

Three documents emerged from the final Rio process: a broad non-binding statement of principles relating to development and the environment, called the Rio Declaration on Environment and Development; a declaration on forest management principles, also non-binding; and a detailed programme of action principles at national and international level, called Agenda 21 (meaning an agenda for the twenty-first century).

An important corollary was Local Agenda 21, an initiative which sought to bring concerns with sustainable development down to the local level. A key feature of Local Agenda 21 is its emphasis on grassroots participation in the process of sustainable development, a theme which is increasingly emphasized in both non-governmental organization and government activity. Analysts confronted with complex multi-dimensional problems, scientific uncertainty and differing value orientations amongst stakeholders, have increasingly been drawn to emphasizing participation as central to a new decision-making framework. Whereas highly technical and sophisticated scientific analysis may be appropriate to problems with clear parameters governed by well-recognized laws, the outcomes of development decisions are governed by large numbers of variables, and many of the relationships are essentially unknown. Although techniques such as environmental impact assessment and multicriteria analysis can go some way to cope with these uncertainties, local people working on problems at their own level of interest and experience are cited as having more rather than less expertise than outside analysts. These themes are developed in more detail in Section 10.7.

2.9 Summary

Economic growth, which is an increase in the scale of an economy, should be distinguished from economic development, which is a change in the structure or form of an economy. Forms of economic organization can be split into four broad types: self-sufficient, communal, feudal, and capitalist or market-based.

The development of economic theory can be divided into classical and neoclassical periods. The classical economists studied the use of the market mechanism in stimulating economic development; Smith was positive about the market's ability to maximize social welfare, whereas Malthus and Ricardo foresaw natural resources as limiting factors. Marx saw the market in terms of a power struggle over the methods of production, with both social and physical aspects. Whereas classical economists had mostly considered supply issues related to the economy, neoclassical economics introduced a fuller notion of consumer demand. Market prices were then seen to arise out of a balance of supply and demand. Environmental problems could be addressed in this framework by adjusting prices or rearranging property rights.

Modern insights in ecological economics hinged around the perception of absolute limits to the size of the economy as a result of natural resource constraints. Models such as Boulding's spaceship Earth and the Club of Rome's *Limits to Growth* emphasized such constraints. Following the Rio conference in 1992, the issue of defining 'sustainable development' and, in particular, the notion of the substitutability between different types of natural and man-made capital, has become central.

Further reading

Cameron, R. (1993) *A Concise Economic History of the World*. Oxford University Press, Oxford.

Daly, H.E. (1992) *Steady-State Economics*, 2nd edn. Earthscan, London.

Kula, E. (1998) *History of Environmental Economic Thought*. Routledge, London.

Meadows, D.H., Meadows, D.L. & Randers, J. (1992) *Beyond the Limits: Confronting Global Collapse, Envisioning a Sustainable Future*. Chelsea Green, Post Mills, VT.

Pearce, D. (1993) *Blueprint 3: Measuring Sustainable Development*. Earthscan, London.

Reid, D. (1995) *Sustainable Development: An Introduction*. Earthscan, London.

3 Economic Principles for Non-Economists

'If economists could manage to get themselves thought of as humble, competent people on a level with dentists, that would be splendid.' [John Maynard Keynes]

3.1 Introduction

In this chapter we introduce some of basic principles of economic theory. The principles of economics given here are all associated with what is known as neoclassical economics. Neoclassicism is currently the dominant school of thought in economic theory, and some of the historical developments leading up to it are given in Chapter 2. Here we outline the foundations of the neoclassical model of economic behaviour related to buying and selling decisions in a market, which leads to the theory of supply and demand. This set of principles, taken together, is both an explanation of how markets work and a justification for their use in organizing economic activity. The issues covered here are therefore central to understanding economics, not simply in terms of a technical explanation of market behaviour, but also to understanding the extent of moral and political justifications for basing decision-making on economic criteria, an important theme which resurfaces throughout this text.

3.2 What is economics?

Economics studies society's production and consumption of commodities, i.e. goods and services. Economic analysis must solve three fundamental problems: *what* commodities to produce, *how* to produce them, and *for whom* to produce them. Economics is thus concerned with the allocation of effort to produce and consume goods and services and the associated *decisions* made by the various elements of society: individuals, firms and the state. In studying the interrelations between production decisions and other factors, such as limited resources, technological development, and demands for commodities, economics attempts to improve the workings of the economic system. Such an improvement is characterized by an overall increase in *social welfare*. In economic terms, social welfare (or well-being) is determined by consumption levels (the quantities of goods being consumed) in society.

Economic decisions are thus concerned predominantly with balancing two things:

1 the efforts spent in producing commodities; and
2 the demand within society for consuming those commodities.

The complexity of an economic system arises because individuals can have an overall larger set of commodities to consume if there is *specialization* in production. As we do not, in general, live lives of self-sufficiency, we consume a whole host of commodities which we have not produced ourselves. The economic system enables individuals to exchange their labour for commodities, using money as the medium of exchange. It also determines how much individuals receive in return for their employment (their salaries) and therefore how much they can consume.

3.2.1 Scarcity

An economic system has a certain endowment of resources, such as land, manufacturing machinery (man-made capital), raw materials and labour. These resources are generally *scarce*. In economic terms, scarcity implies two important factors:

1 that there is a finite supply of the resource; and
2 that there are alternative competing uses for the resource.

Alternative uses means that there is then an *opportunity cost* entailed in using a resource for one use rather than for another competing one. The opportunity cost of any choice is the cost of the best alternative which must be forgone in order to satisfy that choice. To take a simple example: if I can spend my afternoon either sunbathing or making a wooden chair, deciding to go sunbathing entails an opportunity cost of the value of the chair that I might have produced. In this example my time is the scarce resource; using it for sunbathing entails an obvious lost opportunity for chair-making.

Scarcity therefore also implies that an individual's wants cannot all be satisfied simultaneously, so in the example above, I would clearly like *both* to sunbathe *and* to have the chair at the end of the day. If a commodity is scarce, then the economic system must determine some way of *allocating* that commodity between competing end uses. This is the very definition of the 'economic problem'. If there is no scarcity associated with the consumption of a commodity then it is termed a *free good*. Whether a commodity is a free good or not may well be context-dependent; spring water in remote hillside locations may be consumed 'freely' because supply is abundant, whereas (bottled) spring water in town takes resources to deliver and is 'scarce'.

3.2.2 The circular flow

As well as providing the labour input for production, individuals generally own the firms which operate the production processes, either as shareholders or entrepreneurs. Firms, like individuals, can be thought of as the centres of activity between which goods and services flow, and which both consume some goods (labour, raw materials) and produce others (finished goods, waste). The system of production of commodities within any single firm is decided by the managers of that firm, acting in response to the prices that they have to pay for their inputs, and the revenue from their output. Firms exist because they can carry out production more efficiently than individuals acting alone. The economic system determines the payments made to the owners of firms in return for providing land, machinery and/or materials: the payments are for the goods that the firm sells. Thus, conventional economics depicts the economy as a circular flow system, as represented in Fig. 3.1.

A useful summary of the discussion above in a formal definition of the discipline of 'economics' is given by Begg *et al.* (1998):

> 'Economics is the study of how society decides what, how, and for whom to produce.'

The central concern of economic analysis is *human* behaviour. Economists are concerned essentially with developing hypotheses about the relationship between important variables, such as demand for goods, knowledge about markets, purchasing power and many other factors.

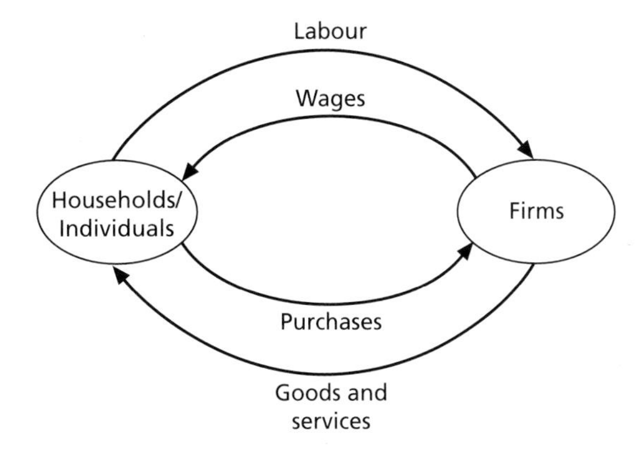

Fig. 3.1 The circular flow of the economy. Households provide the labour input and own the means of production. Firms produce the goods and services which are then distributed to the households.

3.3 Modelling

Economic hypotheses predict market behaviour on the basis of assumptions about individual behaviour. Such assumptions form the basis for building economic models, from which predictions can be made about the operation of the economy; for example, the way market prices move in response to goods shortages. A model is a simplified representation of reality. By making assumptions which simplify the details of real life complexity, trends and tendencies can be drawn out from data which would otherwise remain hidden. Models provide a framework for organizing the way a problem is to be addressed. The modeller isolates important features of the system and specifies relationships between these features on the basis of the stated assumptions; these can subsequently be tested against real market data.

One problem facing modelling and hypothesizing in the social sciences is the impossibility of undertaking controlled experiments. The subject under study for economists is actual human behaviour in a social environment, and thus while economists may be interested in one set of relationships—say, investment decisions and the interest rate—a large number of other factors will be influencing individuals' choices in the real world. Economists do try to establish what are termed *ceteris paribus* ('other things being equal') conditions—keeping all variables except one constant, and then predicting the response to a change in that given variable. However, in practice 'other things' are rarely, if ever, constant, and thus predicting the outcome of a number of interrelated factors is a highly problematic task.

An important feature of modelling is reliance on the *law of large numbers*. This 'law' is not a statistical law, but rather a methodological guideline which requires economists to study trends in *collective* rather than *individual* behaviour. Individuals are not robots, and therefore do not maintain the same behaviour and make the same choices from one day to the next, even if the external conditions are controlled. However, if a large enough group is analysed, the law of large numbers allows any one individual's variations in behaviour to be counterbalanced by equal but opposite variations by another individual. Economists then can synthesize data about a large number of choices to identify 'representative' or 'typical' individual behaviour. This variability means that the nature of economic laws is different from that of 'natural' laws. Natural laws apply in all circumstances, whereas economic laws are, strictly speaking, only tendencies derived from aggregating large numbers of observations.

If economists recognize the inherent limitations implied by assumption-formation and the law of large numbers then modelling can be a productive exercise. Correspondingly, modelling is clearly open to error if fundamental relationships are missed out of the original assumptions of the model, or are not captured by its variables. Economics has been criticized from this perspective for being prone to the *fallacy of misplaced concreteness* (Whitehead 1929). This refers to modellers forgetting the degrees of abstraction applied within their models, leading to unwarranted conclusions about reality.

Daly and Cobb (1991) have referred to this problem as one in which the fundamental structure of problem-solving leads to a misconception of the nature of the problem itself. Schumpeter's notion of a 'preanalytic vision' is pertinent: the preanalytic vision is the development of the set of concepts that are used to define a problem before we actually begin to solve the problem. As noted above, models provide a means for structuring the way in which a problem is addressed. If we do not recognize some aspect of reality within the initial model, such as an absolute limit to physical resources, the model will continue to provide guidance on ways of increasing consumption without ever recognizing this fundamental limitation (for further aspects of this issue see Chapters 2 and 11). Georgescu-Roegen (1971), speaking from the perspective of an energy analyst concerned with physical limits to growth, suggests that 'it is beyond dispute that the sin of standard economics is the fallacy of misplaced concreteness'.

3.4 Positive, prescriptive and normative statements

The kind of advice that economists give can be split into 'engineering' and 'ethical' issues (Sen 1987). Engineering issues are to do with the mechanics of the economic system and are concerned with specific relationships between variables. Statements related to engineering issues are called *positive*: they are based solely on logical deductions, starting with assumptions that define a model, and working through to conclusions based on established economic laws or tendencies. Such statements should be *value-free* or *objective*. If two economists use an identical model then they should arrive at the same positive conclusions.

Prescriptive statements concern the model's predictions of what can be done to achieve a particular defined aim. Such statements are indirectly value-laden in that the defined aim is based on some person's value judgement of what is important. However, a prescriptive statement does not make claim to one situation being better than another, only that in order to achieve a given aim a certain action should be undertaken.

In contrast, *normative* statements directly involve ethical issues and are explicitly value-laden. They concern what *should* be done on the basis of the model's outcomes. In order to illustrate this distinction, consider the following three statements concerning the state's taxation of cigarettes:

1 *If the state increases the tax which is applied upon cigarettes, then total consumption will fall.* This is a positive statement. The underpinning assumption is that cigarettes are 'normal' goods, and therefore obey the conventional laws of supply and demand (see Section 3.14).

2 *If the objective of the state is to reduce the consumption of cigarettes below current levels, then it should increase the tax applied to them.* This statement is prescriptive in that there is no inference that the state ought to set this objective, but only that a particular mechanism will help to achieve it.

3 *The state should increase the tax on cigarettes.* This statement (in isolation) is normative because it makes an explicit value judgement, namely that one state of affairs (higher tax) is better than another.

Conventional economics attempts to steer away from making normative statements, and indeed a highly popular textbook (Lipsey 1995) draws attention to this feature by defining itself as a text on 'positive economics'. It aims to be a 'mechanics' guide to the economy without engaging in much discussion of values.

Some socioeconomists, notably Polyani (1965) and Myrdal (1958) have denied that positive economics is in fact as value-free or objective as its proponents claim. They suggest that the choice of concepts that economists use, and the subjects they study, are themselves promoting particular visions of the way things should be. A focus on rates of wages in economic literature, for example, can thus be seen as an implicit promotion of the employee/wage-earner system of organization of labour. In this regard Polyani refers to the concept of a 'commodity fiction' in positive economics. The 'fiction', in his view, is the assumption that all goods can, in theory, be bought and sold in a market, even those that are not traded in markets, and therefore that everything has an implicit market price. He points out that other, non-Western cultures do not perceive many goods in these terms of ownership, buying and selling—for example, Native American Indian views on land ownership—and that economic analysis that rests on these assumptions is therefore promoting strongly value-laden views of how these goods should be handled.

Some of these issues are developed further in Chapters 4 and 5. Here we concentrate on outlining the basic features of conventional economics: the 'mechanics'.

3.5 Microeconomics and macroeconomics

The discipline of economics can be divided into two sections: *microeconomics* and *macroeconomics*. Microeconomics considers the allocative decisions taken by individual agents: consumers and firms. Microeconomists tend to focus on one particular aspect of these individuals. They examine the variables which determine why an individual demands any one commodity and why a firm supplies it. Microeconomics thus considers the circular flow in the economy at a microscopic level.

Macroeconomics takes a different perspective. It considers interactions in the economy as a whole. Macroeconomics considers such variables as aggregate employment, the average rate of price increases (inflation) and the foreign exchange rate. These are concepts which are familiar to the lay person from the daily news. Macroeconomists model the economy's behaviour at this macroscopic scale.

3.6 The allocative system

There is more than one way in which commodities can be produced and distributed in an economy. The two principal alternatives are the *market mechanism* and the *command economy*. A *mixed economy* brings elements of both together.

3.6.1 The market mechanism

A market need not be a geographical location. A market can be defined as *the interaction of individuals in a freely operating exchange of goods and services*. Crucially, it is characterized by the use of *prices* as a signal to the different groups acting within it. For individuals, price determines commodity consumption and, correspondingly, their decisions as workers about how much employment to substitute for leisure time. For firms, decisions about what to produce and how to produce it are also governed by price. The prices of resources are adjusted, through changes in *demand and supply*, so that resources are employed to produce those goods and services which society desires. The operations of the laws of demand and supply are analysed in Section 3.14.

How does price determine allocation within a market? In the market mechanism, *consumer sovereignty* exists, which means that consumers can freely determine for themselves how to spend their money—they are sovereigns over their own decisions regarding what to consume. Consider spring water. If an individual chooses to purchase a bottle of one particular brand of spring water from a shop, then he or she is *revealing a preference* for that commodity purchased from that shop. The market offers competing products which can quench thirst. The choice of that branded spring water over, say, lemonade *signals* to the respective drinks producers information about the individual's tastes and spending power. The purchasing decision answers the 'what' and 'for whom' production decisions.

The chosen shop has certain characteristics: the level of service, convenience, price competitiveness, etc. In the act of purchasing from that shop, the individual signals his or her preference for that particular balance of attributes over the ones offered by competing shops. If a sufficient amount of sovereign consumers' money signals a similar preference, then the competing stores will be bankrupted. Furthermore, the attractiveness of the chosen store's attributes stimulates other entrepreneurs to establish similar but competing stores, owing to the desire to make profits, profits being simply the difference between total costs (shop establishment, wages) and total revenues (sales) from the production process. The store is providing, or 'producing', the service of 'retailing', and the individual's choice answers the 'how' question for this type of production.

The store's staff member who serves the customer purchasing the spring water has chosen to be an employee of a firm, i.e. that store. The offer of employment was made by the firm's owner(s) because, given the 'price' of the employee (the wage), profits were expected to be higher as a consequence of employment. Conventional economic theory assumes that the single objective of firms is *and should be* to maximize their profits. The job offer was accepted because, given the wages and conditions offered by other competing employment opportunities, this one was the most preferable from the point of view of the new employee.

The spring water purchase demonstrates the operation of the market. Individuals are sovereign over their decisions, and prices alone determine the production process. Further, entry and exit of firms to the market is what the process of *competition* in capitalist society is all about. If competition exists, then firms which do not make sufficient profits are driven out of the market. This forced exit arises because the firm has not answered the fundamental 'what, how and for whom' to produce question sufficiently well when confronted with sovereign consumers who allocate their own money to select the commodities they prefer.

3.6.2 The command economy

In the command economy, the *government* takes responsibility for production and distribution. The former Soviet Union and Eastern bloc adopted this allocative system. State planning offices determine 'what, how and for whom' to produce. Detailed instructions are then issued to firms and workers. In its purest form, this allocation is achieved without *entrepreneurship* (the activity of individual firm managers to solve allocative decisions) and without the private ownership of property. Under capitalism, the organization and mobilization of resources to achieve and regulate production are left to entrepreneurs who are driven by the profit motive to solve these supply problems.

Solving every allocative decision at state level is an immense organizational task. What differentiates the two systems is the incentive structure. Under the market mechanism, a *hard budget constraint* exists. This means that a firm's costs of supplying a particular commodity cannot exceed its revenues in the long term; if they do, it goes bankrupt. Competition implies that both inefficient production practices and the production of commodities which sovereign consumers do not desire are weeded out.

Under the command economy, this consumer sovereignty is absent. Thus, in the case of spring water, the state administrator decides what quantity society should consume. This may or may not approximate to consumer preferences. Further, operating a command economy implies the need to *monitor* both the quality and quantity of what is actually produced. This costly, and imperfect, surveillance is unnecessary under capitalism: if a commodity is distinguished as substandard then it attracts a relatively lower price in the market; if the firm produces too much then there is a glut on the market and the price falls. Too low or too high a production level can imply a relatively high cost per unit output, and consequently an unattractive price to consumers. Thus, unlike their command economy counterparts, capitalist entrepreneurs must continually optimize their production processes, reducing costs and improving quality, as they are subject to the forces of competition.

3.6.3 The mixed economy

The market mechanism promotes consumer sovereignty with no government intervention, and the command system state control with no individual economic liberty. Between these two poles lies the *mixed economy* in which the state and private sectors interact in solving the economic problem. In reality, it is hard to find applications of either pure capitalism or pure command-and-control by nation states. Indeed, it is generally optimal in terms of *efficiency* to have a mixed economy, in that some tasks are better performed by the market and others by central planning authorities. Efficiency in economics refers to allocating resources so that a given objective is realized at least cost.

Consider the command economy first. What is the role of entrepreneurship under this allocative system? The central planning authorities can only cope with 'brush stroke' management, that is *estimating* the level of production quotas adequate to meet needs. If a product is an *intermediate* one needed to produce the *final* commodity, e.g. flour for bread, then its non-availability can halt the larger production process. Therefore there needs to be some slack in the system to compensate for imprecise estimation. In the former Eastern bloc's implementation of command economics, consumers were allowed some choice in that private markets selling agricultural produce coexisted with centrally planned agricultural cooperatives. Further, there were market-based incentives for factory controllers to achieve or better output targets. Thus, the former Eastern bloc nations operated in a mixed economy, but with a strong command element.

Under capitalism, there are three instances where state intervention is advisable: public goods, antitrust regulation and externalities. Each of these are positivist arguments based on the assumptions of market operation and aimed at correcting misfunctionings in the mechanics of those markets. State intervention might also be based on the normative concept of fairness, e.g. the reallocation of resources from the rich to the poor and the provision of certain services, such as health and education, irrespective of individuals' income levels. This is considered in more detail in Sections 5.2 and 5.3.

PUBLIC GOODS

Public goods are defined as goods for which consumption by any one individual does not detract from the ability of others to consume them. They are characterized by *indivisibility* ('all or nothing' provision) and *collective consumption* (individuals cannot be excluded from consumption). Examples of such commodities are street lighting, national defence and internal law and order. With regard to national defence, for instance, it is either provided for everyone in that state, or it is not provided for anyone. There is no opportunity for citizens individually to opt out of whatever collective decision is taken.

Importantly, if individual citizens were to be given the option of opting out, then there would be an incentive for individuals not to pay for national defence owing to the *free rider* effect. The free rider effect occurs when individuals can enjoy the benefits of a good while relying on others to pay for that good. In relation to defence, the contribution made by each individual citizen towards the national budget is insignificantly small in relation to the total amount needed to provide the service. Whether he or she contributes or not is unlikely to be the determining factor. Thus, regardless of individuals' preferences, it is in their *self-interest* to declare that they do not wish to contribute to a fund for national defence. If each individual follows this line of reasoning then national defence is not provided, even though the majority of the populace might actually desire its consumption. This is a socially suboptimal outcome, thereby justifying state correction through intervention.

ANTITRUST LEGISLATION

In order for the price mechanism to function efficiently, conditions must permit competition. *Perfect competition* is an economic 'ideal' state characterized by several factors, most importantly: a large number of buyers and sellers; 'perfect information' about goods and prices amongst these buyers and sellers; and costless entry and exit for firms into and out of the industry. This kind of situation is ideal because it ensures that no firm can then manipulate prices by controlling the market. Instead, competition acts to drive prices down to their lowest levels. The absence of perfect competition is the reason why, even in countries such as the USA which vigorously espouse the free market, the state enforces *antitrust legislation* to prevent large firms from manipulating their market power to suppress competition.

The antitrust argument applies in particular to industries which are *natural monopolies*. A monopoly is an industry in which there is only one firm operating: in effect, the firm simply is the industry. In the case of a natural monopoly, the efficiency of the firm's operation increases perpetually as its size increases: the larger the output level, the lower the cost per unit output. As an example, consider a railway network. The cost to a potential entrant to this market of establishing the physical infrastructure of rail tracks is prohibitively large. It is much cheaper to add a new station to an existing line than to construct a new line. In such cases, there are *barriers to entry* for new firms (that is, very high start-up costs) and thus no effective competition. Without state intervention the dominant firm could set prices as it wished, leading to high profits for the firm but a total provision of rail travel less than the socially optimal level. This arises from the lack of competition characteristic of a monopoly industry structure, and government regulation can help to correct it.

EXTERNALITIES

A third form of optimal state intervention is to counteract the existence of *externalities*. The state of externality exists when firms or individuals impose costs or benefits on other agents in society and when these costs or benefits are not compensated. If this happens then there is a chronic overproduction of those commodities which impose costs, and a chronic underproduction of those that accrue third-party benefits. The term 'chronic' implies that, without a change in circumstances (intervention), the situation will persist. The term 'externality' refers to the fact that the costs and benefits are outside or 'external' to the market.

Many of the tools of environmental economics are applied to correct this form of market failure, when prices do not reflect full production costs to society—there is a need to *internalize the externality* (this is discussed further in Chapter 14). An example of such a failure is pollution. If unregulated, firms will use the environment as a waste sink in which to dump the non-valuable outputs from production processes. This behaviour minimizes a firm's production costs. However, the capacity of the environment to assimilate this waste is limited (assimilation being the process of converting harmful substances into relatively benign ones over time). If the total flow of pollution exceeds the assimilative capacity then toxicity levels will rise, ultimately reducing the assimilative capacity itself, and the welfare of people is then affected. The firm is not compensating such people, even though it is affecting their welfare. Thus, there is a discrepancy between the private cost to the firm of dumping, which approximates to zero, and the social cost to society which is positive and potentially large. Societal welfare is increased if this discrepancy is addressed by making the firm pay in some way for whatever pollution damage it is causing.

There are also positive externalities. In the case of a positive externality it is the affected parties enjoying the benefit who compensate the producer in order to internalize the externality. Consider the example of a vaccine that has no negative side effects and immunizes against a communicable virus. The probability of someone who is not immunized catching this disease goes down when the number of people who are immunized goes up, because there are then less potential hosts for the virus. The reduction in numbers catching the disease is a benefit to society, but a non-monetarized one which is not 'internalized' into individuals' choices as to whether or not to pay for vaccination. Thus, there is an economically valid argument for the free state provision (or subsidization) of such services.

DISTRIBUTIONAL ISSUES

Even if functioning perfectly, markets may produce a level of *inequality* in consumption and income distributions which is considered unacceptable. The 'for whom' question is answered in the market mechanism by consumer sovereignty. Commodities follow those with money, not necessarily those with the greatest needs. Thus, the state might intervene in order to redistribute income through the taxation system, providing *transfer payments* as a social safety net for the state's citizens. Transfer payments are flows of money that do not form part of an exchange of goods and services; they simply shift income between groups in the interests of improving equity. These issues are considered in Section 5.3.

3.7 The assumptions of economic modelling: rational economic behaviour

The assumptions that economists make about how individuals make choices between commodities are important in understanding the application of economic theory to real world problems. This section sets out the conventional microeconomic analysis of individual behaviour.

3.7.1 The concept of the margin

Conventional economics assumes that rational economic individuals allocate their resources in order to maximize their overall *utility*; that is, the satisfaction derived from consumption. This is the basis of individual economic behaviour. What makes it particularly powerful is the concept of the *law of diminishing marginal utility*: the more of a commodity that an individual has already consumed to satisfy a want, the less is the extra gain in utility generated from consuming one more unit (the 'marginal' unit). In other words, the level of utility *increases decreasingly* with additional units of the same good.

This law can be applied with reference to the spring water example. The utility from drinking the contents of the first bottle is greater than that from drinking a second, which in turn exceeds that from drinking a third. If the consumption of one more marginal unit actually decreases the individual's utility then a state of *satiation* is said to apply. Consumer choice theory is only applicable before this state is reached; that is, subsatiation. It is assumed that individuals, having a limited budget, do not purchase commodities unless these commodities derive them utility. There are exceptions to this law. Commodities which are *addictive* do not exhibit diminishing marginal utility. If a commodity forms part of a *collection*, then the marginal utility of the item which completes the set is likely to be greater than the one proceeding it. Apart from these exceptions, the law is generally applicable.

It is important to distinguish between the *marginal* unit consumed, the *average* number of units consumed and the *total* number consumed. The average is simply the total number of units consumed divided by the number of consumers. Economic decision-making generally focuses on marginal analysis because the costs and benefits associated with the consumption or production of an extra unit of a commodity vary with the number of units already produced or consumed. Using total or average measures obscure this.

3.7.2 Equimarginal utility per budgetary unit

Marginal analysis is used to determine rational utility-maximizing behaviour for individual agents. Consider an individual who has a fixed budget and faces set market prices for commodities. In other words, the individual has no ability to influence the market prices through purchase decisions. This consumer is sovereign over his or her purchasing choices and can choose any affordable combination, or bundle, of commodities from all those on offer. *The condition for utility-maximization is that the marginal utility of the last unit of income spent on each of the purchased commodities should be identical.*

Perhaps the best way to explain why this condition should apply is to consider the situation were it not to apply. Imagine that the basket of commodities that this individual has purchased contains *m* muffins and *d* doughnuts. The unit of currency is the pound sterling (£). Muffins cost £*M* and doughnuts cost £*D*. Let the marginal utility derived from spending £*M* on the *m*th muffin be given by MU_M. If this amount, £*M*, were to be spent on doughnut(s) instead of muffins, then the marginal utility associated with these doughnut(s) is given by MU_D. Crucially, if MU_D is greater than MU_M then the individual's total utility would be increased by switching consumption away from muffins to doughnuts. Utility is maximized when MU_D is equal to MU_M; that is, identical marginal utilities for the last unit of income spent (£*M*).

We now take the analysis a stage further. If muffins cost twice as much as doughnuts (£*M* = £2*D*) then the marginal utility gained from the last muffin purchased (MU*m*) should be twice that gained from the last doughnut (MU_d). This can be formalized for any number of goods, each with a different marginal utility (MU) and price (P):

$$\frac{MU_{\text{good 1}}}{P_1} = \frac{MU_{\text{good 2}}}{P_2} = \frac{MU_{\text{good 3}}}{P_3} \quad (3.1)$$

$$\ldots = \text{MU per pound.}$$

The utility-maximizing process is one of increasing spending on those commodities which are relatively cheap, in terms of the marginal utility provided per unit pound spent, until their diminishing marginal utility means that they reach parity with the other commodities in the consumption basket. The last penny spent in *any* direction will then yield the same utility. This result has an important moral implication. It suggests that the freedom of individual consumers to allocate any of their resources, not necessarily just money but also time or possessions, as they wish *enables* consumers to maximize their own welfare by equalizing marginal utilities. The issue of moral justification for the market system is taken up at more length in Chapter 5, but it is useful to note this key element of its justification here.

3.8 Indifference curve analysis

In this section, the principles of rational utility maximization are extended to explain the *indifference theory of consumer choice.* Once the principle of indifference theory is understood, it can be used to derive an individual's demand curve, and individual demand curves can be used in turn to produce the overall societal demand curve for a given commodity. This societal or total demand curve provides one side of the supply and demand relationship which, when combined with a theory of supply, demonstrates how economic analysis solves the issues of what, how and for whom to produce.

Analysing the indifference theory of consumer choice requires the setting out of two concepts: the *indifference curve* (IC) and the *budget constraint.*

3.8.1 The concept of an indifference curve

As noted above, rational agents compare the utility derived from different commodities (*A*, *B*, *C* . . .) in order to determine their utility-maximizing consumption bundle. If just two such commodities are selected, say *A* and *B*, then combinations of different quantities of *A* and *B* can be depicted graphically as shown in Fig. 3.2. This figure shows what is known as an indifference curve. It shows combinations of goods, *A* and *B*, such that the individual's utility is the same for all these combinations. In other words, the individual is equally satisfied whether he or she has combination α or β, or indeed any of the other combinations along the curved line.

To consider this another way, assume initially that

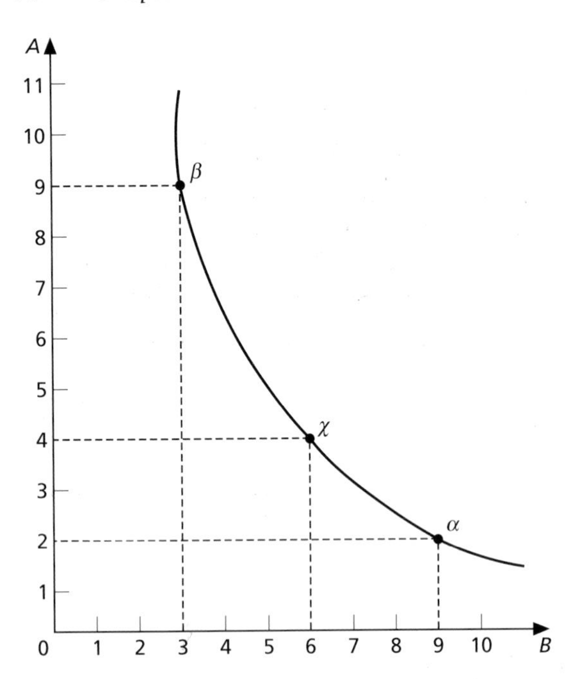

Fig. 3.2 A convex indifference curve in *A*/*B* space. The indifference curve is a locus of bundles, each comprising a different amount of *A* and *B*, between which the individual is indifferent.

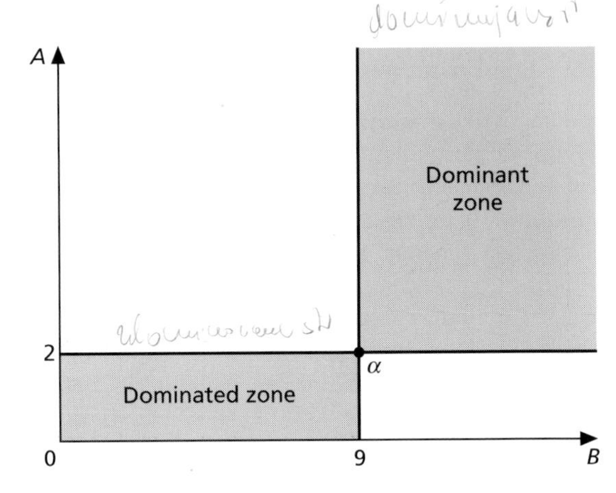

Fig. 3.3 The feasible region for indifference bundles. A point on the indifference curve passing through α cannot lie in either the preferred zone or the dominated zone because of the subsatiation condition.

the individual has a given starting endowment of (2*A*, 9*B*), point α in Fig. 3.2. If six units of *B* are taken away from this individual, then how many extra units of *A* would be required to exactly compensate for this loss? If the answer is seven units, then this corresponds with the point β (9*A*, 3*B*). As this refers to *exact* compensation, the individual is said to be *indifferent* between the bundles α and β. These are two points on one of the individual's indifference curves.

The rest of the points on this IC are mapped out by posing the same question to the individual but with varying initial endowments. Thus, the bundle χ (4*A*, 6*B*) is on this IC because, with reference to the point α, the individual requires two extra units of *A* to compensate for the loss of three units of *B*.

The number of extra units of *A* needed to exactly compensate for a unitary loss in *B*, while holding the utility level constant, is termed the *marginal rate of substitution* (MRS). The MRS changes along the IC. A bundle comprising a relatively high quantity of *A* compared to *B* is likely to have a high MRS. In Fig. 3.2 the MRS increases as the individual's consumption of *B* decreases, and thus decreases as the consumption of *B* increases. The MRS is equal to the slope of the IC.

3.8.2 The shape of the indifference curve

A conventional IC is downward-sloping and convex to the origin. The reason that ICs are downward-sloping is because of the assumption of non-satiation. Consider Fig. 3.3. The initial endowment α is again (2*A*, 9*B*). The location of β, or indeed any point on the IC, cannot be in the 'dominant zone' as the individual has strictly more of both *A* and *B* at every location in this zone as compared with α. Similarly, every bundle in the dominated zone represents strictly less of both *A* and *B*, and thus β cannot be located there. Thus an IC is downward-sloping.

The convexity condition is implied by the law of diminishing marginal utility, discussed in Section 3.7.1. Consider Fig. 3.2 again. It has been established that the individual is indifferent between α and β. The bundle χ (4*A*, 6*B*) has exactly three units less of *B* than α, and exactly three units more of *B* than β. At α the individual has a lot of *B* but relatively little *A*, whereas at β the reverse applies. Thus, the (negative) *marginal* utility of losing three units of *B* is smaller than the (positive) marginal utility of gaining three units of *B* at bundle β. Thus, to maintain a constant level of utility, the individual requires only, say, two units of *A* to compensate for the loss of three units of *B* from α ($4A - 2A = 2A$, $9B - 6B = 3B$), whereas at β three extra units of *B* exactly compensate him or her for loosing five units of *A* ($9A - 4A = 5A$, $6B - 3B = 3B$).

Importantly, the exact location of χ is dependent on the individual's tastes, but it must be located strictly

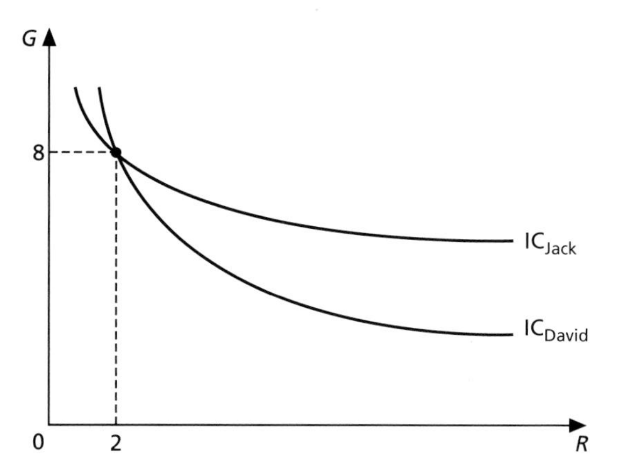

Fig. 3.4 Indifference curves for David (IC_{David}) and Jack (IC_{Jack}). The different slopes represent a difference in tastes.

below the line joining α and β. The negative slope and strict convexity of ICs are logical constraints based on the assumption of rational economic behaviour, whereas the exact curvature and slope of the ICs are determined by the individual's tastes.

How are these tastes reflected in IC analysis? Consider the following scenario. Jack is a fitness enthusiast who likes to maintain a balanced diet to aid his training regime, whereas David is a *bon viveur* who occasionally compensates for his indulgences with exercise. They both have an initial endowment of (8*G*, 2*R*), where *G* represents visits to the gym and *R* outings to the restaurant. Their respective ICs are depicted in Fig. 3.4.

This is not a comparison of utility between David and Jack. Given the initial endowment, each individual's independent preferences are mapped out. The reason that Jack's IC is shallower in *G*/*R* space is that, in comparison to David, he is willing to give up more outings to the restaurant for a given increase in the number of visits to the gym.

3.8.3 The indifference curve map

Each individual IC mapped out thus far has corresponded with a utility level associated with some given initial endowment. There is an IC corresponding with every feasible initial combination of *A* and *B*. Consider the scenario depicted in Fig. 3.2, with α at (2*A*, 9*B*). Let the IC passing through this point be termed $IC_{\alpha1}$ as depicted in Fig. 3.5. If the initial endowment is α_2 at (3*A*, 9*B*), then the corresponding $IC_{\alpha2}$ is shown. If this

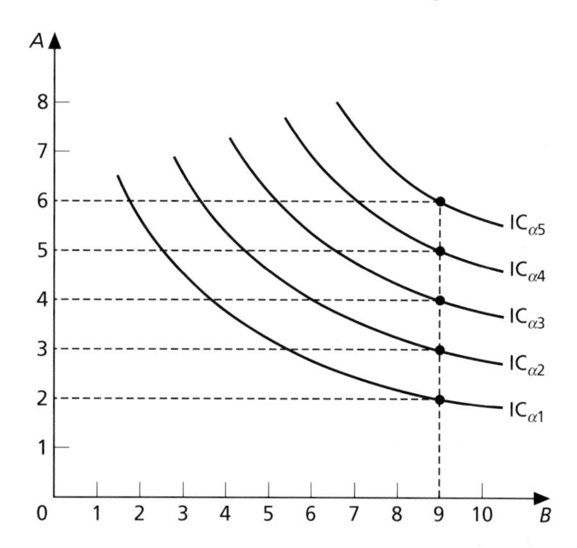

Fig. 3.5 The indifference curve map. There is one, and only one, indifference curve which passes through any given bundle of *A* and *B*.

procedure is repeated for α_3 at (4*A*, 9*B*), α_4 at (5*A*, 9*B*), α_5 at (6*A*, 9*B*), etc., then an *indifference curve map* for the individual can be graphically represented as in Fig. 3.5.

Every point in the *A*/*B* plane is located on one, and only one, IC. As those ICs further from the origin correspond with higher utility levels, the rational individual would want to consume a bundle on the highest possible IC. But what consumption level can he or she afford? We now introduce the *budget constraint*.

3.9 The concept of the budget constraint

A significant assumption of utility theory is that individuals' desires are assumed to be insatiable. In this case, more total consumption is assumed to raise the individual to a higher utility curve. What prevents an ever-increasing level of consumption is his or her budget, the total amount available to spend, which must be allocated between the competing commodities on the market. For the purposes of the model applied, the following assumptions apply.

1 This income is wholly spent on commodities *A* and *B* in the allotted time period, with no savings.
2 Market prices are fixed.
3 The individual's goal is to maximize his or her utility given the available budget.

For this example, let the unit cost of *A* be 2 and that

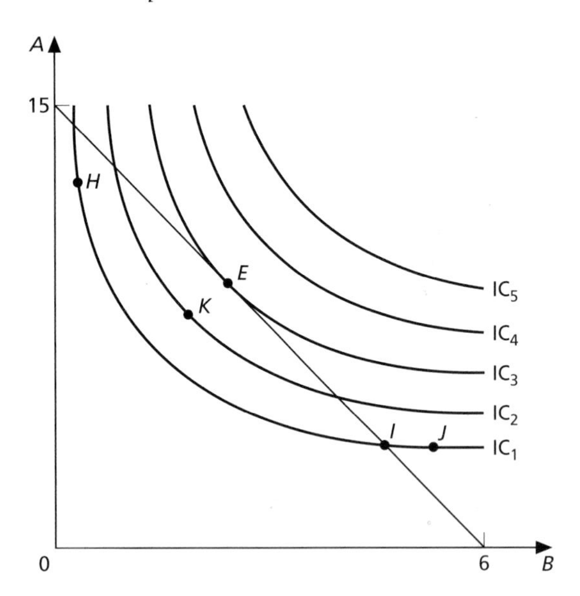

Fig. 3.6 Constrained optimization. The individual is maximizing utility at the bundle where the budget line is tangential with the indifference curve.

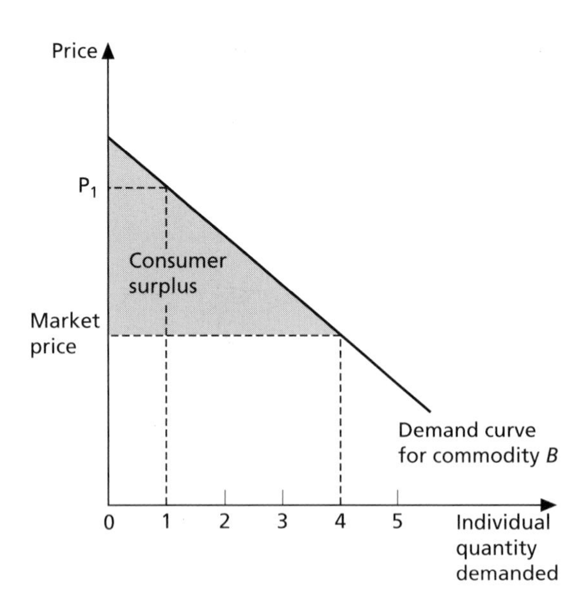

Fig. 3.7 An individual's demand curve for commodity *B* showing quantity demanded at each possible commodity price. The area under the downward-sloping demand curve and above the market price is the consumer surplus.

of *B* be 5, and let the individual's budget be 30 for the given period. If this individual were to spend the total budget on *A* then 15 units would be consumed ($30/2 = 15$), while a total allocation to *B* buys six units ($30/5 = 6$). More generally, the consumption of λ units of *A* and μ units of *B* is feasible if the following condition holds: $2\lambda + 5\mu \leq 30$. For the individual to spend completely the available income, the condition must be an equality: $2\lambda + 5\mu = 30$. Further, as assumption 1 designates zero savings, it is optimal for the individual to spend the entire budget. The budget line is depicted in Fig. 3.6. The slope of the budget line is termed the marginal rate of transformation (MRT) and is determined by the relative prices of the two commodities, *A* and *B*. It represents the number of extra units of *A* that can be purchased by giving up one unit of *B*. Figure 3.6 also has the individual's IC map superimposed on the budget line. *Constrained optimization* means that the individual chooses a bundle which is on the highest possible IC given the budget constraint.

In Fig. 3.6, the consumer can only consume bundles of goods lying on or below the budget line. Thus, no bundle on IC_4 or IC_5 is available. On IC_1 the individual has a sufficient budget to buy bundles *H* and *I* but not *J*. However, bundle *K* on IC_2 is also feasible, and is strictly preferred to both *I* and *H* as it is on a higher IC. The point of *constrained optimization* is given by the bundle *E* where the budget line is tangential to the IC. At this point, the following condition holds: MRS = MRT; that is, the slopes of the IC and budget line are equal. This selection alone maximizes the individual's utility given his or her tastes, budget and the commodity prices.

3.10 The individual demand curve

If this process of constrained optimization is repeated for different prices of *B*, then a *demand schedule* can be constructed for the individual, given his or her tastes. This schedule can be plotted as an *individual demand curve*, as given in Fig. 3.7.

This individual demand curve shows how much of *B* the individual demands at different price points. Demand, in economic terms, means that the individual has a desire for the commodity and is both willing and able to satisfy this desire through payment. A thirsty individual might crave a bottle of spring water, but this desire only becomes economically relevant if it is satisfied through the individual revealing a preference for this commodity by purchasing it. This is linked to the concept of consumer sovereignty, discussed in Section 3.6.1.

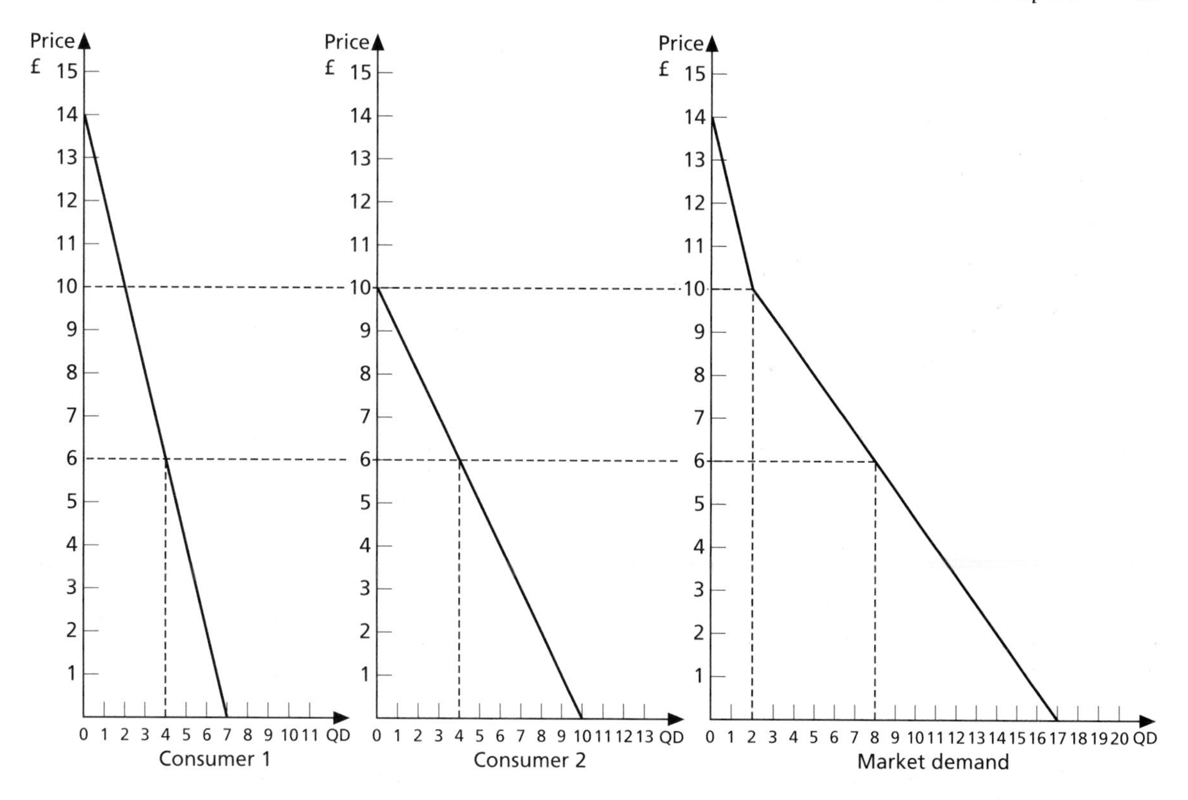

Fig. 3.8 The relationship between individual and market demand. The market demand curve is the summation of all consumers' individual demand curves.

3.11 Consumer surplus

Conventional market transactions generally entail individuals choosing how much of a commodity to consume at a given market price. Reconsider Fig. 3.7. At the prevailing market price, the individual demands four units of *B*, and pays the market price for each of them. Had the market price been at p_1, he or she would still have demanded one unit. The difference between what the individual is willing to pay and actually has to pay is called the *consumer surplus*, shown as the hatched area in Fig. 3.7. Consumer surplus is significant in assessing social welfare, when it can be added to a similar measure called producer surplus (considered below) to give an indicator of society's welfare arising from the market.

3.12 Market demand

Having derived individual demand from first principles, the next logical step is to consider the *market demand* for a commodity. The market demand is simply the aggregation of the individual demand curves for all consumers. To illustrate this, consider an economy in which there are only two consumers of a commodity *B*. The individual demand curves, for commodity *B*, for consumers 1 and 2, are depicted in Fig. 3.8 which shows quantity demanded (QD) on the horizontal axis. In this economy, the market demand curve is found by horizontal summation. At a price of £6 for *B*, consumers 1 and 2 each demand four units, thereby implying a market demand for *B* of eight units at that price. This process is repeated for all prices, and summed across all consumers. This gives the economy-wide demand for each particular commodity.

3.13 The supply side

Thus far we have derived the market demand for a commodity from first principles. In order to analyse how the market mechanism works, it is necessary to input the *supply side*, i.e. production by firms. There are many parallels to be drawn between the behaviour of firms and individuals. Whereas individuals are considered as utility-maximizers, firms are considered *profit-maximizers*. Profit for a firm is given by total production costs minus total sales revenues. In order to maximize profits, firms must optimize the allocation of their inputs into the production process subject to certain constraints.

In conventional economic analysis there are three factors of production which, when employed together and managed by an entrepreneur, achieve output: natural resource inputs (NK), such as wood; labour (L); and plant and machinery which are used to fabricate output, known as man-made capital (MK). Each of these factors has an associated market price per unit: for NK it is their extraction or purchase costs; for labour it is the wage rate; and for MK it is the purchase price or rental charge for the equipment. Associated with each factor is a *marginal physical product* (MPP); this is the extra physical output arising from the addition of one more unit of one of the factors, while holding the input level of the other factors constant.

To clarify this, consider the following example. Two cobblers working in a shoe repair shop mend seven pairs of shoes per hour between them. If one more cobbler is employed in the same shop, sharing the same tools, the total output for the shop rises to nine pairs per hour. The MPP of this extra unit of labour is then two pairs per hour.

Let us assume that the charge to the customer of shoe repairs is £5 per pair. In this case, the marginal revenue product (MRP), which is the *monetary value* of the extra unit of labour in terms of its contribution to the extra output, is £10 per hour. More formally: MRP = (MPP) × (price). If the wage rate, that is, the price of this factor input, is less than the MRP then the profit-maximizing entrepreneur employs this extra unit.

Each factor of production is subject to the law of diminishing returns, wherein the MPP of successive units of each input decreases. In this example, the first cobbler with sole use of the shop's facilities might have an MPP of four pairs per hour, whereas the addition of the second cobbler less than doubles the total output to seven pairs per hour, implying an MPP of three pairs per hour for this extra unit of labour. The MPP of the third cobbler is lower still, at two pairs per hour. An entrepreneur sets the level of employment for *each factor input* such that the MRP of the last unit employed is equal to the cost of employing it. Using this optimization procedure, the entrepreneur can determine the least-cost method for producing any given level of output. These least-cost combinations of factor inputs are described as *technically efficient*.

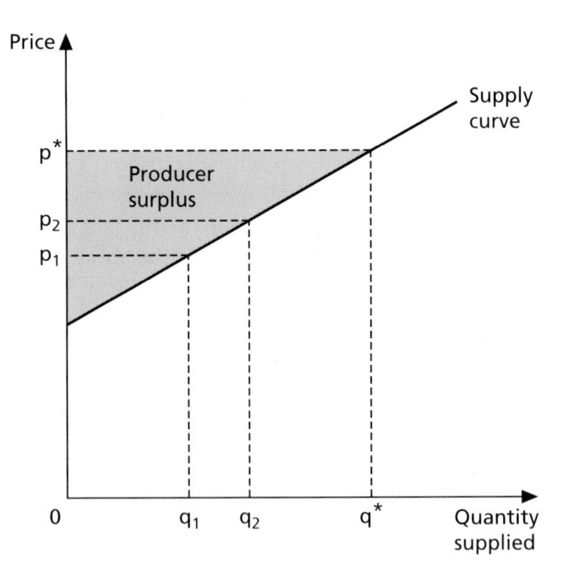

Fig. 3.9 A conventional supply curve showing how much is supplied at different prices. Producer surplus is given by the shaded area.

Technical efficiency describes how a profit-maximizing producer operates, but not how much is produced. Intuition suggests that more is produced when the market price for the output is high, and vice versa. This implies that a conventional supply curve is upward-sloping in price–quantity space, as in Fig. 3.9. An individual supply curve depicts how much output any one producer is willing to supply at each possible price. A market supply curve simply aggregates the individual supply curves of all producers of a given commodity in a market. Individual supply curves will vary depending on the skills and resources of different firms.

Just as the area *under* the demand curve measures the surplus of the demanders of a commodity, the area *above* the supply curve denotes producer surplus, as in Fig. 3.9. Let us assume that the market price for the commodity is given by p^* in Fig. 3.9. The producer would supply unit q_1 for price p_1, but actually receives the market price p^*. The producer surplus (PS) for this unit is thus (p^*-p_1). Similarly, the PS for unit q_2 is given by (p^*-p_2). For unit q^*, the PS is zero as the producer requires price p^* to supply this unit, and that is the exact (market) price obtained. Summing for all the units up to q^*, it is apparent that the aggregate PS for the individual producer is given by the hatched area above the supply curve. If we aggregate individual supply curves—just as we aggregated individual demand curves in Section 3.12—then we construct a market

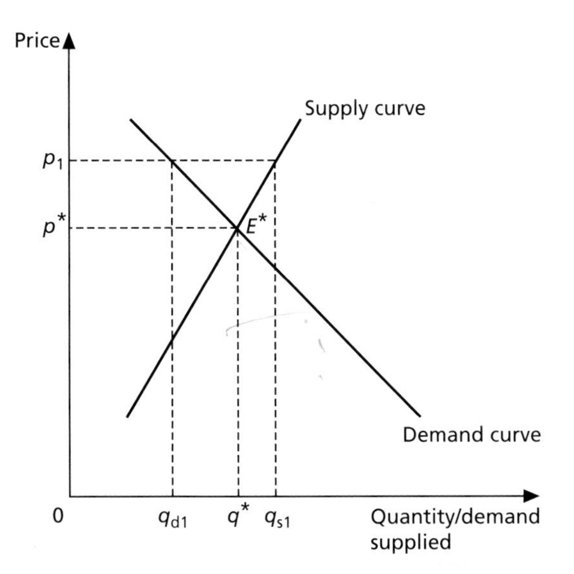

Fig. 3.10 The interaction of market demand and supply curves: equilibrium analysis. The point E^* is a stable equilibrium position, where quantity demanded equals quantity supplied.

supply curve. The total producer surplus for the industry is the area above the market supply curve.

3.14 Supply and demand interactions: equilibrium analysis

We have already considered the *market mechanism* as an allocative system using price as a signal to both consumers and producers (see Section 3.6.1). But how is price determined? The answer is that the *interaction* of demand and supply determines price. This is the realm of *equilibrium analysis*. An economic equilibrium is a situation where all the agents are choosing the best possible actions for themselves, given their preferences and the available opportunities, and where each agent's behaviour is consistent with that of the others.

We have analysed the process of constrained optimization which results in an individual demand or supply curve for each agent in a market. The market demand and supply curves for a given commodity, resulting from the aggregation of these individual curves, are presented in Fig. 3.10.

The point E^* in Fig. 3.10 is the equilibrium point, implying that the equilibrium level of supply/demand is q^* and the market price p^*. Why is this the case, and what exactly does it imply? Well, each agent is assumed to be a *price-taker*, which means no individual supplier or consumer has sufficient influence in isolation to affect the market price. This is the defining characteristic of a *competitive market*. However, although each individual's choice is insignificant in isolation, the actions of all agents together determine this equilibrium price, also termed the *market-clearing price*. This is the only price where the market supply is equal to demand, and geometrically is given at the interception point of the two curves.

The reason that this is an equilibrium is that the curves represent the optimal choices of the agents involved in the market, and these choices are mutually consistent in that the market clears at this price; that is, demand equals supply. At any price/quantity other than p^*/q^*, these two conditions do not both apply. Take, for instance, price p_1 in Fig. 3.10. At this price, the quantity demanded is q_{d1} and the quantity supplied q_{s1}. These quantities are mutually incompatible because supply exceeds demand, and there is thus a glut on the market (q_{d1}–q_{s1}). This excess supply forces down the market price if there is a competitive market.

This is a simple but powerful model of supply and demand interactions. The model can be adapted to accommodate non-competitive markets, wherein one or more agents have some level of market power. However, such adaptations are beyond the scope of this text. What we have analysed is how the market mechanism operates, under certain assumptions, in order to deliver the optimum production and consumption decisions amongst the members of society.

3.15 Summary

Economics concerns itself with the optimal allocation of resources in society. Economic analysis must solve three fundamental problems: *what* commodities to produce, *how* to produce them and *for whom* to produce them. There is a circular flow between firms (which produce commodities) and individuals in households (who consume commodities and own firms). The dominant mechanism for resource allocation is the price mechanism. Price acts as a rationing signal to both firms and households. If demand rises then price rises, and vice versa; if supply rises then price falls, and vice versa. The price mechanism relies on the market functioning properly. But the market fails in certain cases, one being the unregulated emission of pollution. Such failures generally imply the need for state intervention.

The economic model of behaviour assumes that

individuals try to maximize the utility, or satisfaction, derived from consumption, and firms to maximize the profits from production, given the constraints of prices and budgets. This assumes that consumers chose between bundles of goods to consume by comparing the marginal utilities associated with each good. Such consumer behaviour can be modelled using indifference curve analysis.

Economics as a discipline strives to be 'positivist', that is value-free and objective. As economics is concerned with human decisions in social systems, it relies on the law of large numbers to model general trends in behaviour. However, the discipline has been criticized for the 'fallacy of misplaced concreteness', which involves forgetting the assumptions involved in economic model-building, and which can lead to unwarranted conclusions about real-world behaviour.

Further reading

An elementary introduction to economics:

Begg, D., Fischer, S. & Dornbusch, R. (1998) *Economics*, 5th edn. McGraw-Hill, Maidenhead, Berkshire.

A more advanced introduction to microeconomics:

Varian, R.H. (1999) *Intermediate Microeconomics: A Modern Approach*, 5th edn. Norton, London.

A critique of economics discussing systems theory:

Clayton, A.M.H. & Radclife, N.J. (1996) *Sustainability: A Systems Approach*. Earthscan, London.

4 Ethics and Environmental Philosophy

'I plead not for the suppression of reason, but for a due recognition of that in us which sanctifies reason.' [Mahatma Gandhi]

4.1 Introduction

Moral philosophy both pre- and postdates the study of ecology and economics. Palaeolithic hunter-gatherers were both economists, husbanding scarce resources to meet their needs, and ecologists, intimately aware of the natural environment that surrounded them. It seems unlikely, however, that they had much time for moral theorizing. Yet, as an academic study, the discipline of moral philosophy predates both economics and ecology by over 2000 years. The word 'academic' itself stems from the name of the garden, the Academe, where Plato first taught the principles of philosophical investigation.

Why is an understanding of ethics and environmental philosophy important? There are at least three areas in which ethical and philosophical considerations are fundamental to the analysis of environmental problems.

1 *Goals and policy objectives.* In public choice situations, policy-makers must operate with a set of value criteria by which to judge the outcomes of different decisions. How these criteria are decided upon, what they comprise and how they are evaluated are therefore important influences on the direction of policy and planning decisions. Regardless of the many different belief systems prevailing in society, policy-makers must, in many cases, select certain specific values by which to guide their decisions.

2 *Institutions and social values.* All societies operate with some form of shared cultural values, which are apparent in formalized institutions, such as the system of law, the specification of individual rights regarding ownership, political representation, and many others. These institutions define the parameters within which members of society must act, in terms of their freedoms, expectations and responsibilities. Environmental philosophy encourages an examination at a fundamental level of the interrelationship between these kinds of institutions and their effect on both the environment and on environmental attitudes.

3 *Motivation and individual behaviour.* The set of moral beliefs or value systems held by individuals influences the way in which they behave. Some understanding of what values are important to individuals, why they believe those values are important, and how they translate those values into action is therefore critical to developing policies and social institutions that will work effectively in solving real-world problems.

These three areas are obviously related. In a democracy, policy-makers are likely to design policy in response to the values of the people who elected them, although they need not represent them precisely. Conversely, over time, policies may themselves influence values; for example, through the effects of education. In every society, value systems evolve through time in a dialectical process; beliefs are challenged and defended, sometimes tradition is re-established and sometimes new ideas gain acceptance.

This chapter covers four related areas of interest to the development of ecological economics and which can roughly be included under the title ethics, but more properly perhaps be called environmental philosophy. The first is the meaning and practice of ethics, understood as the 'science' of right action or goodness, which is outlined in Section 4.2; a brief history of ethical thought is then outlined in Section 4.3. Section 4.4 reviews the foundations of environmental philosophy, summarized in the historical development of environmental attitudes. As modern philosophical responses to the environment are relatively recent, Section 4.5 treats them as a separate topic which defines some modern philosophical responses to the current environmental debate.

4.2 What is moral philosophy?

It is helpful to make a distinction between ethics (or moral philosophy), which is the study of the foundations of such concepts as 'goodness', 'right' and 'justice'; and morality, or morals, which more correctly refers to the particular social code or system of conduct prevailing in a society. Thus Victorian morality refers to an identifiable set of attitudes towards 'the morally good life' held around the turn of the twentieth century in Britain, encompassing views on work, the role of women, attitudes towards other cultures and other social values. These attitudes are influenced by moral philosophy, but not defined as a coherent and specifically rational moral system. The following section is concerned with moral philosophy and moral systems; what might be called the development of environmental morality (that is, social attitudes to the environment) is reviewed in Sections 4.4 and 4.5.

The concerns of moral philosophy can be divided into three closely related strands. These encompass the fields of *metaethics*, which considers the meaning of moral statements; *substantive ethics*, concerned with developing coherent theories of justice and goodness; and *practical ethics*, which attempts to determine the 'best' or 'just' course of action in specific situations.

4.2.1 Metaethics

Metaethical questions are seldom raised outside the philosophy departments of universities. They ask: what *kind* of statement is a sentence like 'You should do X', or 'John is a good man'? Note that this is a question regarding the type, not the content, of moral statements. It is not looking for a justification or a description, such as 'Do X because it is charitable', 'John is good because he does not tell lies'. Rather it concerns the logical form of moral statements, and whether it is possible to 'justify' any moral judgement at all. For example, are moral judgements relative to cultures or are there absolute standards of right? The answers to metaethical questions will determine the foundations of ethical theories and how moral dilemmas are resolved. Although seemingly very abstract, metaethical issues can be fundamental to the way in which moral debate proceeds. Metaethics is introduced in more detail in philosophies of value surveyed in Chapter 5.

4.2.2 Substantive ethics

Substantive ethics, sometimes referred to as normative ethics, is the study of the contents (the 'substance') of moral theories. It is concerned to see if such theories are coherent, and what general approach they adopt to moral questions, such as what kind of actions are right and what kind of things are good in themselves. As such, most informal ethical reasoning is generally regarding substantive issues.

Two primary categories are generally distinguished in substantive ethics. The first is *deontological* or rights-based theories, which are concerned with the form of moral duties, and the structure of rights (the word 'deontology' comes from the Greek word for duty, *deont*). These theories specify duties and obligations to act in a certain way, or respect certain rights as a matter of moral imperative. A correct or morally justified action is one that fulfils a pre-established duty or obligation, without regard to the results of following that duty.

The alternative string of substantive theories are *teleological*, or consequentialist in outlook (the word 'teleology' also derives from the Greek, *telos* meaning end or purpose). These theories are concerned with judging actions according to their outcomes. Thus, in the teleological ethical theory of utilitarianism, the best or most morally correct action is defined as the one that results in the greatest overall gain in happiness for society. This is a judgement based on the predicted consequences of the action, rather than on fulfilling a duty (e.g. keeping a promise) regardless of its consequences.

It is worth noting in passing a further category—*procedural* theories—because these are sometimes contrasted with substantive theories. Procedural theories derive moral justification for actions from the procedures or processes used to arrive at decisions, leaving open any judgement about the content of the decision itself. The distinction is most clearly seen in the case of democratic judgements; any candidate who secures the most votes in an election is the 'morally right' candidate to take up the post, not because their manifesto was the best in itself but simply because the election process itself—the procedure—makes the winner the 'just' or morally endorsed candidate.

4.2.3 Practical ethics

Although ethics is technically a branch of applied philosophy, in the sense that it concerns the correct way in which to act, or in Plato's famous phrase 'no small matter but the way we ought to live', it is useful to separate the process of determining right or just actions in particular situations from the theories that justify right actions in general. This distinction is useful for two reasons. First, it may be no small task in itself to determine the correct application of a general theory in a particular case. The process of cost–benefit analysis is primarily a procedure for applying utilitarianism in public decisions, but it is hardly a simple task (see Chapter 8). It is, however, in one sense a straightforward exercise in practical moral evaluation, given a number of simplifying starting assumptions.

Secondly, and perhaps more significantly, there may be conflict in practical situations over which of a number of moral theories is the most compelling. This element of 'pluralism' does not entail that because there is debate over which theory is most appropriate that no theory is correct; merely that no one theory will be appropriate in all situations. Pluralism accepts that moral situations and moral justifications are complex, and aims to see which substantive moral theory offers the most compelling arguments in a particular case, while allowing that a number of different theories all present some justification.

Practical ethics has to solve at least three problems, derivative from substantive ethics. It has to identify the moral issues at stake in a decision; it has to determine what moral principles are involved in these issues; and it has to see how those principles should be applied in this particular instance. Fields of applied ethics, such as medical and business ethics, have grown up in response to recurring issues of this kind.

4.2.4 The scope of ethics

Moral philosophy is best considered as an investigative process—a 'science of right action'—rather than as providing a set of predetermined answers to moral questions. In many instances, it may not be possible to say which of a choice of actions seems morally right without the extensive analysis of both natural and social sciences.

It should also be noted that moral philosophy is concerned with matters both of personal action and public policy. Privately, it provides tools with which the individual can analyse the rights, duties and responsibilities involved in any given situation. In the public domain, it concerns the design of just institutions and the procedures used for determining choices in which a number of individuals have a stake. Although these latter issues are ultimately derivative from the former, in that they must be answered by some view of what in a very general sense constitutes a good or desirable life for people, they can usefully be separated.

4.3 A brief history of Western ethics[1]

For convenience, the development of ethical thought within the Western tradition can be split into three broad phases. The first, stretching from the pre-Socratics in around 400 BC to the middle of the Dark Ages at around AD 500, is dominated by Greek thought exemplified by the work of Plato and Aristotle. Their principal concern is with analysis of the virtues and the role of man in society, in particular the Greek *polis* or city state.

The medieval period, by contrast, is dominated by the institutions of the Christian Church and the preoccupation of philosophers with the Divine and its relationship to events on Earth. The distinction between Christian moral philosophy and Christian morality is therefore a fine one; ethical theory is rooted primarily in deductive reasoning anchored in the supreme perfection and rationality of God, and founded on biblical interpretation.

The dominance of the Church and other monumental social and political institutions, and the certainty and rigidity in moral life that they ensured, was steadily eroded from around the end of the sixteenth century, leading into what can be called the modern period. This slowly evolved into a renewed concern with earthly interests from the divine concerns of the medieval period. However, man as an individual, as distinct from man's place in society, became the centre of ethical and political interest.

4.3.1 The Greeks

The cultural background of classical Greece, where Socrates first began his philosophical investigation of morality, was one of long-established political and social hierarchies. The Ancient Greek heroes are seen as defending their authority by appropriate behaviour: strength, courage and cunning as each situation

[1] This text focuses primarily on moral philosophy as it has developed in Western Europe since the Greeks—and within what is known as the 'Anglo-American' tradition. The ideas within this tradition are what currently dominate the moral landscape in advanced Western economies and they are explored for this reason. From an environmental perspective there is increasing interest in alternative philosophical systems, such as Taoism, Buddhism and Hinduism, and the ways in which they perceive of the human–environment relationship. For more details of these approaches see Armstrong and Botzler (1993).

demanded. To be virtuous, in these situations, was to exhibit a state of 'flourishing' or 'living well', encompassing not only good conduct but peace of mind, respect, honour and material prosperity, maintaining a relationship of appropriateness between rank, power and overall bearing.

In the two centuries prior to the birth of Socrates, the sphere of Greek influence had been extending steadily throughout the Mediterranean into Africa and Asia. The encounter with new and diverse cultures revealed moral and political systems wholly at odds with the accepted norms of Athenian civilization. Such diversity questioned the absolute quality of the prevailing social norms of Greek society, and initiated some radical reappraisals of them. Socrates was pre-eminent amongst philosophers in pursuing the issue of fundamental knowledge of moral truth and justification and, by careful questioning of those who were considered to possess virtue, he revealed they could give no satisfactory account of what it was they actually possessed. From such investigations he drew his famous conclusion—'All I know is that I know nothing'—for which the revered oracle at Delphi proclaimed him as the wisest man in Greece.

PLATO

Socrates himself published nothing, but his pupil Plato, who founded the Academy after Socrates' death, recorded many of his dialogues. In *The Republic* (Plato 1987), Plato offers a detailed account through Socrates of what constitutes goodness and justice for the individual. He stresses that true justice must be sought for its own sake; that is, not as a route to power or a means to riches, but as the kind of life that is the best that can be led.

To discover this 'best' kind of life he considers what makes a city or society 'just' (the best it can be), on the basis that it will be easier to discover justice on a macroscale and, by analogy, this principle can then be applied to the individual. His eventual conclusion is that justice lies in the harmonious coexistence of the three classes that constitute the state: the guardians or rulers; the warriors or auxiliaries; and the economic or working class. When each of these fulfils its function—in particular, observing the diktats of the guardians who are trained in philosophy—a virtuous and happy state results. By analogy, the three classes of the state correspond to the three primary elements of the individual personality: intellect (the guardians); spirit or temperament (the auxiliaries); and appetite or desire (the workers). A truly just individual is one in whom these three elements are in balance, with the intellect regulating conduct and the other elements of personality as appropriate.

Three features of this conclusion are worth noting. The first is its emphasis on hierarchy and stability. It is significant that Plato lived through the traumatic events of the Athenian war with Sparta, which saw the humiliating defeat of the state and imposition of a ruthless tyranny in place of the legendary Athenian democracy. Plato himself, in essence related to the ruling class, was profoundly concerned by the excesses of the tyrants. His emphasis on rule by reason and a stable social structure can be seen as an intellectual defence of what was good in the Athenian state.

Secondly, the insistence on intellect and rationality as the prime elements in defining justice follows from Plato's *epistemology*, or theory of knowledge. For Plato, there existed two worlds: the phenomenal, or real world of concrete everyday experience; and the noumenal world of abstract ideas and 'Forms'. True knowledge was of abstract Forms and ideas, not of everyday experience. Whereas things in the real world changed continually, the world of ideas was both perfect and unchanging. The rules governing mathematics are perhaps the clearest example of this. No humanly drawn circle is perfectly circular in the phenomenal world, but a perfect circle, as an idea, can exist in the noumenal world. Defining a good state in relation to an ideal concept of harmony, as discovered through the analysis of *The Republic*, was to make the definition rational, absolute and immutable.

A third feature of note is that, in contemporary terms, Platonic justice resembles in many ways a modern Freudian view of a 'healthy personality' (Kenny 1973). According to Plato, a just or virtuous individual is guaranteed contentment and a happy life because his personality always remains resilient and unperturbed, despite any misfortunes. In contrast, the unjust person who is not in a state of harmony is subject to limitless desires which can never be satisfied, ignorance of the highest forms of happiness, and is condemned to perpetual conflict between his desires and his long-term interests.

To modern eyes Plato's conclusions may seem at best paternalistic, and in political terms they translate into an approach wholly totalitarian (great detail is supplied on how the young should be educated to fit the system). Its significance, however, is greater than this conclusion simply stated. Plato's analysis establishes the critical issues within an active philosophical inquiry into justice. To what extent is it possible to develop an objective or rationally grounded measure of justice or goodness? And what is the relationship between individual morality and the political system? In providing answers to these and other questions, Plato realized his inquiry touched not simply on substantive ethical positions, but on the theory of knowledge, the concept of rationality, the composition of human nature, and the nature of social institutions including the effects of education, custom and political involvement. All theories that subsequently followed thus had to provide at least implicit answers to these issues.

ARISTOTLE

Plato was succeeded at the Academy by Aristotle, a physician's son from Macedonia in northern Greece. Aristotle studied under Plato for 20 years, subsequently founding his own school, the Lyceum. During an academic career of outstanding diversity he literally founded the science of biology, while contributing to virtually every sphere of knowledge. He also tutored the young Alexander the Great, who went on to extend the boundaries of the Greek empire to a position of unparalleled power.

Whereas Plato was an idealist, Aristotle was primarily a natural scientist. His investigative method was therefore empirical, the exact opposite of Plato's abstract theorizing. His principal book of moral philosophy, *The Nicomachean Ethics* (Aristotle 1975), contains detailed analyses of the virtues. Moral philosophy was thus for Aristotle a practical enterprise, concerned with determining judgements about correct actions in particular circumstances.

Two principal threads are developed to this end. In Aristotle's epistemological system there are four causes of all events: material, efficient, formal and final. Thus, a house has a material cause (its composition of brick, timber, tiles, etc.); an efficient cause (the workers who build it); a formal cause (the plans by which it is constructed and laid out); and a final cause (to provide shelter). Whereas the first three concern the substance, process and structure that explain why things are as they are, the final cause is teleological, or purpose-driven. In a contentious step, Aristotle suggests that all objects have a characteristic 'function' or 'quality', which personifies the kind of thing they are: for fish, it may be swimming; for birds, flying. In this case, goodness or final cause lies in exemplifying this characteristic activity: for man, the activity that singles him out is rationality. Thus, exercise of the intellect is a supremely good activity. To act according to rational deliberation is to be virtuous.

The second related element of Aristotelian practical moral philosophy is the concept of virtue defined as the 'golden mean' between two vices. This suggests, for example, that the virtue of courage stands midway between cowardice and foolhardiness; magnanimity stands midway between selfishness and liberality. This view can be interpreted as a rather placid doctrine of 'moderation in all things', but this is not fully accurate: Aristotle puts emphasis on acting in a way *appropriate to the situation at hand*. This may call for great courage in certain situations, or justifiable anger or love. These things can only be judged in a particular situation, when the golden mean can be determined through a combination of experience, including emotional reactions, and intellect. Although there are evidently a number of problems with acting on this doctrine, and with its coherence with all virtues (is it possible, for example, to possess too much compassion?), its influence on medieval thought was highly significant.

In summary, Greek ethics is primarily concerned with the individual as an active member of the city, a social being imbued with a rational sense that acts as a guide towards appropriate conduct. Although Aristotle suggests at the end of the *Ethics* that the philosopher can be supremely happy in solitary contemplation, the principal emphasis is on the meaning given to virtue as an aspect of social relations.

In conclusion, it is interesting to note a distinction made by Aristotle between two forms of 'economic enterprise'. Economic activity in general, as shown in Plato's treatment of the 'labouring class', was not considered with high regard in Greek or Roman society. Aristotle, however, distinguishes between *oikonomia*, which is the careful management of the household's resources so as to increase its value over time, and *chrematistics*, which is the manipulation of property and wealth to maximize exchange value. Although neither of these activities is the highest good that humanity seeks, the virtuous citizen will practise *oikonomia*, and scorn *chrematistics*, because the latter is 'unproductive' and contributes nothing to the health of society as a whole (see Daly and Cobb 1989).

4.3.2 The medieval period

The thousand years from AD 500 to 1500 may, in Western ethical terms, be equated with an almost absolute dominance of the Christian Church and moral issues equated with the correct interpretation of God's will on Earth. It is frequently considered a period of relatively little innovation in ethical theory, and shown in its least favourable light by the Jesuit practice of casuistry—the detailed and finicky tracing out of the implications of scripture, often to justify a desired result rather than uncover an unknown truth. Thus, whereas the Greeks were concerned with life within the *polis*, treating ethics as a species of political philosophy, medieval ethics turned its attention both upward, towards God, and inward, to examine the proper motivation and conscience of the individual.

The intellectual work of early Christian moralists was primarily a response to Greek thought. St Augustine (AD 354–430) utilized Plato's distinction between the real world and the world of Forms to distinguish between the realm of earthly desires and the divine kingdom of Heaven. The work of the late medieval period's more famous intellectual, St Thomas Aquinas (1225–74), emphasizes the twin concerns of scriptural interpretation and the correct motivation of man, deriving his system from Aristotelean notions of final cause. In Greek cosmology, man was situated in a 'great chain of being' in which plants were used by animals, and animals by humans; Christian doctrine under Aquinas extended the chain upwards to angels and ultimately God.

As a whole, the period is perhaps more significant for the pervasive influence of Christian morality, particularly the virtues of humility, selflessness and devotion, in maintaining a relatively unchallenged and stable social and political hierarchy. The Renaissance and the Reformation were both pivotal points in the transformation of this broad ethic. The scientific flowering of the Renaissance, extending human potential and intellect, and the extensive geographical explorations of the sixteenth century, opened up new worlds of possibilities. The rigid moral guidelines of the Church, particularly regarding wealth and humility, gave way to a view of human endeavour as a celebration of God's work on Earth. This new-found confidence flowered in the Protestant work ethic, and national economic policies of wealth creation through domination and trade.

4.3.3 The modern period

The status of moral guidelines was set under immense strain by the unprecedented changes of the seventeenth century in political, religious and social circles. The confidence in moral teachings, established by the Church, dissolved, or at least receded, to usher in a period of renewed moral theorizing. Ethical thought was open to a much greater individualism than was possible under the domination of traditional Church authority. Furthermore, advances in scientific thinking offered new models of thought to those concerned with moral issues.

The arch-sceptic David Hume (1711–76), a Scottish academic, took the analytical procedures of the natural sciences (he considered Newton the greatest mind of the modern period), and applied them to moral philosophy. His unique 'discovery' was that the source of moral values lay not in the properties of any particular object or action itself, but in the reaction of the observer to those properties. Moral judgements thus became relativistic: things were no longer good or bad in themselves, but merely as judged by the human reaction to them. To call something 'good' was simply to say it aroused in the observer a feeling of approval. This feeling was the source of moral value, not any objective element of what was being observed.

Hume's (1978/1777) position (treated at length in Section 5.2) was both radical and, in the course of time, immensely influential. The change in status of moral judgements from rationally derived truths, as they had been since Plato, to simple emotional reactions, entirely undermined conventional beliefs. This undermining was only partly softened by Hume's claim that the moral feelings were 'universal dispositions', which meant that in general all humans responded with similar feelings to actions that had moral significance. This meant that there remained some criteria on which universal moral judgements could rest, but they no longer had any *rational* foundations.

KANT AND THE ETHICS OF DUTY

Hume's influence was slow to build, and two further developments are central to establishing the forum for moral debate in the twentieth century. The first is Immanuel Kant's (1724–1804) development of the formal nature of moral rules or guidelines. For Kant (1949/1785), morality can be more or less summarized into a notion of fulfilling duties or obligations to an unwritten, but rationally perceived 'moral law'. The law is formulated as what Kant calls the 'categorical imperative', a binding command which identifies the just course of action. It is therefore a deontological ethical system which, in contrast to the Greek identification of just actions with 'the good life', suggests that moral actions, though compelling without exception (categorically imperative), are frequently contrary to personal interests.

The categorical imperative appears in essentially two formulations:

1 act as if you are acting in accordance with a universal law; and

2 act as if you are a member of the 'kingdom of ends'.

The concept of a 'universal law' means that any moral judgement must be impartial, impersonal and equitable. In deciding if an action is right under the Kantian system, it must be possible to formulate the principle of the action as a universal guideline to follow in all circumstances.

Kant suggests that in purely logical form the categorical imperative can offer guidance if interpreted properly. For example, lying cannot ever be defended, as it might be under a consequentialist ethic to avoid perhaps unnecessary suffering, because to formulate some variation of 'telling lies when convenient' as a universal law entails a logical contradiction. That is, if 'convenient lie-telling' is formulated as a universal law, it denies its own possibility because 'convenient lie-telling' can only function on the assumption that everyone tells the truth. If it is known that everyone lies when it is convenient to do so, the framework within which truth and falsity operate collapses. It is not therefore possible to will 'telling lies' as a universal law, and so such activities are always immoral. This is ingenious logic, but the popular influence of Kant's doctrine lies in his second formulation of the categorical imperative which is, fortunately, more straightforward.

The idea of legislating for a 'kingdom of ends' can be summarized in the simple diktat: 'Treat others as ends rather than means'. The emphasis on treating people as ends not means stems from Kant's concern with the 'dignity' of moral agents. While recognizing that in society people frequently 'use' each other—to obtain services, and by buying in labour—in a mutually respectful society this use actually enables other people to pursue their own ends (this theme is apparent in the economist Adam's Smith's work; see Section 2.4.1). An institution such as slavery, however, is unacceptable because it requires some people to have no ends themselves, but only to act as means for their masters. Kant states:

> 'In the kingdom of ends [i.e. a moral world] everything has either a price or a dignity. Whatever has a price can be replaced by something else as its equivalent . . . whatever is above all prices, and therefore admits of no equivalent, has a dignity.' (Paton 1948)

For Kant, humans alone possess this unequivocal dignity or ultimate value. Kantian ethics, by stressing the ultimate dignity or value of human agents, has thus become fundamental to the development of democratic constitutions which specify the minimum rights compatible with securing the dignity of citizens.

UTILITARIANISM AND THE ETHICS OF WELFARE

In contrast to Kant's emphasis on duties and rights irrespective of personal inclinations, the utilitarian school of ethics that developed in the mid-1700s centred entirely on the happiness of the individual. Whereas Kant's doctrine invoked universal laws that had to be followed regardless of circumstances, thus respecting human dignity was paramount whatever the costs of accepting that position might be, the utilitarians advocated examining the details of each individual situation to see what action would increase overall the total sum of happiness in society. Good actions could only be determined by examining their consequences.

The utilitarian ethic was given its clearest statement by Jeremy Bentham (1748–1832), who proposed in the *Introduction to the Principles of Morals and Legislation* (Bentham 1970/1789) that 'nature has placed mankind under the governance of two sovereign masters, pleasure and pain', and that a 'hedonic calculus' should add up the sum of pain and pleasure resulting from any action and judge its value accordingly. The foundation of moral action was thus derived from the principle of utility: 'that principle which approves or disapproves of every action whatsoever according to the tendency which it appears to have to augment or diminish the happiness of the party whose interest is in question' (Bentham 1970/1789). In terms of social policy, the happiness of a community was simply given by the sum of the happiness of its individual members.

A more sophisticated utilitarianism was developed by John Stuart Mill (1806–73), also an influential economist, who made an important, and contentious, distinction between the *quality* of pleasurable experiences. Although retaining the basic 'calculus' approach ('actions are right in proportion as they tend to promote happiness, wrong as they tend to promote the "reverse of happiness" ' (Mill 1962)), he also suggested that certain kinds of pleasure—in particular, intellectual pursuits—were inherently more satisfying than others. This is not a theme adopted by modern utilitarians, but it has some connections with environmental ethics explored further in Section 5.5.2.

Utilitarianism was highly significant, not simply as a moral theory but as a tool of social reform. In contrast to the medieval period when ethical teaching was dominated by the clergy and other educated classes who thus wielded considerable power over individuals, utilitarians were in the vanguard of a social reform movement that championed highly democratic and humanitarian principles. In a consequentialist ethic, the king's happiness counted for no more than the pauper's; institutions, such as the land ownership system, had no ethical underpinning beyond the happiness which they supplied. If other systems offered greater potential gains for happiness, utilitarianism put moral weight behind the need for reform.

Having traced some of the prominent developments in ethical thought, central to modern political, legal and economic thinking, we now consider as a separate theme the evolution of Western attitudes towards the environment. These are attitudes or a sense of environmental morality which define the human relationship to the natural world.

4.4 Cultural attitudes and the context for Western environmentalism

The development of Western cultural attitudes towards the natural world is not in any sense monolithic. In all ages, a complex web of religious, social and economic influences have affected different groups and encouraged different belief systems, all of which have to some degree shaped the views of the present. A cultural overview therefore can do no more than suggest what, with hindsight, seem to be the predominant movements and hint at their complexity. Such an overview, however, is extremely valuable in both expanding the range of issues for debate within the current analysis of environmental issues, and in representing them in some brief context of the development of Western attitudes to nature. For introductions to individual views see Merchant (1980), Passmore (1974) and Pepper (1996).

4.4.1 The pre-Christian perspective

The pre-Christian relationship to the natural world in Europe can be summarized by two countervailing forces. The predominant attitudes to nature in pre-Hellenic (Ancient Greek) societies appear to be a superstitious or a religious reverence for nature. This was expressed in forms of *animism*—natural things each inhabited by their own spirits or gods. The early Greek myths represent gods as frequently inhabiting natural forms, and sacred areas, such as groves and springs, were identified as sites of worship. These animist spirits needed to be respected in order to prevent disruption to the business of life.

In many ways counter to this world view, the early philosophers of Ancient Greece initiated a philosophical tradition of metaphysics that was concerned with 'essences' of matter and wholly disinterested in the perceived phenomena of the real world. Metaphysics is the branch of philosophy concerned with questions regarding the nature of existence. In this field, a radical presentation was given by Plato in his theory of Forms, which separated the 'real' world of theoretical truths from the illusions of mere 'physical' reality (an inversion of the normal view, which is that the physical world is real and the mental world somehow less real). Although religious matters remained nature-orientated (the Roman system, with its passion for order, had a divine bureaucracy of major and minor gods governing all aspects of agriculture from plough to harvest), the highest intellectual respect was reserved for theoretical subjects, such as mathematics, dealing with 'real' truths, and practical and naturalistic studies were considered relatively inferior.

4.4.2 The Christian tradition

In many ways the most significant shift in cultural attitudes towards the natural environment can be considered that occurring between pre- and post-Christian values, though the issue is contentious. The Christian religion embraced a critical change by shifting the religious emphasis away from nature-related spirits to a God who was *transcendant* from nature. Nature itself no longer had any intrinsic spiritual significance; it was only as an example of God's splendour in creation that it retained a divine connection. This was a fundamental shift in spiritual emphasis, away from the animistic beliefs held towards nature in earlier periods, and the complex nature-entwined gods of Greek mythology.

The impact of the transition to Christian values built slowly throughout the Middle Ages and its influence was pervasive. Although essentially animistic beliefs undoubtedly remained an active influence (even nineteenth-century foresters in Germany are recorded as asking the forgiveness of trees before felling them), in social terms Christian ideology determined not simply moral values but economic policy (on, for example, usury and fair prices), and attitudes to science and technological innovation.

Two central doctrines regarding the natural world, and derived from the Book of Genesis, were predominant in the medieval period. The first, and most significant, was the concept of the rule of nature: 'Thou shall have dominion over the fish of the sea, and over the fowl of the air, and over the cattle, and over all the earth and over every creeping thing that creepeth upon the earth' (1 : 26). Because God had created the Earth for human use—essentially human dominion—then humans were free to do with it as they liked. Although other sections of the Old Testament can be interpreted as advocating a role of stewardship rather than domination for humanity (the animal-loving St Francis took up this tradition), the established medieval position was firmly in favour of absolute dominion.

The second doctrine, supporting the established political and economic hierarchies of the feudal system, concerned the human predicament since the Fall. Whereas in Eden the Earth had spontaneously provided all the necessities of life, now as a result of original sin mankind could only secure comfort through continual labour. The image of people as fallen from grace and thus trapped in their situations was appropriate to the rigid hierarchy of the feudal system, in which the Bible itself had to be mediated by the Church's teaching. The notion that human labour with the Earth was a necessary consequence of original sin resulted in the early distrust of labour-saving machinery. New inventions from Asia, such as the waterwheel and clock, were initially viewed with deep suspicion. For Genesis insisted that 'through thy act the ground is under a curse. All the days of thy life thou shalt win food from it with toil; thorns and thistles it shall yield thee, this ground from which thou dost win thy food. Still thou shalt earn thy bread with the sweat of thy brow. . .'.

4.4.3 The Scientific Revolution

The rebirth of learning that started with the Renaissance in Italy around AD 1500 and developed into the Scientific Revolution, brought into question these established medieval values in two areas. In science, new discoveries by Galileo (1564–1642), Bacon (1561–1626) and later Newton (1642–1717) revised important aspects of the traditional world view. Bacon asserted, 'For man by the fall fell at the same time from his state of innocency and from his dominion over creation. Both of these losses however, can even in this life be some part repaired; the former by religion and faith, the latter by arts and sciences'. The 'sweat of the body' was to be replaced by 'the sweat of the brow', applied so that 'the human race recover that right over nature which belongs to it by divine bequest'.

In philosophical circles the French philosopher, Rene Descartes (1596–1650), began his experiments with radical doubt, finding secure knowledge, like Plato, not in the physical world but only in mental processes. He drew attention to the 'primary' qualities of matter—mass and length—as the only true properties they possessed, whereas the secondary qualities that made them attractive and wonderful—colour and texture—were ignored. The division between reason and natural processes initiated by Plato became a central aspect of philosophical reflection. Knowledge, both religious and secular, was a matter of intellectual theory and 'seeing past' appearances. Descartes went so far as to suggest that animals themselves were essentially only 'bio-mechanisms', thus incapable of pain or consciousness. Of the scientific endeavour, he intended 'a practical philosophy by means of which . . . we can . . . render ourselves the masters and possessors of nature' (Descartes 1996/1641).

The scientific developments most clearly characterized by Bacon and Robert Boyle (1627–1691) opened up the possibility that nature could be adapted to serve mankind's will more effectively. This new optimism, which might have been seen as sinful pride in earlier thinking, coincided with a 100-year period which saw unprecedented change in political structures with the English Civil War, in religious teaching with the Reformation, and in economics with the increase in overseas trade and nation states. The new-found confidence and optimism changed people from mere adjuncts to nature, unable to develop it beyond subsistence, to individuals as masters and controllers of it. This mastery was given a further religious angle by assuming that it was in fact human destiny to *perfect* nature, which was in a waste state and needed to be refined by human design. The formal gardens of the eighteenth-century Age of Enlightenment presented this ideal state of nature in which geometrical perfection replaced natural variation. Far from being secular in outlook, however, the work of scientists, such as Bacon and Newton, retained a religious emphasis; it was an act of devotion to God to reveal the workings of His creation, in particular the beautiful simplicity of the first principles of physics upon which all else relied.

Simultaneously, Calvin's development of Protestantism can be seen as producing a climate more favourable to economic expansion in several ways. The Protestant 'work ethic' reinterpreted wealth generation as a glorification of God's work in man, rather than as a sinful exercise in material acquisition. The emphasis on individual conscience paved the way to the spirit of free enterprise on which economic prosperity flourished. Although respect for the virtue of charity retained caring for the community as a social responsibility, the pursuit of wealth was no longer restricted either by religious teaching or social status. The emphasis was instead on an individual path to salvation. The first works of political economy, such as Adam Smith's *The Wealth of Nations*, appeared.

4.4.4 Romanticism and the secular return to nature

Throughout the development towards the Industrial Revolution, nature had been represented in primarily two forms: agricultural lands, symbolizing the need to work since the Fall; and wilderness, either as royal estates or as uncultivated land, symbolizing threat, darkness and evil. At around the turn of the nineteenth century, however, the sense of victory in the 'war with nature' appeared to engender a sense of regret at what was being lost in the process. Those who had received the full benefits of the period of agricultural expansion had the leisure to regret the opportunities for appreciating the wilder aspects of nature.

The spiritual aspect of nature therefore returned in a secularized form in the Romantic movement which swept Britain and Germany at the turn of the nineteenth century. With the conditions of the newly industrializing mill towns degenerating, the relationship to nature was given renewed significance by the artists and writers who criticized the loss of spiritual significance derivative from the contemplation of nature.

Although confined to a small and privileged elite, and insignificant in its impact on the political or economic imperatives of colonial expansion and domestic industrialization, the Romantic movement represents the forerunner of more active environmentalism exemplified by writers such as William Emerson, Henry Thoreau and John Muir. It is at this point that many themes which are still current in modern environmental debate developed, and some of these are introduced in the following sections.

4.5 Ethics and modern environmentalism

The Industrial Revolution saw Western environmental concern, which had previously been preoccupied with taming wilderness and improving agriculture, given two new emphases. One was resource conservation and preservation, aimed at either rationalizing or reducing the human impact on natural resources now that industrial development was expanding so rapidly. The other was the improvement of urban environmental quality, particularly air and water quality, as the supply of both had been shrinking with the Industrial Revolution.

Whereas the environmental quality problem was essentially one of pathology, where human well-being was directly affected by poor conditions as a result of human activity, the preservationist issue was more problematic. The problems of urban air quality were 'brown–green' issues, where the need for action was obvious and quickly gathered political support. Preservation, on the other hand, was a 'green–green' issue, where two different basic philosophies—economic progress and nature preservation—essentially collided. For areas of the Western world, development was reducing one kind of scarcity (availability of cheap consumer goods) but initiating another (loss of wilderness and biodiversity). As the technical problems of improving air and water quality were tackled, they were accompanied by philosophical and institutional problems, involving how to trade-off between production of tangible consumer goods and destruction of intangible environmental ones.

In the following sections we outline some of the main ideologies regarding the use of the environment that arose in response to these issues.

4.5.1 Environmental ideologies

Two distinctive themes can be found in early stages of the environmental movement's concern with the natural heritage: conservation and preservation. Both ideologies were entirely human-centred, but in many cases they drew radically different conclusions on resource use.

CONSERVATIONISTS

Early conservationists like Gifford Pinchot (1865–1946), founder of the US Forestry Service, emphasized the responsible utilization of natural resources for the social good. In this sense, conservation was a response to what he perceived as the irresponsible development practices of unrestrained business enterprise. He stated that:

> 'The first great fact about conservation is that it stands for development . . . Conservation does mean provision for the future, but it means also and first of all the recognition of the right of the present generation to the fullest necessary use of all the resources with which this country is so abundantly blessed.' (Pinchot 1910)

Pinchot went on to state explicitly that 'the object of our forest policy is not to preserve the forests because they are beautiful . . . or because they are refuges for the wild creatures of the wilderness . . . but . . . the making of prosperous homes'. (Pinchot 1910).

PRESERVATIONISTS

In contrast, preservationists, such as John Muir (1838–1914), Ralph Waldo Emerson (1803–82) and Henry Thoreau (1817–62), favoured the complete protection of unique landscapes and ecosystems by emphasizing the spiritual and cultural value they held for mankind. Muir offered as a defence of wilderness on cultural and aesthetic grounds his belief that 'thousands of tired, nerve-shaken, over-civilized people are beginning to find out that going to the mountains is going home; that wilderness is a necessity and that mountain parks and reservations are useful not only as fountains of timber and irrigating rivers but as fountains of life' (Muir 1898). The prime directive for preservation was thus the value of the 'wilderness experience'. In political terms, however, this was difficult to equate with the economic value of timber. Pinchot himself deplored what he saw as the waste of natural assets represented by designations of national parks.

It is evident in the efforts to establish the first national parks in America that the balance between pioneers on the American frontier and preservation was difficult to establish. The system of rights to land acquisition that Thomas Jefferson had envisaged a hundred years prior to the great Western expansion in North America required the granting of freeholdings to all those who took up the challenge of turning the undeveloped wastes of the Western frontier into cultivated land (a system that was itself founded on Jefferson's view of small-scale farming as nurturing core moral values for civil society). The advocates for Yellowstone National Park had to declare that the park lands were of no agricultural merit in an attempt to appease this view, stating that 'the withdrawal of this tract therefore from sale or settlement takes nothing from the value of the public domain, and is no pecuniary loss to the government, but will be regarded by the entire civilized world as a step of progress and honor to Congress and the nation' (Hargrove 1989).

Similar sentiments were expressed by the Romantic poet William Wordsworth in Britain, calling for the Lake District to be recognized as a 'national property'. Critics in America responded that if the land was uncultivatable, it stood in no need of protection from would-be settlers and official designation was unnecessary. The debate was resolved in part because although the preservationists' economic argument was overstated, the argument from an agricultural perspective was not compelling either.

ECOCENTRISTS: LEOPOLD'S LAND ETHICS

The debate between responsible management and absolute protection was given a different perspective in the writings of Aldo Leopold, a forest manager who published *A Sand County Almanac* in 1949. Leopold is commonly regarded as the founder of an ecologically orientated ethic which 'changes the role of *Homo sapiens* from conqueror of the land community to plain member and citizen of it'. In this sense, he suggested that 'all ethics so far evolved rest upon a single premise: that the individual is a member of a community of interdependent parts . . . The land ethic simply enlarges the boundaries of the community to include soils, waters, plants, and animals, or collectively: the land' (Leopold 1949).

The 'land ethic' thus replaced an anthropocentric (human-centred), rights-based ethical approach with an ecocentric, community-based one. Leopold expresses its central tenet as: 'A thing is right when it tends to preserve the integrity, stability and beauty of the biotic community. It is wrong when it tends otherwise.' Expressed in this way, the land ethic is a simple substantive theory, directly applicable to human interaction with the environment.

Some recent commentators have suggested that the land ethic is a clear example of a pragmatic approach to land management, rather than a fully substantive ethical theory. Given the difficulties encountered by managers in controlling complex systems, the land ethic emphasizes the importance of working with nature and managing for system function rather than specific objectives. As a game manager, Leopold developed a strategy aimed at maximizing the number of deer within his area, principally by eradicating wolves. 'I later realised', he wrote, after seeing the impact of over-grazing on forest regeneration, 'that just as the deer fear the wolf, so perhaps a mountain may come to fear its deer' (Leopold 1949).

ANIMAL RIGHTS

In contrast to the holistic emphasis of the land ethic, the animal rights movement which gathered momentum during the 1960s was underpinned by an extension of traditional utilitarian theory to sentient animals. As Bentham recognized 200 years earlier, 'The day may come when the rest of animal creation may acquire those rights which never could have been withholden from them but by the hand of tyranny. The question is not, Can they *reason*? nor Can they *talk*? but, Can they *suffer*?' (Bentham 1970/1789). Proponents saw the protection of sentient animals as the logical conclusion of a hedonic calculus ethics which had already been expanding its constituency from property owners, to slaves, women and now everything capable of feeling pain. More recent advocates have also considered the extension of Kant's notion of dignity to animals as important. The animal rights movement has established itself as a separate branch of applied ethics, and it is probably fair to say that its impact on the wider environmental movement has been limited by its single issue focus.

4.5.2 Philosophical environmentalism

The academic development of environmental ethics, which established its own journal, *Environmental Ethics*, in the mid-1970s, was spurred on by two key publications, both of which summarized the nature of the environmental dilemma by representing it as a fusion of conflicting ecological, economic, political and social forces. Subsequent arguments can be traced back to answering issues raised in these two articles, focusing on institutional and cultural impediments to establishing a more sustainable society.

THE TRAGEDY OF THE COMMONS

Garret Hardin's 'The tragedy of the commons' (1968) represented the human predicament as one of the selfish interests of individuals predominating over the collective interest of society, to the detriment of all. Why should the shepherd grazing common land, he asks, refrain from adding yet one more sheep to his own flock to increase his personal profits, even if this activity when carried out by all the shepherds will easily surpass the carrying capacity of the common lands? Individually each shepherd stands to benefit by adding an extra sheep and if anyone voluntarily abstains he will lose the potential for profit while others will continue to gain by abusing the system. This is a case, Hardin argues, of ecological degradation that has no 'technical solution', unless the rules governing the situation are changed to ensure carrying capacity is not exceeded.

The modern environmental problem as conceived by Hardin is thus ecological collapse, driven by short-sighted economic rationality, unaffected by ethical constraints, requiring a political solution. The institution that is at fault is what Hardin calls 'common property', or what economists term 'open access'—resources which have no ownership and are therefore subject to no control (ocean fisheries are the clearest modern example; see Section 12.5). The emphasis of Hardin's essay is that in a competitive environment moral or benevolent behaviour selects for its own extinction, and society must accept a guiding principle of 'mutual coercion mutually agreed upon' that forcibly constrains freedoms in many aspects of life—for example, numbers of children and levels of consumption—just to ensure survival.

THE HISTORICAL ROOTS OF OUR ENVIRONMENTAL CRISIS

In 'The historical roots of our environmental crisis', Lynn White (1967) analysed the historical and cultural influences that led to environmental problems. His conclusion was, in a very close parallel to Hardin, that the modern predicament was a conflict between social freedoms and ecological constraints: 'Our ecological crisis is the product of an emerging, entirely novel, democratic culture. The issue is whether a democratized world can survive its own implications.' In White's analysis, the principal problem is the loss of a significant *cultural* relationship to the natural world, traced back to the teaching of Christianity. Whereas Hardin suggests institutional reform to *forcibly* control an economic free-for-all which he sees as untenable, White suggests rediscovering an environmentally benign cultural attitude that will encourage *voluntary* restraint and shift the direction of democratic development onto a sustainable path.

White thus suggests that 'we shall continue to have a worsening ecological crisis until we reject the Christian axiom that nature has no reason for existence save to serve man'. In a direct echo of Hardin's view that there is 'no technical solution', he states: 'What we do about ecology depends on our ideas of the man–nature relationship. More science and more technology are not going to get us out of the present ecological crisis until we find a new religion, or rethink our old one.' However, whereas Hardin's proposed solution of 'mutual coercion' is an appeal to human *self-interest* in the interests of survival, White argues that it is only by *transcending* self-interest and promoting an alternative interest—that of the well-being of nature itself—that the problem can be resolved.

ECOFEMINISM

The term 'ecofeminism' was coined in the mid-1970s to identify a movement in feminist philosophy which drew parallels between the perceived social and political domination of women in society, and the 'domination' and destruction of nature. Environmental problems could be seen as the result of stereotypically male characteristics, in particular brute strength and rationality exemplified in hard technology and reductionist science, dominating those associated with nature and the feminine, such as love, intuition, reverence and emotion. Reversing the trend of ecological damage involves redressing this imbalance in attitudes both to women and to nature; this change can be taken forward through reform of social relations both between men and women, and between all humans and the natural world, advocated by ecofeminism.

This very brief summary masks a wide range of views that have since developed within ecofeminist thought. Perhaps the strongest division is between those ecofeminists who emphasize the distinctions between men and women, and identify women as in some ways 'closer' to nature through experiences such as childbirth, and thereby more able to promote an ecologically responsible ethic. Others reject what they see as this 'women/natural vs. male/rational' stereotyping, and instead emphasize the need for a more 'gender neutral' approach to social development in general.

Within both these approaches, there is a strong emphasis on the importance of what can be called holistic rather than dualistic views of human personality. That is, the traditional division between body and mind, with the physical life as essentially separate from the mental life (a view associated with Descartes), is seen as a damaging and false division; these aspects of personality are much more integrated and interactive than is generally assumed. Ecofeminism thus provides a psychological and social critique of what are considered traditional domineering attitudes to nature and to women. The rationalist individualistic behaviour assumed in economic models and theory (Section 3.3) is seen as a classic example of a male-orientated methodology which abstracts only certain aspects of the human personality (predominantly rationality and self-interest), thereby implicitly reinforcing these values and thereby belittling alternative more holistic ones, such as altruism and community relationships. For fuller statements see Merchant (1980) and Plumwood (1993).

DEEP ECOLOGY

The phrase Deep Ecology was first applied by the Norwegian philosopher Arne Naess (1973) to identify a style of environmental philosophy which questioned responses to the natural world at a fundamental level. In contrast to what he calls a 'shallow ecology' ethic which relies on conventional ethical theories, Naess interprets ecological problems as arising not simply out of an incorrect *ethical* position regarding responsibilities to the environment, but an incorrect *metaphysical* view of the universe which misunderstands the human relationship to nature. Whereas for Descartes the mind and the body had been essentially separated, and individual humans separated from other things in the world, Deep Ecology tries to forge a greater sense of unity between isolated individuals and the environment within which they exist.

Two principal elements of Deep Ecology are *self-realization* and *biocentric egality*. Self-realization is the personal impact of Deep Ecological thinking, which leads a person to understand the complexity and interconnectedness of the human field of experience in relation to the natural environment. Essentially, individuals come to perceive themselves as more integrated into a giant web of interrelationships with their surroundings than as isolated self-interested beings. Biocentric egality ('equality of biological lifeforms') is the direct implication of this understanding, recognizing that as humans are not isolated but subsumed into a wider whole—the whole universe—the only appropriate response is to treat all living things as of equal importance. The position has strong Eastern connections with the Taoist view of interconnectedness and unity. In practical terms, it is associated with the most radical environmentalism, requiring humans to minimize their impacts on the environment wherever possible.

SOCIAL ECOLOGY

The social ecology movement, identified with the American, Murray Bookchin, sees ecological problems as essentially caused by failures in institutional and social relations rather than personal philosophies. Bookchin (1980) criticizes Naess heavily for his 'anti-human bias' and stresses that social change, primarily through local participatory democracy and decentralized economic systems, is necessary to bring about a transition to a more ecologically sustainable lifestyle.

Social ecologists emphasize the primary importance of human relationships, and see all kinds of discrimination and minority exclusion as contributing to a collective problem: a failure to develop genuinely humane institutions and social conditions within Western economic and political systems. Bookchin suggests:

> 'We need to create an ecologically orientated society out of the present antiecological one. If we can change the direction of our civilization's social evolution, human beings can assist in the creation of a truly "free nature", where all of our human traits—intellectual, communicative, and social—are placed at the service of natural evolution to consciously increase biotic diversity, [and] diminish suffering.' (Bookchin & Foreman 1991)

Social ecology stresses the need for public participation in decision-making, adequate representation for interest groups, devolution of decisions to the lowest feasible level of authority, protection and empowerment of minorities, and effective communication and cooperation in the interests of building vibrant and effective local communities.

INTRINSIC VALUE THEORISTS

Aside from positions requiring major social, political or metaphysical changes, the environmental debate in philosophical circles has centred on the understanding of the concept of intrinsic or inherent value in the natural world. Theories of intrinsic value are reviewed in more detail in Chapter 5. Here it is worthwhile to note that many of these theorists have generally prescribed to a moral *monism*. That is, they have sought a single major principle or source of value that establishes the requirements for something to be given moral consideration; often, that it is in some way 'natural'. By applying this principle in any given situation, the morally right course of action can be identified. These kinds of monistic theories can be contrasted with the pluralistic approach introduced in Section 4.2, which accepts that several alternative values systems all warrant some degree of consideration.

4.5.3 Conclusions on philosophical environmentalism

Philosophers are divided as to whether the ecological crisis calls for a new environmental ethic, or whether current traditions simply need to be modified to take more account of an environmental dimension. The traditionalists argue that a new ethic cannot simply be created, but environmental concern must emerge out of a meaningful modification of current ideas and traditions. In opposition, other schools of thought see no way of progress without a new ethic, which substantially redraws concepts of moral responsibility, and even personal identity.

These two broad ideologies can be summarized as representing either *ecocentric* or *anthropocentric* values. Ecocentric philosophies place the maintenance of, and respect for, natural systems at the centre of their value systems. Actions are thus justified or condemned according to their impacts on the ecosystem. Although many animal rights movements focus on only certain species within ecosystems, without a holistic conception of natural values, their recognition of non-human rights allies them closely with these philosophies. Ecocentric schools of thought see humans as directly involved in responsibilities to the natural world, independently of their own interests.

Anthropocentric ideologies see values deriving solely from human needs and interests. As such, an economic analysis of costs and benefits is an anthropocentric approach to resolving resource use issues, as monetary valuations, by default, represent only human interests. There is no reason why anthropocentric philosophies should necessarily cause environmental degradation, as it is entirely possible that human interests are served by conservation and high environmental quality. The important distinction is that whereas under an anthropocentric system environmental protection is dependent upon a particular set of interests and preferences held by society, under an ecocentric philosophy preservation is an absolute moral duty.

4.6 Summary

Ethics is concerned with policy objectives, social systems and individual behaviour. Three branches—metaethics, substantive and practical ethics—can be distinguished. Historically, Greek ethics is concerned with an analysis of virtue; medieval ethics with scripture and Christian principles; and modern ethics with two competing types of theory—Kantian notions of dignity and human rights, and utilitarian notions of maximizing welfare.

Attitudes towards the environment have interrelated with both ethical theories and the general cultural evolution of Western society. An important change can be seen in the shift to Christianity from earlier traditions of nature worship. Christianity itself influenced scientific attitudes and development, and hence resource use; early Christianity emphasized humility, whereas later periods saw scientific improvement (and corresponding economic growth) as celebratory of Christian work. The value of nature in these periods was instrumental; that is, valuable only for human purposes. The Romantic movement was a response to this situation, which emphasized spiritual values in nature.

Recent environmentalism can be seen as a conflict between resource conservation, which maintains the instrumental view of nature and aims at the husbanding of resources, and preservation, which sees intrinsic value in nature and aims to preserve untouched wilderness. These views divide roughly into anthropocentric and ecocentric outlooks. A number of philosophical positions have become established around this debate, most notably Deep Ecology, ecofeminism, social ecology, animal rights and Aldo Leopold's land ethics.

Further reading

Armstrong, S.J. & Botzler, R.G. (1993) *Environmental Ethics: Divergence and Convergence*. McGraw-Hill, New York.

Clarke, J.J. (1993) *Nature in Question: An Anthology of Ideas and Arguments*. Earthscan, London.

Hargrove, E.C. (1989) *Foundations of Environmental Ethics*. Prentice Hall, Englewood Cliffs, NJ.

Norman, R. (1983) *The Moral Philosophers: An Introduction to Ethics*. Oxford University Press, Oxford.

Pepper, D. (1996) *Modern Environmentalism: An Introduction*. Routledge, London.

Thoreau, H.D. (1974) *Walden*. Collier Books, New York.

Part 2
Value and Valuation Tools

5 The Concept of Value

'The triumph of economic growth is not a triumph of humanity over material wants; rather, it is the triumph of material wants over humanity.' [Richard Easterlin]

5.1 Introduction

The subject of this chapter is 'value'. It is concerned with the process of making value judgements, the justification of moral beliefs, value systems, and ideas about justice, goodness and rights. Although such a subject seems very theoretical, issues concerning value do have very practical relevance. Put simply, values influence action—hence the study of ethics is known as *applied* philosophy, concerned with behaviour in the real world.

Here we focus on four aspects of value. In Section 5.2, we begin with a metaethical inquiry into value; that is, the justifications that can be made in favour of holding any values at all. In Sections 5.3–5 we then review three more specific areas of value judgements: views of 'justice' in the economic system (Section 5.3); the value components that contribute to human welfare (Section 5.4); and value issues arising out of concern for the natural world (Section 5.5).

5.2 Value and justification

Before we explore some of the different value systems that operate in society, and the justifications for making particular value judgements, we can ask a prior question: how can we justify any value judgements at all? In other words, what are we actually doing when we make a value judgement?

5.2.1 Categories of value

As a working definition, we can start by defining a value as a framework for identifying positive or negative qualities in events, objects or situations. Building on this very broad definition, we can make a further distinction between three fundamental types of value: functional, aesthetic and moral. If we claim that an object, action or belief is in some way 'valuable' or represents a value, we are claiming that it possesses at least one of these three kinds of value.

Functional values arise when any object or action serves a recognizable purpose, though this purpose does not need to be consciously chosen. Functional values are commonly referred to as instrumental values, because they identify a functionally valuable thing as an instrument for achieving an objective. Functional values are therefore relational: they value something *relative* to its purpose.

Aesthetic values are concerned with beauty. A beautiful object is not valued as a means to an end, but as an end in itself. In this case, an aesthetic value is an *intrinsic* value. Whereas instrumental values derive their value from their relationship to something else, aesthetic values arise from the qualities of the object in itself.

Moral values are values concerned with goodness. They are judgements of virtue, rightness of actions and justice. Like aesthetic values, moral values are definite ends in themselves. No further justification is needed of an action than that it is moral. By performing moral actions we aim to do the right thing simply because it is the right thing and not to achieve any further objective.

Do these three categories cover all the kinds of evaluative judgements we wish to make? It seems fair that they do. When we use general evaluative phrases—a good team, a good car, a good view—we may not be consciously separating our judgements in this way. Indeed, we may be making two different value judgements at the same time—the league winners may also be highly virtuous—but it is clear that our judgements will always refer to one of the three kinds of value identified above.

5.2.2 Making value judgements

So far we have taken for granted the ability to distinguish values. But what are our criteria for making such value judgements? In general, we rely on a notion of a shared value system when we discuss values. Thus, in the simple case of functional values, the 'shared' criteria

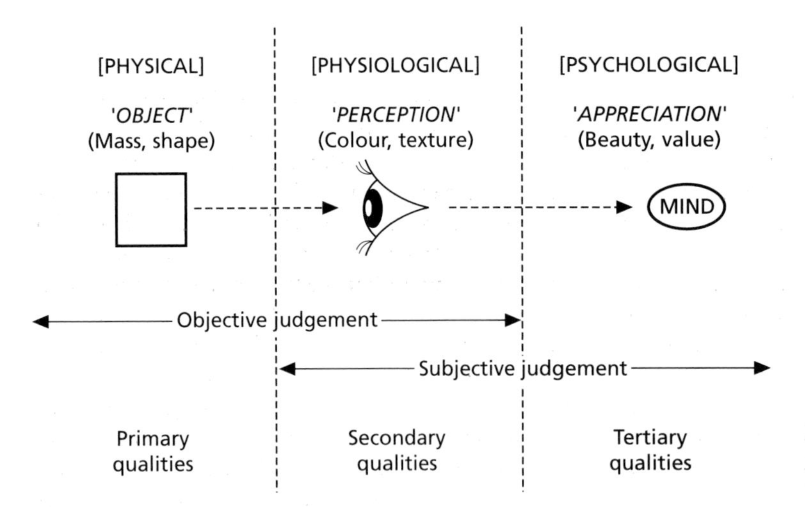

Fig. 5.1 The distinction between primary, secondary and tertiary qualities of objects.

for judgement relates to the function of the object. We know it is the function of a knife to cut, so we understand that a good knife is one that fulfils this function well. Furthermore, we can test if any particular knife cuts well by trying it out. Quite simply, the presence of a function means that anyone who understands that function can make a judgement of functional value.

Is there a similar test for aesthetic values? What is the justification for calling a picture or a piece of music 'beautiful'? It is clear that in these cases we cannot point to a further purpose or aim towards which the music or painting is directed. There is something intrinsic to the object which makes it valuable. But how can we describe or define this value? If two people disagree over the aesthetic quality of a painting, what is the basis of their argument? In other words, what exactly is the source of the value argument—the object, or their reactions to it?

One way of answering this question is to differentiate between the kinds of 'qualities' or characteristics that objects possess. At the most basic level, objects have a certain mass and shape. These qualities are *primary*. *Secondary* qualities are sensory qualities, such as colour and texture. Secondary qualities are obviously dependent on primary qualities, but they are also dependent on the senses of an observer. A colour-blind person cannot perceive certain secondary qualities correctly. That is, he or she cannot make evaluative judgements about certain colours in the way that others—normal observers—can.

Should judgements about these secondary qualities be considered objective or subjective? They are clearly dependent on the person sensing the object, which suggests that they are subjective. On the other hand, all normal people who view the object under the same conditions perceive the same qualities. The quality is thus universal; in this sense, it seems fair to call it objective. Certainly in everyday speech, secondary qualities, like 'redness', are taken to be objective properties of objects, not simply experiences of observers.

This distinction is more significant if we introduce third-level, or tertiary, qualities (Alexander 1968). An example of a tertiary quality is beauty. The experience of beauty is simultaneous with the act of seeing, but it is not the same as seeing. The observation of beauty gives a sensation in the way that simple sense-perception does not. We can meaningfully discuss why we find some things beautiful, whereas we cannot meaningfully discuss why we perceive certain colours. We simply see them. Secondary qualities are in this sense simple, whereas tertiary qualities are complex, or 'twice removed from reality' (Rolston 1982). This division is represented in Fig. 5.1. It seems that in order to recognize tertiary qualities, we require some sense of *appreciation* in addition to basic sensory perception.

5.2.3 Judgements of moral values

Up to this point we have been considering aesthetic value judgements. Is there any similarity between these and judgements of moral value? The idea that moral and aesthetic values are both sorts of 'tertiary' qualities leads to what Gaus (1990) calls the cognitive-affective view of moral and aesthetic judgements:

THE COGNITIVE-AFFECTIVE VIEW

The cognitive-affective view suggests that all value judgements are actually reflections of the *feelings* of observers, caused by the properties of objects or actions. In this sense, to call something 'valuable' is to say that it has a particular effect on people: that they like it or approve of it in some way. But, according to the cognitive-affective view, it is not possible to go further than this, to ask 'Why does the object affect people like this?'. It is simply a fact that this reaction happens, and calling something valuable is simply to identify this reaction. An art critic may be able to point out the features of a painting that provoke a positive reaction, but he or she cannot explain why this reaction *itself* should arise, other than referring to what Hume calls 'the general taste of mankind'. This can be contrasted with the 'objective property' view.

THE OBJECTIVE PROPERTY VIEW

The objective property view suggests that 'being of value' is a property that some objects have *in themselves*. The objective property view of value suggests that values are not just human reactions but are in some sense real properties of objects. In this sense, value judgements are claims to *identify* sources of intrinsic value in the world. Value, in this case, is not just *assigned* to objects but *originates* in objects. These properties or moral values can be identified by observers either through a moral sense—what we might call moral intuition—or through some process of rational judgement.

5.2.4 Metaethical positions

The two basic theories about how value judgements arise define the central problem of metaethics: what does it mean to make a claim that an action is morally correct? Is it a claim to some form of truth, or is it a statement about feelings and preferences? Or, in fact, are these two positions much closer than we might think in terms of their influence on behaviour? We can note in passing that it is not possible to resolve value disputes through experiment or research. We can conduct research to find out what moral values people do hold, of course, but it is one of the problems of ethics that a simple majority opinion does not in itself constitute proof of right judgement.

In response to the questions posed above, there are four basic positions on metaethical justification.

EMOTIVISM

There are no moral truths. Moral judgements are based on emotional reactions, thus moral values are no different to 'tastes' for particular kinds of actions. Emotivism is a position endorsed powerfully by the great twentieth-century philosopher Bertrand Russell (1935), who claimed that 'questions as to "values" lie wholly outside the domain of knowledge'. In this, he meant that the truth or falsity of value judgements was something that was incapable of proof. The original proponent of the theory was David Hume (1978/1777). In his account, a commonality of feeling amongst people—for example, cold-blooded murder always arouses a feeling of revulsion—gave rise to moral judgements, and thus to moral standards or values. In this sense, moral values arise from this commonality of feeling, which may make them in a genuine sense universal.

INTUITIONISM

There are moral truths and intrinsically good things, and all normal people can intuitively recognize them when they see them. The position asserted by the Cambridge philosopher, G.E. Moore, who presented it in his *Principia Ethica* (Moore 1971/1903). He considered moral value to be a kind of simple quality of objects and actions, different in kind to other qualities but which none the less could be appreciated through a moral sense or basic intuition.

RATIONALISM

There are moral truths which can be rationally defended if we think through the implications of them clearly enough. Immanuel Kant (1949/1785) maintained that the diktats of the 'moral law' were compelling on rational grounds. His categorical imperative claimed that it was impossible to form the intention to act immorally without involving oneself in a position that would prove ultimately self-contradictory. As rational arguments are logically true or false, moral values are capable of proof in Kant's system.

RELIGIOUS JUSTIFICATION

There are moral truths, established by a Divinity. In the case of religious justification, moral goodness is derivative from a higher power, and the challenge to humans is to identify it and act accordingly.

Each of these positions has been defended vigorously by philosophers, theologians and scientists over the last century. Their deliberations, however, have had little impact on the kind of moral discussions that animate legal, political and policy debates and which are a familiar part of life in a democratic society. These debates are always regarding substantive moral issues. How much liberty is acceptable? What are the duties of the individual to the state? Should species be protected at any cost? Such arguments take it for granted that the underlying values of liberty, duty and care can be justified.

This lack of interest in metaethical issues is really not surprising. The important issue in value theory, as Hume identifies, is not the philosophical justification of the entire system, but simply what you do believe in. Thus, even moral sceptics, such as Russell, who hold that there can be no solid justification for holding particular moral beliefs, do still hold beliefs. They simply cannot rationally justify why they hold them. Moral beliefs still regulate their behaviour, because they are based on feelings and feelings are the basic determinants of action.

5.2.5 Conclusions on value and justification

The division between 'matters of fact', which are scientifically testable, and 'matters of value', which are based on feeling or intuition and incapable of proof, is known as Hume's 'guillotine' (Black 1970), or simply Hume's Law. The guillotine severs any relationship between moral judgements about what we ought to do, and any statement of the basic facts. Hume's Law suggests it is only by introducing an explicit moral value into an argument—which will itself be based implicitly on a feeling or intuition about what is right—that any moral conclusions can be drawn from matters of fact.

However, this guillotine between facts and value judgements does not mean that moral judgements are simply 'added on' to the real world facts. Hume's thesis is that value judgements are made *simultaneously* with a grasping of the facts. The moral beliefs that we do hold are clearly the product of numerous factors, including upbringing, personal experience, self-reflection, rational criticism and acquired knowledge. These act together in combination when considering any issue to produce a value judgement. Moral debate over this judgement remains very necessary for three reasons.

First, any moral discussion involves what Blaug (1992) calls a mixture of 'pure' and 'impure' value judgements. Pure value judgements are the basic kinds of judgements that an emotivist claims simply cannot be justified rationally, for example, 'malicious cruelty is always wrong'. Impure value judgements mix pure value judgements with beliefs about the facts of the case, and about how the world works. In any argument involving impure value judgements, discussion can certainly persuade us to change our moral judgements if it shows some aspects of a situation to be other than we thought they were.

Secondly, an increase in our knowledge of 'facts' in general can also change our moral attitudes, because they can change the orientation of our feelings. Moral judgements may be dependent on feelings, but feelings change with knowledge and experience. A scientific theory, such as evolution, or a philosophical theory, such as Deep Ecology, could change our moral reactions to the lower order animals by suggesting that we share a common evolutionary line. This knowledge may generate certain emotional reactions, which mean we make certain moral judgements in a different way than if we are ignorant of evolutionary theory. In this sense, moral arguments hinge on the way in which we come to *perceive* issues.

Thirdly, moral argument can show us to be inconsistent. In this sense, we may start with pure value judgements and work through their implications. This may lead us to discover that we are not actually behaving in a way consistent with our more fundamental beliefs. Such a discovery would again be an important influence on our behaviour. Animal welfarists, such as Singer (1975), for example, argue that unnecessary suffering is generally considered a pure moral wrong. It may seem in many ways inconsistent to differentiate between human and animal pain once this judgement has been made.

PRACTICAL IMPLICATIONS

We noted in Section 4.1 that moral values have three important practical implications: they directly motivate individual behaviour; they guide public policy; and they influence, and are influenced by, the design of systems of legal rights and other social and economic institutions. The focus of the rest of this chapter is therefore on substantive ethical judgements: the reasons given in support of particular moral judgements and sources of value. A distinguishing feature of many of these moral debates is not the direct opposition of values, but the

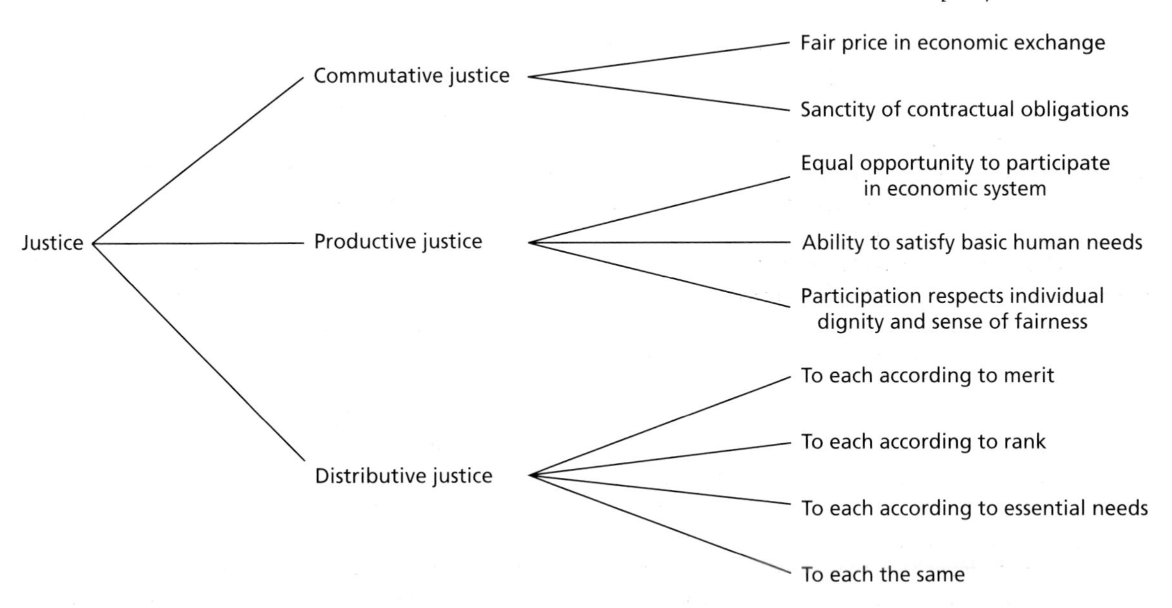

Fig. 5.2 Three branches of socioeconomic justice and their possible interpretations. (Source: Wilson 1992.)

question of the *degree* of importance that each value should have. Thus, the moral dilemma in these situations is finding the correct balance between competing claims of value, or deciding on the relative strength of different notions of value, rather than specifying one absolute criterion of right judgement.

5.3 Moral value and the economic system

In Chapter 3 we explained the neoclassical economic theory of value. Economic values depend on the 'preferred choices' made by individuals in markets. These choices collectively determine the price of commodities when the consumption possibilities open to consumers and the production possibilities open to suppliers are both limited. The price mechanism balances the consumption desires of consumers with the rewards earned by suppliers, thereby providing a relative theory of value of goods for all those participating in the market system.

Moral judgements are apparent in this system in two forms. Internal to the system itself, they are expressed through the choices made by individuals or firms, for example in purchasing environmentally friendly products or producing goods with 'clean' technologies. Externally, they are apparent in much broader notions, such as the right to private property, freedom of contractual exchange, wage–labour and other social conventions that establish the institution of the 'free market', and also including modifications to the pure free market, such as income taxation. These broader ethical notions both create the preconditions for a market system itself, in the basic right of ownership to goods, and circumscribe the extent of that system, for example allowing the ownership of land but not people. This *external* interaction between moral values and the market system is the subject of this section. In the words of Lord Robbins (1935), we are here concerned with ethics and economics set 'in juxtaposition', with moral value systems acting as real constraints on the operations of actual markets.

This juxtaposition can be clearly seen in the concept of 'economic justice'. Theories of economic justice are concerned with the fairness of the economic system according to a number of moral criteria. Here we consider three branches identified by Wilson (1992): commutative, distributive and productive justice, which together cover many of the moral issues arising out of the operation of market economies. A summary of the three branches is shown in Fig. 5.2.

5.3.1 Commutative justice

Commutative justice concerns fairness in exchange. If one set of goods are exchanged for another set of equal worth, then commutative justice has been done. The principle is outlined by Aristotle, who distinguished between profits generated through production, increasing the real physical wealth of the community, and profits derived from financial speculation. These speculative profits, which do not come from increasing the overall stock of goods available within society, were considered in Aristotle's view to be an inferior form of wealth creation. Whereas the pursuit of production-based profits was clearly beneficial for the whole community, activities that made money simply from speculation appeared to distort the real purposes for which money existed.

Two moral injunctions arose out of Aristotle's analysis of commutative justice, both of which were highly influential in the medieval period. They provide an interesting insight into a highly integrated socioeconomic system that considered intervention in markets to be a moral imperative. The first was the banning of usury, or lending money at interest, and the second was direct market intervention which involved setting a 'just price' for goods and services.

USURY

In simple terms, if commutative justice involved the exchange of like for like, then a pound borrowed should be a pound returned. The 'breeding' of money through lending at interest is, on the Aristotelian view, contrary to the nature of money, which has no role other than to facilitate exchange.

The objection to money-lending lay in the fact that the lender had put money to no active use himself or herself, and yet received an increase in its value. It was different to the practice of financial investment, which bore an outward similarity to a money-for-nothing activity. The investor bears the risk of failure, and so it is just that if the venture is successful he or she should have an agreed share in the profits in proportion to the amount he or she has ventured in order to make it happen. The injustice in usury stems from returns gained without risk.

The prohibition against usury had some influence on the medieval financial system. Aquinas endorsed the Aristotelian position that charging interest was unjust and, in addition, considered it a clear example of the sin of avarice. Christians were therefore barred from engaging in it. It was left to foreigners, typically Jews, to handle loans, on the basis that they were non-Christian and their souls were therefore beyond saving.

There were also consequential moral considerations supporting the ban on usury. It was calculated to hit the poor hardest, because they were often those who needed loans most strongly yet could not negotiate good terms and had little means by which to repay high rates of interest. Usury thus contributed to an increasing injustice in distribution (see below) and to the exploitation of the weak. The welfare of the poor was therefore explicitly protected at the expense of entrepreneurial opportunities for the rich.

THE JUST PRICE

Under the just price system, an official body regulated the prices charged for goods and services according to a notion of the 'intrinsic' value of the good in question, set through a combination of religious and historical precedents.

Two elements are of interest. First, the idea of a price set in advance by a superior authority was in keeping with the highly paternalistic approach towards government that was one of the defining characteristics of the medieval age. The wages boards of the city guilds, which organized the skilled trades, involved a representative of the Church in addition to the prominent master craftsmen of the guild. In this way the interests of the wider community were again represented, particularly the poorer sections of it.

Secondly, the just price system was clearly a deterrent to entrepreneurship, in so far as any individual's ability to prosper was restricted by the limited opportunities for profit provided under the system. Rewards for effort were set at what the authorities considered a socially acceptable level. If there appeared to be a demand for carpenters, then more would be taken into training, but current craftsmen could not enrich themselves at the expense of the community while this adjustment was taking place. As Vickery (1953) notes, 'as long as circumstances changed slowly, this idea did much to prevent exploitation in potentially monopolistic situations'. Correspondingly, as the economy diversified beyond all recognition from the stable hierarchies and limited outputs of the feudal system, the rigidity of the just price mechanism simply became unworkable.

The principles of commutative justice in a modern economy are maintained in aspects of the economic system, such as laws on misrepresentative advertising, contractual obligations and statutory rights. Thus, they are of concern only in so far as they smooth the operation of the market, enabling consumers to make judgements of value within an institutional framework that guarantees against failure to receive what was paid for in good faith. The control of monopolies through such bodies as the Monopolies and Mergers Commission in the UK does, however, recognize that unfair competition can lead to consumers paying unjustly high prices.

5.3.2 Distributive justice

Ethical concerns in Western economies over the last century have focused largely on the issue of distributive justice. Distributive justice is simply the fairness of the distribution of the goods and services of the economic system between all the members of the community involved in or affected by that system.

What defines a distributively just system? We must be careful to distinguish between judgements regarding just *processes*, and judgements regarding just *outcomes* (the distinction made between a procedural and a substantive ethic; see Section 4.2.2). In considering just processes, a just distribution might be any arrangement that is the outcome of commutatively just exchanges, regardless of what this end distribution is. Libertarian philosophers (for example, Nozick 1974) often support this view. They stress returns to individual effort, strong rights to personal property derived from that effort, and as few limitations on free exchange as possible. Clearly, the outcomes of such systems may well be very inequitable, but this inequality is reached in a just (meaning essentially honest) way. As an alternative, we may wish to disregard the causes of the current distribution and judge its fairness against some independent standard, such as equality of possessions or income.

It is clear that a large number of criteria might have a bearing on distributive justice. How should a pie be divided between four people to be distributively just? Equal shares would perhaps be the obvious answer. But is there a concern over who provided the ingredients? Or who baked it? Is anyone starving? Has another one just been divided? If so, how? In order to judge a just outcome, we may want considerable information on the context of the decision.

In fact, we are likely to operate with several ideas of what is distributively just, dependent precisely on the context. 'First come, first served' may be considered a distributively just way to allocate film tickets, and US companies tend to operate a 'last hired, first fired' system for laying off staff. Hospital beds are likely to be allocated on a relative need basis, however, and academic qualifications are based on merit, not arriving at the exam hall before everyone else. Military conscription is run on a random distribution. Few judgements of distributive justice are likely to be entirely independent of context.

Sen (1987) suggests that judgements about distributional justice require agreement on three distinct processes:

1 *Measurement.* How do you measure what people have? For example, do you consider possessive or property characteristics and, if so, is total income, disposable income or total wealth most relevant? Or are opportunities more important (to use services; to gain education)? Or is the actual development of innate capacities more significant still (measuring by potentialities rather than opportunities)?

2 *Aggregation.* In judging the 'justness' of any particular system, how are the measured characteristics aggregated? In other words, what weightings, if any, are given to differences between individuals?

3 *Prioritization.* Do some aspects of welfare lie in a hierarchy of importance? For example, is equality of liberty, perhaps to participate in exchange, always prioritized over equality of final wealth?

The ramifications of all these different possibilities cannot be followed here. However, it is worth noting a final consideration. If some efforts towards increasing justice in distribution are to be made, at what stage of the economic system should they be targeted? Free marketeers suggest that the market should be allowed to function without interference, and any distributional changes should be made by directly transferring wealth through taxation at the end of the process. Those with more communistic leanings favour more control over the means of production, partly because they consider the means of production itself to be an important influence on what people have, through the kind of work they do, not simply the rewards they receive for that work.

In favour of redistributive taxation, keeping justice issues external to the market, are arguments that stress:

1 maintaining incentives for individual entrepreneurship/hard work, assuming taxation is not too punitive;
2 corresponding efficiency of allocative effort, leading to universal enrichment; and
3 trickle down benefits of wealth.

In favour of more direct state control are concerns with:

1 greater social cohesion through cooperative effort;
2 valuing work in social as well as wage terms; and
3 allocation according to need rather than ability.

Some aspects of these divisions are taken up in the next section on productive justice.

5.3.3 Productive justice

The concept of productive justice is much less familiar in the context of developed economies than distributive issues. Strongly influenced by the involvement of the Catholic Church in developing world economies (Wilson 1992), productive justice lies in the ability to participate in the economic system and satisfy basic needs through meaningful work. In this sense it has elements of both commutative justice, in receiving fair returns for labour, and distributive justice, in mitigating inequalities of consumption and opportunity. It is recognizably distinct, however, also stressing the human needs satisfied through the process of work itself, rather than simply through the wages that result from that work.

Productive justice has a long history. Aquinas considered the activity of work to be necessary to human development. The need to engage in cooperative activity was one element of its value; the opportunity to engage in creative activity another. The labour theory of value of Karl Marx has a strong emphasis on productive as well as commutative justice. If the worker's labour is the source of value in commodities, clearly commutative justice would demand that value is returned to its owner through the wage mechanism. In addition, however, the communist ideal has a clear productive emphasis, stressing the need for the cooperative participation of workers unified in a common goal.

Under productive justice the economic system does not exist simply to respond to the consumer's desire for commodities and the producer's desire for profits, but as a social arrangement that develops human capacities and provides citizens with a sense of dignity or value through their work. This arrangement need not lead to a communist system of social organization. 'Green' economist Fritz Schumacher (1979) emphasizes the importance of meaningful work, whatever the organizing principle, as important to human development. He identified smaller scale, particularly craft methods of production as the ideal example.

The issue stressed by Schumacher is one of striking a balance—in essence, reconciling the demand-driven side of the market system with the human needs of those involved on the supply side of the system. A productively just system might therefore be less efficient when measured according to economic criteria, such as value added per capita or net profit, but proponents claim it would be more efficient at satisfying human needs that are not measured economically, such as self-respect, creative involvement and job satisfaction.

5.3.4 Conclusions on values and economics

The free market economies of the West have been undeniably successful in the production of commodities and the generation of wealth. Operating within appropriate institutional restraints, the three key values of the market emphasized by free-marketeers are that it is:

1 liberal, impinging as little as possible on individual freedom;

2 democratic, providing a way in which the market is very literally controlled through the individual choices made by all those involved in the system; and

3 efficient, in rewarding enterprise and encouraging wealth creation which eventually benefits everyone.

As the market is a self-organizing system, alternatives seem to require a greater degree of state control, which is less directly democratic and threatens to edge ever closer towards the excesses of communism.

In opposition, ethical and ecological critiques do seek greater intervention in market interactions based on a different interpretation of moral values. Social critics stress the damaging effects of unrestrained competition on public goods, such as social harmony and mutual aid; they question the democratic assumption of markets with regard to inequalities of control over wealth, stressing the relative exclusion of the poor and the centralized power of large corporations; and they may place higher values on intangible goods, such as community structure and job security, than on simple consumer goods. Environmentalists point out the exclusion of the interests of future generations in decisions made in markets (see Section 8.4 for a discussion); and incidences of environmental damages that can occur outside the market system and therefore remain unvalued. The more radical amongst them also point out the exclusion of the interests of other life-forms, at the extreme leading to irreversible loss of ecosystems and species.

In its simplest form, this controversy concerns the ultimate ends of the economic system. The free marketeer stresses the production of goods and services, measured as an output from the system, and the liberty of the individual within that system. Environmental and social critics stress the intangible and therefore unmeasured values that arise both within and outside the system of production and exchange, thereby treating the physical outputs as only one subset of the system, albeit a valuable one. In the rest of this chapter we look at some further issues arising out of this broader conception of the creation and measurement of value.

5.4 The concept of welfare

In this section, we concentrate on human welfare. In particular, we look at some of the components that make up human welfare—the aspects of experience that make an individual's life 'go well' (Griffin 1986). In this sense we are concerned with examining what is of value in human experience.

Conventional welfare economics is concerned with human welfare issues, but it is not much concerned with the *contents* of individual utilities. Rather, it is concerned with maximizing the *opportunities* for the satisfaction of individual desires, without investigating in any detail what those desires are. Such desires are taken as given, or exogenous to economic analysis. Welfare economic theory does not try to distinguish between the concepts of *needs* and *wants*. It restricts itself to the assumption that greater opportunities for consumption, *ceteris paribus*, lead to greater personal satisfaction.

The ideas reviewed in this section do try to make some comments on the contents of individual utilities. In general, they focus on the more intangible goods arising from the social and economic interactions taking place in society, such aspects as feelings of security or job satisfaction. These 'intangible' sources of welfare are mostly ignored in conventional economic analysis, and yet they may have a potentially far greater influence on welfare than the production and exchange of market goods. Furthermore, even though these intangible goods are not analysed in conventional economic theory, they are strongly affected by the workings of the market system. Their importance is perhaps particularly evident in situations where previously subsistence or community-based economies have undergone transformations into cash-based market economies (see, for example, Norberg Hodge 1992).

Aside from issues of analytic methodology, however, economists have also resisted using ideas such as wants and needs because of a justifiable rejection of *paternalism*: making decisions about consumption priorities on behalf of individuals which in a liberal society ought to be made by the individuals themselves. Defining a need can be an infringement of individualism, by assuming that everyone has this need, thereby promoting policies that force everyone to have it satisfied even if in fact they value something else more strongly.

What looks like a strong division of opinions over values here is really a question of degree. Every public decision involves applying some idea of individual welfare, even if it is simply more freedom to do as the individual likes. Anything short of absolute anarchy assumes regulation based on an idea of what conditions are good for society's members. Investigating more fully what contributes to welfare simply expands the set of criteria available for consideration when approaching a planning or policy decision.

5.4.1 Needs and wants

Mallmann (1973) defines 'needs' as those requirements that are always found when 'the behaviour of human beings is analysed irrespective of culture, race, language, creed, colour, sex or age'. They are thus not conditioned by social structures, natural environments or technical advances. The United Nations Human Development Index, developed in 1984, is perhaps one of the best-known definitions of human needs and lists 32 indicators of development, ranging from education and life expectancy to security and home life.

The models outlined below adopt a less formal approach to needs analysis. They are concerned with identifying the fundamental categories of value that contribute to human welfare. This concern with categories recognizes that the means for satisfying any particular category are diverse. Kamenetzky (1992) identifies four levels of human needs, arising from biological, biopsychological, psychological and sociocultural needs. This simple hierarchy is represented in Fig. 5.3. One of the best-known models of human needs is that of psychologist Abraham Maslow (see Box 5.1).

MAX-NEEF: NEEDS AND SATISFIERS

In response to the development problems he encountered in Latin America, Chilean human ecologist Manfred Max-Neef developed a typology of fundamental human *needs* and *satisfiers*. One distinguishing feature of Max-Neef's (1992) approach is his insistence on needs analysis as a participatory and exploratory process. Investigating needs is a way that people can discover and develop their own potential, particularly those who have traditionally been excluded from the political or economic system through lack of education or democratic rights.

While accepting that the perception of a need is always a subjective judgement, Max-Neef distinguishes between *subjective-universals*, which are true or valid for all people; and *subjective-particulars*, which are the domain of traditional welfare economics and deal

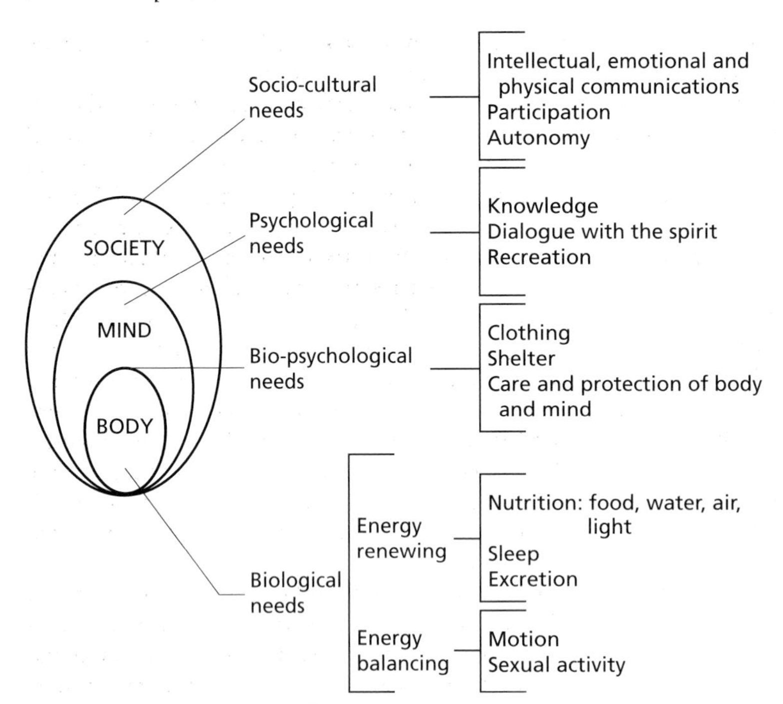

Fig. 5.3 Kamenetzky's four levels of needs relating to both individual and social aspects of human welfare. (Source: Kamenetzky 1992.)

Box 5.1 Maslow's model of human needs

In the 1950s, psychologist Abraham Maslow (1954) developed a simple model of human needs based on a division into two kinds: either essentially negative 'deficiencies', or essentially positive aspects of 'self-realization'. These he set in a rough hierarchy, differentiating material, social and ethical needs, and the values that represented and fulfilled each one.

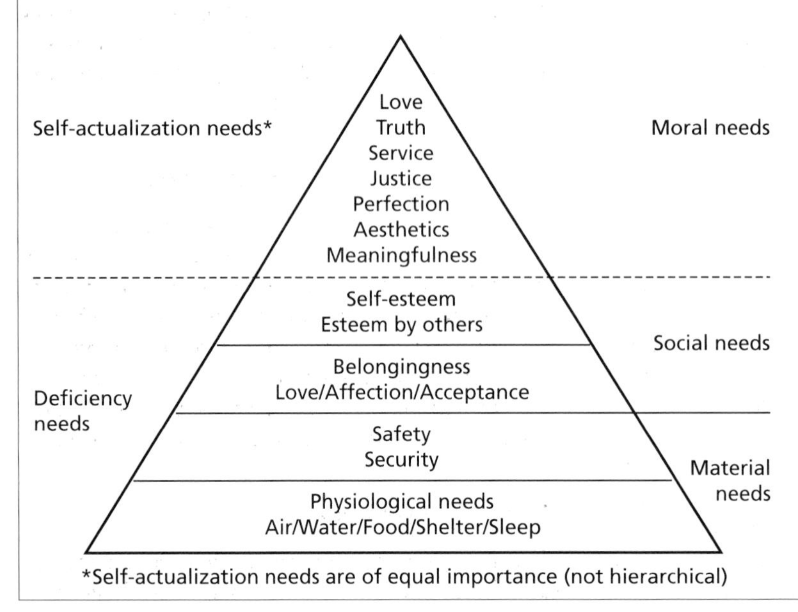

Fig. 5.4 Maslow's (1954) needs model.

with the specific preferences and choices of individuals. The subjective-universal aspects of needs are defined by nine basic categories. The universal particulars are the innumerable ways in which these nine categories of needs can be satisfied in practice.

It is important to note that the analysis of needs under this system is not simply a question of measuring physical deficiencies but is bound up with political, social and anthropological elements. The expanded notion of needs that Max-Neef attempts to define is, in his words, an attempt to rediscover the 'molecular composition of the social fabric', as opposed to the atomic model of independent rational individuals assumed within conventional welfare approaches. In his view therefore, 'satisfiers are not the available economic goods'. Satisfiers can include 'forms of organization, political structures, social practices, subjective conditions, values and norms, spaces, contexts, modes, types of behaviour and attitudes, all of which are in a state of tension between consolidation and change'. Under this analysis, needs and desires cannot be analysed in isolation from the context which both generates and helps to satisfy them.

The needs/satisfiers matrix (Table 5.1) developed by Max-Neef distinguishes two aspects of needs: existential categories, and axiological categories. Existentialist categories are the four so-called modes of operation of any human being: being, having, doing and interacting. The axiological categories are divided into nine types: subsistence, protection, affection, understanding, participation, leisure, creation, identity and freedom.

From the basic matrix, the cells of which are only intended to be suggestive and will always be culturally specific, Max-Neef draws out a further five categories of possible satisfiers. These comprise violators/destroyers, pseudo-satisfiers, inhibiting satisfiers, singular satisfiers and synergic satisfiers. Examples of these categories are given in Tables 5.2 and 5.3.

The needs/satisfiers model of welfare was designed as a tool in development work. It has value from an ecological economics perspective for the way in which it restructures the conceptual framework through which development and resource use planning proceeds. In particular, it can be valuable for the following:

1 By replacing linear thinking by systematic thinking. In a linear model, development planning proceeds by establishing priorities and tackling them sequentially. In general, this can lead to a focus on the most obvious, observed poverty of 'subsistence'. Satisfiers devoted to this end may well be singular, and in some cases inhibit the development of satisfiers for other needs that are less readily observed or measured.

2 By replacing the emphasis on the explicit production and consumption of artefacts and services with a model of the implicit needs being met by that production and consumption. This opens up a wider array of decision criteria for planners and managers by reorientating the issues towards the categories of need rather than the most obvious physical aspects of deprivation. These physical aspects form part of, but not the whole problem.

3 In planning terms, it offers a framework for participatory models of developmental planning. A similar philosophy underlies techniques such as participatory rural appraisal (PRA) (see Section 10.7). Needs analysis thus becomes an educational exercise which expands the range of possibilities open to both planners and participants by actively engaging participants in developing solutions.

In conclusion, Max-Neef emphasizes three essential differences between his 'barefoot' economics and conventional theory. First, he maintains that a meaningful separation can be made between needs and wants. In his view this can justify intervention in markets, if the needs of some are being sacrificed to the wants of others. Secondly, he assumes that human nature is in some senses fixed, but in other ways capable and, indeed, desiring of continued development and evolution. Such developmental needs may actually be inhibited by relying solely on the transactional system of the market to supply needs. Thirdly, he stresses that ecological limits are now impressing on the economic and social systems at an ever-increasing rate. Adaptation of the economic and social system in response is, in his view, therefore inevitable.

5.4.2 The social aspect of consumption

Max-Neef addresses the question of value through a grass roots analysis of needs, developed in a context of extreme poverty. A very different perspective is provided by economist Fred Hirsch, who focused on what he saw as the increasingly *social* function of consumption in affluent Western economies. In *The Social Limits to Growth* (Hirsch 1976), he describes his view of three related problems of these advanced capitalist systems:

The paradox of affluence

Why has economic advance become and remained so compelling a goal to all of us as individuals, even

Table 5.1 Max-Neef's matrix of needs and satisfiers. (Source: Max Neef 1992.)

Needs according to axiological categories	Needs according to existential categories			
	Being	Having	Doing	Interacting
Subsistence	1. Physical health, mental health, equilibrium, sense of humour, adaptability	2. Food, shelter, work	3. Feed, procreate, rest, work	4. Living environment, social setting
Protection	5. Care, adaptability, autonomy, equilibrium, solidarity	6. Insurance systems, savings, social security, health systems, rights, family, work	7. Cooperate, prevent, plan, take care of, cure, help	8. Living space, social environment, dwelling
Affection	9. Self-esteem, solidarity, respect, tolerance, generosity, receptiveness, passion, determination, sensuality, sense of humour	10. Friendships, family, partnerships, relationships with nature	11. Make love, caress, express emotions, share, take care of, cultivate, appreciate	12. Privacy, intimacy, home, spaces of togetherness
Understanding	13. Critical conscience, receptiveness, curiosity, astonishment, discipline, intuition, rationality	14. Literature, teachers, method, educational policies, communication policies	15. Investigate, study, experiment, educate, analyse, meditate	16. Settings of formative interaction, schools, universities, academies, groups, communities, family
Participation	17. Adaptability, receptiveness, solidarity, willingness, determination, dedication, respect, passion, sense of humour	18. Rights, responsibilities, duties, privileges, work	19. Become affiliated, cooperate, propose, share, dissent, obey, interact, agree on, express opinions	20. Settings of participative interaction, parties, associations, churches, communities, neighbourhoods, family
Leisure	21. Curiosity, receptiveness, imagination, recklessness, sense of humour, tranquility, sensuality	22. Games, spectacles, clubs, parties, peace of mind	23. Day-dream, brood, dream, recall old times, give way to fantasies, remember, relax, have fun, play	24. Privacy, intimacy, spaces of closeness, free time, surroundings, landscapes
Creation	25. Passion, determination, intuition, imagination, boldness, rationality, autonomy, inventiveness, curiosity	26. Abilities, skills, method, work	27. Work, invent, build, design, compose, interpret	28. Productive and feedback settings, workshops, cultural groups, audiences, spaces for expression, temporal freedom
Identity	29. Sense of belonging, consistency, differentiation, self-esteem, assertiveness	30. Symbols, language, religions, habits, customs, reference groups, sexuality, values, norms, historical memory, work	31. Commit oneself, integrate oneself, confront, decide on, get to know oneself, recognize oneself, actualize oneself, grow	32. Social rhythms, everyday settings, settings which one belongs to, maturation stages
Freedom	33. Autonomy, self-esteem, determination, passion, assertiveness, open-mindedness, boldness, rebelliousness, tolerance	34. Equal rights	35. Dissent, choose, be different from, run risks, develop awareness, commit oneself, disobey	36. Temporal/spatial plasticity

The column of 'being' registers *attributes*, personal or collective, that are expressed as nouns. The column of 'having' registers *institutions*, *norms*, *mechanisms*, *tools* (not in a material sense), *laws*, etc. that can be expressed in one or more words. The column of 'doing' registers *actions*, personal or collective, that can be expressed as verbs. The column of 'interacting' registers *locations* and *milieus* (as times and spaces). It stands for the Spanish *estar* or the German *befinden*, in the sense of time and space. There is no corresponding word in English.

Table 5.2 Max-Neef's matrix: violators and destructors.* (Source: Max Neef 1992.)

Supposed satisfier	Need to be supposedly satisfied	Needs whose satisfaction it impairs
1. Arms race	Protection	Subsistence, affection, participation, freedom
2. Exile	Protection	Affection, participation, identity, freedom
3. National security doctrine	Protection	Subsistence, identity, affection, understanding, participation, freedom
4. Censorship	Protection	Understanding, participation, leisure, creation, identity, freedom
5. Bureaucracy	Protection	Understanding, affection, participation, creation, identity, freedom
6. Authoritarianism	Protection	Affection, understanding, participation, creation, identity, freedom

* Violators or destructors are elements of a paradoxical effect. Applied under the pretext of satisfying a given need, they not only annihilate the possibility of its satisfaction, but they also render the adequate satisfaction of other needs impossible. They seem to be especially related to the need for protection.

Table 5.3 Max-Neef's matrix: singular satisfiers.* (Source: Max Neef 1992.)

Satisfier	Need which it satisfies
1. Programmes to provide food	Subsistence
2. Welfare programmes to provide dwelling	Subsistence
3. Curative medicine	Subsistence
4. Insurance systems	Protection
5. Professional armies	Protection
6. Ballot	Participation
7. Sports spectacles	Leisure
8. Nationality	Identity
9. Guided tours	Leisure
10. Gifts	Affection

* Singular satisfiers are those which aim at the satisfaction of a single need and are, therefore, neutral as regards the satisfaction of other needs. They are very characteristic of development and cooperation schemes and programmes.

though it yields disappointing fruits when most, if not all of us, achieve it?

The distribution compulsion

Why has modern society become so concerned with distribution—with division of the pie—when it is clear that the great majority of people can raise their living standards only through production of a larger pie?

The reluctant collectivism

Why has the twentieth century seen a universal predominant trend toward collective provision and state regulation in economic areas at a time when individual freedom of action is especially extolled and is given unprecedented reign in non-economic areas, such as sexual and aesthetic standards?

Hirsch identifies what he considers a structural aspect of the problems outlined above, namely that 'as the level of average consumption rises, an increasing proportion of that consumption takes on a social as well as an individual aspect. That is to say, the satisfaction that individuals derive from goods and services depends in increasing measure not on their own consumption but on consumption by others as well.'

The limiting factor to economic growth, which is increasing access to commodities for all sectors of the population, is set in Hirsch's view not by the 'uncertain, distant physical limits' suggested by the Club of Rome (material scarcity), but by the concept of social scarcity. Social scarcity occurs when the satisfactions that individuals can obtain from further material consumption become constrained by the actions of others.

In Hirsch's terminology, economic goods are increasingly seen as *positional goods*—valued not for their intrinsic qualities that provide a private benefit, but for their relational value which determines an individual's position in the social hierarchy. There can then be no connection between individual and aggregate advancement of welfare, because individual advancement must increase beyond the aggregate in order for it to have relative value for the individual, and this is clearly logically possible only for a minority.

As it stands, Hirsch denies that the liberal approach to market organization founded on Smith's 'invisible hand' can therefore deliver welfare increases for all. He suggests that 'without a supporting ethos of social obligation . . . the principle of self-interest is incomplete as a social organizing device'. His conclusion is that the liberal free market is a 'transitional' case.

> 'A major adjustment needs to be made in the legitimate scope for individual economic striving. Individual economic freedom still has to be adjusted to the demands of majority participation. [T]he prime economic problem now facing the economically advanced societies is a structural need to pull back the bounds of economic self-advancement.' (Hirsch 1976)

Hirsch's thesis is certainly open to criticism. Most obviously, his three basic propositions can be questioned. Moreover, the extent of the problem may be exaggerated. The positional aspects of goods may be relatively insignificant in many areas. And if serious congestion is occurring at the top of the consumption pyramid, this may in itself encourage a change in the values attributed to this consumption. What remains significant is his observation that a focus on private objectives regardless of social conditions seems likely to undermine those conditions and ultimately make those objectives less attainable.

5.4.3 Community values

The theme of a supporting social morality, or the importance of social institutions to economic life is central to the work of Daly and Cobb (1989). In *For The Common Good*, subtitled *Redirecting the Economy Towards Community, the Environment and a Sustainable Future*, they emphasize the importance of the social aspect of human welfare. They suggest:

> 'We believe human beings are fundamentally social and that economics should be refounded on the recognition of this reality. We call for rethinking economics on the basis of a new concept of Homo economicus as person-in-community. [T]his fact does not preclude an element of individualism. The persons in question are individuals. [T]hese individuals are without doubt interested in acquiring commodities, and much of their behaviour expresses just the rational self-interest attributed to Homo economicus in the dominant economics. Hence, many principles of classical and neoclassical economics, with proper historical qualifications, will function in an economics based on the different model of Homo economicus as person-in-community.
>
> But what is equally important for the new model—and absent in the traditional one—is the recognition that the well-being of a community as a whole is constitutive of each person's welfare. This is because each human being is constituted by relationships to others, and this pattern of relationships is at least as important as the possession of commodities. These relationships cannot be exchanged in a market. They can, nevertheless, be affected by the market, and when the market grows out of the control of a community, the effects are almost always destructive. Hence this model of person-in-community calls not only for provision of goods and services to individuals, but also for an economic order that supports the pattern of personal relationships that make up the community.' (Daly & Cobb 1989)

The emphasis on community rather than on production and consumption expands the sources of value that must be given consideration in economic analysis and policy. It may be difficult to quantify these relationships, because as Daly and Cobb suggest, they lie outside the market system. In many cases, it is this which adds to or even generates their value. In terms of increasing welfare, however, the impact of these elements of value is as real as that of tangible commodities.

The suggestion that the good of the community is symbiotic with the good of the individual has a long history in political theory. The Swiss political philosopher Jean-Jacques Rousseau (1973/1762) distinguished what he called the General Will—the community's collective decision of its best course of action—and contrasted it with the Will of All, simply the aggregation of individual decisions based on individual preferences. In his view a good society, meaning one that is good to live in, was one that fostered this sense of communal enterprise through the General Will in public decision-making. A sense of community was thus for Rousseau an important source of value to the individual. This source of value disappeared if the individual was viewed in isolation.

Etzioni (1988) has recently described life in society as a state of constant tension between two 'selves' arising out of this sense of both a common and an individual interest. Etzioni terms this the I/WE division between self-orientated and community-orientated behaviour. What is critical for Etzioni is how we resolve the tension between these aspects. At one extreme, complete liberalism—anarchy—maximizes apparent freedom to do as the 'I' pleases, but in fact it tends to limit opportunities in many ways through, for example, fear and isolation. Correspondingly, stringent social control in the service of the 'WE' leads quite clearly to an unacceptable loss of personal freedom. To identify the poles of this spectrum is easy; it is finding an acceptable middle way between them that forces society into a continuous moral debate.

These issues come to a head in many environmental planning decisions, precisely because of the public nature of many aspects of the natural world. Questions of appropriate governance structures, personal freedoms and perceived responsibilities are recurring issues in the debate over natural resources. Some of the moral attitudes towards nature that underpin these issues are covered in the following section.

5.5 Moral values and the natural world

We saw in Section 4.5.2 that the field of environmental ethics has only a short history, covering the last 30 years and only producing a specialist journal in 1979. In this section we outline some of the arguments that have developed over this period.

Environmental issues clearly involve all three of the value categories introduced in Section 5.2.1. That is, natural systems have a vast array of functional values, through the supply of economically valuable materials in addition to maintaining vital life-support functions. Clearly, many natural elements also have great aesthetic value, ranging from majestic landscapes to the beauty of an individual flower. Both these sources of value are readily acknowledged, and economic values (prices) are themselves a reflection of the functional value of such environmental goods to people (see Chapter 6 for attempts to establish these values). Such values arise when environmental goods raise the sense of well-being, happiness or welfare of an individual, and individuals are willing to pay for them. In economic theory, it is this sense of well-being, or welfare, located in individuals, that alone has intrinsic value, and price is the means of quantifying its importance.

The main focus of this section is on arguments about the appropriate moral attitudes to hold regarding natural entities. Thus we are asking in what way the natural world involves us in specifically moral rather than scientific, aesthetic or economic issues. These concerns encompass both anthropocentric and ecocentric ideologies (see Section 4.5.3). In terms of environmental economics, such discussions are essentially asking the question: is the free market the appropriate way of deciding how we should use environmental goods? In other words, what ethical arguments are there for treating environmental goods—biota, habitats, landscapes—as different in kind from typical market commodities?

5.5.1 Moral considerability and moral significance

According to Goodpaster (1978), moral problems in the natural world require the definition of two related concepts: moral considerability and moral significance. If a thing is morally considerable, then consideration of it must play some part in determining the right course of action. Moral significance determines the relative importance of competing claims to moral considerability. All morally considerable things thus have some moral significance, but some warrant more significance than others.

The presence of moral considerability sets the ultimate ends of moral action. Thus all those things that have moral considerability are 'good in themselves', not needing a further justification as to why they are important. Moral considerability can be equated with the possession of intrinsic or non-relational value. Other things that are not morally considerable in themselves may, however, still be morally relevant because they are instrumentally valuable for things that are morally considerable.

One distinction to make initially is that judgements of moral considerability *made* by humans do not need to be *centred* on humans. This is the difference between anthropogenic and anthropocentric values. An anthropogenic view holds that humans are the only source of *value judgements*; but they do not have to be the only source of value itself. It is not a very contentious position that humans are the only source of value judgements —certainly they seem to be the only source that we are likely to be able to understand. However, humans may still perceive other, perhaps non-conscious things to be valuable in themselves. Under anthropogenism, humans are the only source of valuing activities, but the centre of value can be found in non-human entities.

5.5.2 Valuing nature

By direct implication, arguments that claim there are sources of moral value in nature are arguments for nature conservation. In fact, in the case where renewable natural assets have a recognizable market value, economic arguments are themselves one set of arguments for ecosystem conservation. Furthermore, a number of techniques for valuing non-market goods have been developed which try to extend economic arguments beyond direct market values. These arguments try to capture all the possible ways in which the simple existence of particular natural assets raises human welfare (see Chapter 6).

Here we leave those arguments to one side. The arguments outlined below are concerned with the *moral* consideration of nature in the absence of market values. These are split into four kinds: *welfare* arguments, and *virtue* arguments, both of which are anthropocentric; and *biocentric* and *ecocentric* arguments, both of which recognize intrinsic value in non-human things.

HUMAN WELFARE ARGUMENTS

Human welfare arguments for conservation point out the ways in which human welfare is dependent on preserving natural assets in their current form. We have already noted the value of ecosystem services, such as flood control, and aesthetic/recreation experiences which add to human enjoyment. Below, we note in addition to these direct uses two arguments for conservation which reflect the uncertainty surrounding the full measure of these benefits.

The argument from ignorance

The properties of ecosystems and the biota they contain are in many cases still only poorly understood by ecologists. This means that the potential welfare gains from preserving these biota for further study may outweigh the immediate benefits from consumption. The medicinal properties of as yet unexamined plant species, for example, support arguments for ecosystem preservation.

The precautionary principle

Given that understanding of ecological processes is poor, the precautionary principle suggests that the use of the environment should be moderated by an awareness of risk. Thus, the dangers inherent in damaging ecological processes on which humans rely should be explicitly acknowledged. The moral aspect of this argument is similar to that for culpable negligence. It is considered a moral wrong to allow injury to occur in cases where the causes of possible injury can be anticipated and are therefore avoidable. In the case of exercising due caution in ecosystem terms, the judgement must be to avoid taking unnecessary risks.

HUMAN VIRTUE ARGUMENTS

Whereas the arguments above concern either life-support functions or directly experienced enjoyment of the environment, arguments about virtue are concerned with how the natural environment influences human development. This is a classical view of ethics which suggests that there are certain kinds of human understanding, appreciation and experience that contribute to an overall richer, more fulfilling life. These arguments differ from a pure welfare approach because they suggest that some kinds of life and experience are therefore 'better' or more valuable for an individual than others.

Human development

Many philosophers, including John Stuart Mill (1962/1861), have suggested that human personal fulfillment is dependent on a number of levels of awareness and to reduce utilitarian moral arguments to a 'pleasure principle' is too limiting. The welfare of an individual is therefore dependent on a wide range of possible experiences and activities (see Section 5.4.1 above). Access to and experience of the natural world may be two important aspects of this developmental process. On this argument, it is a moral imperative to maintain the possibility for people to experience the natural world because such experience is central to the full, 'authentic' development of the individual.

Such arguments are not conventionally utilitarian, as they focus not on raising utility *per se*, but on developing capacities for experience. The difference is most clearly seen in attitudes to unhappiness. A utilitarian considers human unhappiness to be a moral bad, to be avoided wherever possible. A 'human virtue' view might consider some elements of unhappiness to be important in developing as an individual. The virtue view does not therefore judge an experience solely by how happy it makes people, but by what it contributes to the development of a personality. Such a view is rejected by liberals as unacceptably prescriptive of standards. They refuse to accept that any particular kind of life or experience is better than any other kind: no judgements should be made on what kind of life or experiences individuals choose to have.

Transformative values

In a similar vein to developmentalists, Norton (1991) has argued that nature has great transformative value; that is, experience of nature can radically alter the preferences of individuals in its favour *once* they have experienced it. If they have not, or cannot experience nature, then the option for transformation is clearly lost. In this sense, Norton would argue that only those who have reasonable experience of natural environments are in a position to judge their value.

The argument is difficult to accept from an economic point of view—many decisions have to be made by individuals under so-called 'imperfect' information. Norton's argument rests on the claim that the transformative effect of nature is substantially different in kind from the kind of transformation that a consumer undergoes when switching between different brands of washing powder. If individuals do experience nature, he suggests they can come to realize a highly significant switch in preferences from this kind of experience.

Behavioural attitudes

Immanuel Kant (1949/1785) advocated the benevolent treatment of animals on the basis that practising cruelty to animals could stimulate cruelty to humans, which would be a moral evil. Banning cruelty to animals was thus justified instrumentally on the basis that it would reduce cruelty to humans overall.

While Kant's particular assumptions are obviously questionable, many philosophers have drawn attention to the connection between attitudes towards nature and attitudes to other social classes, nations and races. The way in which a culture defines its attitudes towards natural resource exploitation may therefore be a significant influence on its wider moral standards. Social ecologists (for example, Bookchin 1980) and ecological feminists in particular have suggested links between the 'control of nature' exemplified by modern capitalism on the one hand, and prevailing attitudes towards certain social groups on the other. The behavioural argument suggests that the development of a more considerate attitude towards natural resources would contribute to a morally better society in general.

Cultural symbolism

The American philosopher Mark Sagoff (1988) has suggested that certain natural features, such as 'untamed' wilderness, are symbolic of important cultural values in Western democracies. The 'freedom' of a wild river is thus an icon of this virtue in a modern democracy—the freedom to express oneself and maintain independence from the control of others. Through these kinds of natural resources, Sagoff suggests that the environment can in effect have a positive moral influence on society. It is therefore important to retain these symbols.

It is important to note that this view is not suggesting that examples for moral guidance can be found in nature. Clearly, human civilization distinguishes itself from nature precisely because it orders itself on moral rather than biological principles. The symbolic argument only suggests that we can find appropriate expressions of our moral beliefs in aspects of the natural world. If we do, it may be prudent to preserve them to help strengthen these attitudes.

EXTENSIONIST OR BIOCENTRIC ARGUMENTS

The arguments above are all anthropocentric. They perceive in the natural environment a number of influences on humans, and it is as a result of these influences that the environment acquires value. In the arguments below, this anthropocentrism is replaced by arguments which identify sources of intrinsic value in non-human individuals.

Animal welfarism

Animal welfarism is a direct extension of human welfare arguments, and grants moral considerability to all other species that demonstrate the capacity to experience pain. As Peter Singer (1985) suggests, animal welfarism is a liberation movement, in so far as it advocates rights for animals commensurate with their status as morally considerable beings. It is important to note that this entails *apportioning rights* to individuals, on the basis that they are *sentient*, or experience pain.

The granting of rights to sentient creatures gives them the status of moral *patients*, who are affected by the actions of moral *agents*. Agents are self-conscious beings who can exert judgement and control over their actions. They are thus responsible for their actions towards themselves, each other and to patients. Patients have no moral responsibilities but they are morally considerable (human babies are the most obvious example). Singer points out that the requirement for consideration does not entail absolute equality. There are instances where the suffering of an individual from one species will outweigh that of another. These kinds of decisions have to be weighed on an individual case basis.

Teleological biocentrism

Whereas animal welfarism defines moral considerability according to sentience, biocentric arguments define it according to life. Anything that demonstrates an organic development path, such as a seed developing into a tree or an egg to a fish, therefore satisfies the condition of moral considerability. Welfarism considers that animals can have desires and interests, at least in so far as they aim at the avoidance of pain. Biocentrism suggests that all livings things can have 'interests' in that they have a tendency to develop in a particular way. That development path indicates the 'unconscious interests' of the thing. If it is possible to identify these 'interests', it is possible that they can be helped or hindered. It is considered a moral wrong to hinder them, at least without good reason.

Paul Taylor (1986) has developed an ethic of 'respect for nature' based on biocentric egalitarianism. In it, he distinguishes all biological organisms as having basic interests (essential for survival) and non-basic interests (the equivalent of human wants). As a rule of thumb, human basic interests overrule basic and non-basic interests of other species; but basic interests of non-human species overrule non-basic interests of humans. From this position he develops several more detailed rules that represent in his view a respectful attitude towards nature (Box 5.2).

Such teleological arguments have been criticized from two sides. Ecologists have pointed out that natural systems involve webs of interrelationships, and the concept of respecting 'unhindered development' is virtually meaningless in these circumstances. At the individual organism level competition is the common state, not cooperation. Almost everything 'hinders' everything else. Questions regarding the moral significance of other entities' interests compared to the significance of humans' interests then become very tortured.

On the other hand, some philosophers criticize biocentric arguments for not going far enough. They suggest that arguments based on individual rights fail to protect two critically important entities: species and ecosystems. Under a biocentric 'calculus', members of common species are just as morally significant as members of endangered species, as both have recognizable interests as individuals. Similarly, the focus on individuals may give insufficient emphasis to the ecosystem that supports them. Extending rights away from specific individuals towards collectivities or groups, such as species, is seen by some as a critical step that environmental ethics needs to make away from traditional rights-based moralities.

ECOCENTRIC ARGUMENTS

These final arguments try to justify placing the focus of moral concern on collections or communities of individuals: either whole ecosystems, or species.

Teleological ecocentrism: systemic development

In addition to the interest of individuals, some ecocentrics identify moral concern with natural processes and properties apparent in collectivities and systems. Stalactites develop in a certain pattern, and ecosystems develop through successive stages, interrupted or continually perturbed by other natural climactic processes. The teleological argument defends the moral considerability of these aspects, so that a mountain environment can be said to have interests in that it develops in a particular way under the forces of nature, which have shaped it and continue to shape it. Impeding or redirecting these processes is therefore a matter of moral concern. Similarly, species can be taken to represent evolutionary processes and potentialities, which confer value upon the species as a collective group.

The problems with deriving moral considerability from a developmental tendency are clear. Successional or evolutionary stages appear to be the entirely unplanned uncoordinated result of activities at the individual level. Whereas an individual can be seen as a centre of life, it is difficult to see a system or process as a focus of direct concern. Some biocentrists have therefore endorsed ecosystem considerability but only in support of an individualistic ethic: the best way to observe a moral principle to preserve individuals is simply by preserving their ecosystem.

Box 5.2 The ethic of respect for nature (Taylor 1986, with permission)

Taylor's ethic of respect for nature suggests that humans ought to observe four rules or duties in their relationship to natural, living things.

1 *The rule of non-maleficence* The duty to do no harm to any entity in the natural environment that has a good of its own.

2 *The rule of non-interference* This includes two duties: (i) a general hands-off policy regarding natural systems, and (ii) avoiding placing restrictions on the freedom of any particular organism.

3 *The rule of fidelity* The duty not to deceive or mislead wild animals (thereby outlawing activities such as hunting and fishing).

4 *The rule of restitutive justice* This requires that we make some form of reparation to morally considerable entities that have been harmed.

For Taylor, the duty of non-maleficience is most important, with the priority of other rules being determined in individual cases. He then defines five principles for the resolution of competing claims for moral significance.

1 *Principle of self-defence* It is permissible for moral agents to protect themselves against dangerous or harmful organisms by destroying them.

2 *Principle of proportionality* In essence, this suggests that basic needs of any species should always override non-basic needs of any other species, though care is needed in cases where needs are substantial but still non-basic.

3 *Principle of minimum wrong* If needs are being pursued by humans which will damage the needs of other species, these human needs must be pursued in the way that is least damaging to other living entities.

4 *Principle of distributive justice* All organisms are of equal inherent worth, and should therefore receive equal consideration in deciding any course of action, although certain priorities will undoubtedly have to be determined in favour of certain species.

5 *Principle of restitutive justice* Harms done to organisms should be compensated in some way.

In Taylor's view, these rules and principles combine to make an 'attitude of respect for nature', which is the appropriate attitude for a moral agent to hold. A schema for utilizing the principles with respect to basic interests and non-basic interests is given below.

		Harmless to humans (or their harmfulness can reasonably be avoided)		
Wild animals and plants . . . in conflict with . . .	Harmful to humans	Basic interests in conflict with . . . Non-basic interests		Basic interests
Humans		Intrinsically incompatible with respect for nature	Intrinsically compatible with respect for nature, but extrinsically detrimental to wildlife and natural ecosystems	
Priority principles	(1) Self-defence	(2) Proportionality	(3) Minimum wrong	(4) Distributive justice
			. . . when (3) or (4) have been applied . . . (5) Restitutive justice	

The land ethic: community

The Leopoldian land ethic (Leopold 1949) is probably the best known, and most radical, of ecocentric arguments. This invokes a concept of ecosystem health as a proper focus of moral concern, similar to Plato's identification of social harmony as a moral value in a political system. Both these philosophies use this principle of health to identify a 'unitary being'—ecosystem or state—composed from individuals. In these systems the rights and responsibilities of individuals are then *derivative* from the whole community. In political democracies, in contrast, rights are *constitutive* of a community. That is, democratic organizations derive their legitimacy from the rights they grant their members; such rights are established for the mutual benefit of all citizens, not drawn up in service of the state itself. To adapt a famous phrase from US President Kennedy, the land ethic asks not what your ecosystem can do for you, but what you must do for your ecosystem.

Liberal philosophers have, on this basis, often rejected ecocentric philosophies as fundamentally totalitarian in their attitude to human rights. Certainly there are some Deep Ecologists who have stated categorically that they value endangered wildlife more than individual human life. However, most ecocentric philosophies do not go nearly so far. The land ethic simply suggests that humankind should see itself as engaged in a wider set of moral duties than just those to other humans. Ecosystem considerability is an additional, rather than alternative, set of moral considerations. The needs of humans may remain paramount but they become more closely identified with the components of the ecosystem through its functional, aesthetic and spiritual significance. Maintenance of ecosystem health then becomes an extension of moral duties beyond the traditional boundary of human society.

5.5.3 Conclusions on moral values and nature

The appropriate moral attitude towards the non-human world continues to be a focus for intense ethical debate. In summary, the arguments can be distilled down to two central questions.

1 What is the appropriate moral attitude to have towards natural resources?

2 What is the appropriate way to make decisions about the use of natural resources?

These questions are related but clearly distinct. The issue of an appropriate moral attitude is obviously prior to deciding on action. Defining the moral considerability of nature is both an individual exercise and a social activity. It is individual because ultimately all moral judgements arise at the level of individual responsibility. It is also a social activity because environmental assets are public goods. The way in which they are used thus impacts on all members of society; environmental attitudes of whatever kind thus entail support for particular social policies.

The question of environmental decision-making itself arises out of two aspects of the 'attitude' question. The first issue is how any specific moral attitude can be put into practice. The second is how decisions can be reached when there are differences of basic attitudes between people.

Determining appropriate action in the light of a particular environmental attitude seems at first glance relatively straightforward. There are situations, for example, where species are endangered, when an obvious moral principle supporting species preservation is applied through a particular recovery or restoration programme. However, complications even with this case are apparent. How much effort should be put into the recovery programme? What is the precise definition of endangered? In these cases, as ecologists and policy-makers are aware, the devil is in the detail. Relatively 'pure' value judgements about the intrinsic value of a species may lead to very 'impure' moral reasoning in a genuine effort to determine appropriate behaviour in a real-world situation.

This kind of exercise in practical ethics is complicated even further in the presence of clear disagreements about pure value judgements. Here, there are two basic responses. One is to accept the prevailing majority attitude, if there is one. The second is to try to compromise and accommodate as many attitudes as possible. This may mean satisfying elements of each in proportion to how widely they are held or defining some 'minimum' level of representation for different beliefs.

The search for an appropriate compromise in the case of these seemingly fundamental differences in value judgements is an exercise in moral *pragmatism*. This stresses finding workable solutions to real-world problems, assisted by philosophical thinking which can help to clarify and draw out the range of possible alternatives. In the words of Sagoff (1988):

> '[W]e have to get along without certainty; we have to solve practical not theoretical problems; and we must adjust the ends we pursue to the means

available to accomplish them. Otherwise, method becomes an obstacle to morality, dogma the foe of deliberation, and the ideal society we aspire to in theory will become a formidable enemy of the good society we can achieve in fact.'

Debate over moral attitudes towards the natural world is an integral part of this pragmatic approach to environmental decision-making which encompasses, in addition, insights from ecology and economics. These three disciplines work in combination to illuminate the central issues of natural resource use.

5.6 Summary

Examining the justification for making value judgements is the purpose of metaethics. The basic debate regards whether values are based on *feelings* of observers, or whether they are in some sense *properties* that belong to objects. Whichever position is held, it remains evident that moral values still at a fundamental level determine action.

Moral values interrelate with economic systems in three ways, evident in notions of justice. Commutative justice entails equality or fairness of exchange. Distributional justice involves the equity of economic systems and the division of goods amongst society's members. Productive justice concerns the end purpose of work itself and its contribution to a 'good life'.

The economic notion of welfare, related only to material or consumption opportunities, is acknowledged to be limited. Other factors that can be considered are needs and wants, and the value of social systems and interaction in maintaining welfare as evident in the social limits to growth debate and notions of community well-being. The economic system may inhibit these social sources of value by focusing too exclusively on the utility-maximizing individual as the centre of economic analysis.

Moral duties to the natural world can be considered in terms of defining the concepts of moral considerability and moral significance. Four broad attitudes regarding nature preservation are evident. Human welfare arguments value only human welfare, but accept that the natural environment is still instrumentally valuable for such welfare. Virtue-based arguments see welfare in terms of a whole personality and not (only) enjoyment or utility. Biocentric views consider all biological entities to have a moral value of some kind. Finally, ecocentric positions see responsibilities relating to entire systems or species.

Further reading

Ekins, P. & Max Neef, M. (eds) (1992) *Real-Life Economics: Understanding Wealth Creation*. Routledge, London.

Elliot, R. (ed.) (1995) *Environmental Ethics*. Oxford University Press, Oxford.

Foster, R. (1997) *Valuing Nature?* Routledge, London.

O'Neill, J. (1998) *The Market: Ethics, Knowledge and Politics*. Routledge, London.

Sagoff, M. (1988) *The Economy of the Earth*. Cambridge University Press, Cambridge.

6 The Economic Approach to Environmental Valuation

'A cynic is a man who knows the price of everything and the value of nothing.' [Oscar Wilde]

6.1 Introduction

This chapter is concerned with how economists value particular aspects of the natural environment and changes that take place within it. The central question posed is simply how to place a monetary value on the goods and bads arising from these changes, which may affect environmental quality or the available stocks of some resource. A wide range of economic valuation techniques have been developed to try and answer this key question and they are reviewed here.

6.2 The rationale for economic valuation

Before introducing the techniques themselves it is useful to recap the assumptions that underlie the rationale for such economic valuation in the first place. Recalling the outline of neoclassical economics presented in Chapter 3, the economic valuation of environmental resources assumes that these resources somehow impact on the utility or well-being of individuals. Furthermore, money measures can act, albeit imperfectly, as measures of the extent to which the utilities of individuals are affected. This is because when faced with a choice between two goods, say 'money' and 'air quality', individuals can (implicitly or explicitly) identify a satisfactory trade-off between the quantities they want of these two goods. Such a trade-off represents a point where the individual is indifferent between the two levels of the alternatives on offer. In other words, there is some amount of money that can exactly substitute for some given decline in air quality, leaving the individual no worse off after the decline than before, because the sum of money substitutes or compensates for that decline.

Faced with limited budgets and with sets of conflicting uses for scarce natural resources, most obviously a choice between the conservation or the consumption of such resources, decision-makers seek guidance on how to trade-off between those possible uses so as to maximize welfare or utility overall; in other words, to allocate those resources so that the overall benefits are maximized. For an individual decision-maker this choice can be made with a direct knowledge of personal goals and preferences, whereas democratic governments must operate on behalf of all their citizens in determining how to achieve this improvement in overall welfare.

As making an informed choice seems to entail in some way comparing the goods and bads associated with the alternatives, economists have attempted to help decision-makers by finding ways to measure the wide range of effects of environmental changes on a single monetary scale.[1] Comparisons between the effects of different actions are thereby made very straightforward and the entire decision-making process can be represented as a cost–benefit analysis (CBA), considered separately in Chapter 8. Much confusion can arise in this process from conflating what might be called philosophical or ethical notions of value, and technical or economic ones. The derivation of a monetary value for goods that do not have a market value is basically an attempt to extend the utilitarian and democratic principle of the free market into environmental decision-making. From the market economic perspective, the problem is that, for various reasons, environmental goods cannot be traded in the market, where individuals themselves could decide how much of them they wanted. Valuation techniques seek to compensate for this lack of market values, recalling that by themselves market values indicate the demand for such goods, by deriving implicit prices.

Underlying this valuation approach are therefore several important assumptions.

[1] An interesting study of the philosophical issues entailed in making comparisons between different kinds of goods can be found in Foster (1997).

1 Environmental change, in order to have a (non-zero) monetary value, must impact in some way on the utility, or well-being, of individuals.
2 The total value of a change in environmental goods is the sum of the values of its effects on individuals; that is, 'society' is understood to be simply all individuals added together.
3 Different kinds of impact must be *commensurable*; that is, they can be compared and some quantity of money can always act as a substitute for some quantity of an environmental good.
4 Similarly, environmental goods of equal value can be substituted for each other with no loss of welfare.

In Chapters 4 and 5 some possible philosophical objections to these assumptions were raised and so they will not be repeated here; further aspects of monetarized decision-making are considered in Chapter 8 (CBA). Throughout the rest of this chapter particular objections or limitations are noted where they are relevant to particular techniques, but we concentrate principally on understanding the operation of the techniques themselves.

6.3 Total economic value

Environmental goods clearly impact on human welfare in a wide variety of ways. An important step in valuing the overall effect of such goods is therefore to determine the range of these impacts, or the different categories of value that need to be captured by valuation techniques. These can then be summed to provide a total economic value (TEV), which should indicate the total value of the resource in so far as it affects human welfare.

A taxonomy of such economic values associated with natural resources has been proposed by a number of environmental economists (Weisbrod 1964; Krutilla 1967). Although there is not yet complete agreement on this taxonomy, it is widely accepted that environmental values can be split into two broad categories: *use values* and *non-use values*.

Use values are associated with the benefits that come as a result of direct contact with the natural resource in some way. This might be from direct consumption—for example, from extracting timber from forests—or from so-called secondary or non-consumptive uses, such as soil stabilization and water retention from a forest; values which indirectly benefit humans through their wider effects on the ecosystem. Direct consumptive values are also known as primary values or marketed goods and services (things that can be directly paid for); whereas secondary values are correspondingly non-marketed goods and services, and non-consumptive (the resource itself does not have to be consumed or affected in any way in order for the value to derive from it) (Pearce & Turner 1990; OECD 1989a; Bateman 1993). In addition to these values, environmental economists have introduced an *option value*,[2] which is defined as the value placed on environmental assets by those people who want to secure the use of the good or service in the future.

Non-use values correspond to those benefits which do not imply contact between the consumers and the good. People do not need to use the good in any way, either directly or indirectly, in order to derive value from it. Such values are generally known as *existence values*: values that derive simply from the knowledge that a particular good exists. Existence values can themselves be further subdivided into *intrinsic values* and *bequest values*. Intrinsic values relate to the utility derived from simple knowledge of the existence of a landscape or a particular habitat, such as the satisfaction of knowing a forest is preserved 'for itself' and not as a function of any human use. Bequest values involve altruism towards others as part of the value, such as the satisfaction of preserving forests for the enjoyment of other people both now (intragenerational) and in the future (intergenerational). Table 6.1 provides an illustrative but necessarily incomplete list of the values or benefits usually attributed to forests.

The total economic value of an environmental asset is obtained by summing up the three value components: use value, option value and existence value. A schematic illustration of how to derive such a TEV is given in Table 6.2. Two errors should be avoided in this process.
1 It is important to check that the components are not mutually exclusive. So, for example, the primary use values of clear-felling for timber clearly cannot be added to non-consumptive values, such as recreation, or secondary values, such as soil protection.
2 Values should not be double counted. This creates challenges for valuation techniques in determining the scope of the values captured by any one technique. Environmental goods are rich composites and decomposing them into constituent value strands to be assessed by individual techniques can be problematic.

[2] Other authors regard option value as a non-use value.

Table 6.1 Total economic value (TEV) of forests.

Use values				Non-use values	
Direct and marketable	Direct and non-marketable	Indirect	Option	Bequest	Existence
Timber	Scenery	Climate mitigation	Biodiversity	Biodiversity	Biodiversity
Fruits, nuts, herbs, latex, gum arabic, litter, etc.	Recreation	Air quality	Wildlife	Scenery	Wildlife
Fuelwood	Community integrity	Soil quality	Community integrity	Recreation	
Forage and fodder	Wildlife	Water cycle	Scenery recreation	Wildlife	
Developed recreation and hunting		Biodiversity	Air, soil and water quality	Air, soil and water quality	

Table 6.2 Framework for total economic value (TEV) computation. x indicates the category of value provided by each good, xx and xxx in the final column indicate that the overall economic value of a good is composed of two or three different individual values.

Goods	Use value			Option value	Non-use value	Economic value
	Direct and marketable	Direct and non-marketable	Indirect			
Timber	x					x
Fruits	x					x
Nuts	x					x
Herbs	x					x
Fuelwood	x					x
Forage	x					x
Developed recreation	x					x
Hunting	x					x
Wildlife		x		x	x	xxx
Scenery		x		x	x	xxx
Non-developed recreation		x		x	x	xxx
Biodiversity			x	x	x	xxx
Community integrity		x		x		xx
Climate mitigation			x	x		x
Air quality			x	x	x	xxx
Soil quality			x	x	x	xxx
Water quality			x	x	x	xxx
	Total direct and marketable	Total direct and non-marketable	Total indirect	Total option	Total non-use	TEV

6.4 Economic valuation techniques

The techniques available for environmental valuation have been classified in many different ways. The approach used here (Table 6.3) is adapted from the classification proposed by Munasinghe (1993). It distinguishes between four broad groups: conventional market approaches; implicit market approaches; constructed market approaches; and non-economic methods. All these groups are further subdivided into actual behaviour-based techniques, potential behaviour-based techniques and other techniques. The techniques within each of these groups are now discussed in turn.

6.4.1 Conventional market approaches

The principle behind conventional market approaches is to establish a link between an environmental impact and some other good *that already has a market value.*

Table 6.3 Economic valuation techniques classification. (Adapted from Munasinghe 1993.)

Technique	Conventional market	Implicit market	Constructed market	Non-economic methods
Actual behaviour	Productivity change Opportunity cost Dose–response Preventive expenditures	Travel cost method Wage differential Hedonic pricing	Artificial market	
Potential behaviour	Shadow projects Substitute costs		Contingent valuation	
Other	Cost-effectiveness			
				Multicriteria analysis Delphi techniques Environmental impact assessment Linear programme

PRODUCTIVITY CHANGE OR PRODUCTION FUNCTION APPROACH

Under the production function approach, the background environment is considered as a factor of production. In other words, it contributes to the production of goods just as inputs of labour and capital do. Changes in environmental quality then lead to changes in productivity and production costs, which in turn lead to changes in the prices and outputs of goods. Under the production function approach, the value of an environmental change, such as an increase in soil stability, would be the value of *additional* crop resulting from the environmental improvement, as compared to the level of production without such an improvement.

This method is certainly the most popular of valuation techniques and has been widely applied in real project analysis. For example, Barrow *et al.* (1986) carried out a study on the impacts of a forestry development project on usable run-off from water catchments. They estimated the value of the impact on the basis of the forgone net benefit of crop production downstream as a result of decreased water yields. Anderson (1987) also studied afforestation in Nigeria and showed that shelter-belts increased agricultural yields by 10–30%. Hodgson and Dixon (1988, 1992) used this approach to estimate the value of the impacts of logging on fishery and tourism on a coastal area. It was estimated that a logging ban yielded larger revenues from increased tourism and fishing than the lost revenues from logging.

The technique is popular partly because it is relatively simple to understand, though not necessarily to implement, and non-contentious. It relies on two factors:

1 estimating the physical effects of a project on the output of specific marketable commodities; and
2 pricing these commodities.

Significant problems can arise in specifying the physical effects on production, most notably where there are interactions between a number of forces; isolating the effects on production of a project that influences only a subset of these forces can be very difficult.

Although the pricing of the output appears straightforward, it can also be problematic. In some instances, the going market price may not reflect the 'true' market price (this issue is discussed in more detail in Section 8.3.3), which distorts the estimation of benefits. In addition, if the effect on productivity is large and positive, this will in itself deflate the market price for the commodity by increasing supply. Not only will this affect the final value ascribed to benefits, but it may have wider effects across the whole locality by changing the behaviour of consumers and producers in response to a fall in price. Some understanding of these possible effects should inform the judgement of overall benefits deriving from a project.

OPPORTUNITY COST APPROACH

In many ways a direct corollary of the production function approach, the opportunity cost method simply uses market prices to estimate the value derived from using a resource in one particular way—often, simply preserving it—by examining the *values of alternative uses*. In other words, it estimates unpriced goods and services from preservation by measuring the benefits forgone by not using the same resource for other more consumptive uses. Thus, the costs of preserving forests for a national park rather than harvesting them for timber would be measured by using the *forgone income* from selling timber. This approach measures what has to be given up for the sake of preservation and thereby takes this as at least a minimum value for the preserved resource if it continues to be preserved.

The opportunity cost approach is a very useful technique when benefits of certain uses, such as preservation, protection of habitats, cultural or historical sites and aesthetics, cannot be directly estimated. Krutilla (1969) and Krutilla and Fisher (1985) used this technique in a famous study aimed at valuing the viability of a dam project in Hell's Canyon, USA. Rather than trying to value all the benefits of the canyon in its natural state, the authors carried out a conventional cost–benefit analysis of the proposed project and of its next cheapest alternative (a nuclear powerplant). They showed that the benefits of the dam project were not considered large enough to justify the irreversible loss of a unique natural area, given the costs of the other opportunities available to generate the same power output.

DOSE–RESPONSE APPROACH

The dose–response approach can be considered as a further variation on the production function principle, which is primarily used to estimate pollution effects on health, materials and vegetation. The procedure is as follows: first, scientific factors relating the cause (pollution) to effects (illness) are established in quantitative terms, requiring a dose–response function of the following form to be constructed:

$R = f(P, \text{other variables})$,
where R = response and P = cause (pollution).

Once these relationships have been established, coefficients of the type R/P are established and a demand curve is used to assess the economic value of effects. For example, a study reported by Dixon *et al.* (1994) on the use of the cost-of-illness approach in Mexico in 1991 proceeded to the economic evaluation in three steps:

1 quantify the ambient concentrations of various pollutants;
2 determine the dose–response relationship to infer incidence of disease, including both morbidity and mortality, in the population; and
3 estimate the costs of the increase in morbidity and mortality on the basis of treatment costs (replacement or defensive expenditures), loss of wages and loss of life.[3]

Importantly, benefits and costs estimated with this technique should be considered as lower limit values because it only accounts for out-of-pocket expenses (in the case of pollution-related illness, such costs as loss of earnings resulting from illness, medical costs, etc.) but disregards other values, such as suffering, and other costs incurred for preventing illness (flu shots, immunizations, etc.). It is clearly best suited to situations where direct cause–effect relationships can be established, and where related expenses are clearly identifiable.

DEFENSIVE OR PREVENTIVE EXPENDITURES

Individuals and communities often spend money mitigating or eliminating damages caused by adverse environmental impacts. This is the case, for example, when double-glazed windows are installed for reducing traffic noise, or extra-filtration installed for purifying water. Under the 'defensive' expenditure valuation method, these defensive expenses are themselves considered as minimum estimates of the benefits of environmental improvements, because it is assumed that individuals spend money on these expenditures to try and gain an improvement in environmental quality at least equal to that which has been lost. In other words, individuals are signalling their preferences for a rise in environmental quality through their willingness to pay for 'defensive goods' that raise their personal level of environmental quality. Such an increase in quality must provide a benefit to the individual at least as great as the cost of the defensive equipment, because otherwise the individual would settle for lower quality and avoid spending the money.

The advantage of this technique is that the costs of mitigation are generally far easier to estimate than the direct costs of environmental damage itself. Three options are open to valuers in this regard.

1 They can directly observe expenditures made by individuals and households.
2 They can ask households what they might be willing to pay for such mitigation (an approach close to 'contingent valuation', considered below).
3 They can employ professional estimates of what expenditures would have to be undertaken in order for

[3] Monetary estimations of the value of life have raised some ethical as well as theoretical criticisms (Pearce 1989). A more appropriate and accepted definition of this technique is that it seeks to place a value on changes in the statistical probability of illness or death.

Box 6.1 Defensive expenditure theory

Let P = the householder's subjective value of damage caused by pollution of water, X = the costs incurred for extra filtration, and P′ = residual pollution remaining after mitigation measures.

The rational householder will spend money for damage mitigation when:

P = X + P′ or P − P′ = X.

The economic benefits or the consumer's surplus is the total area under the demand curve.

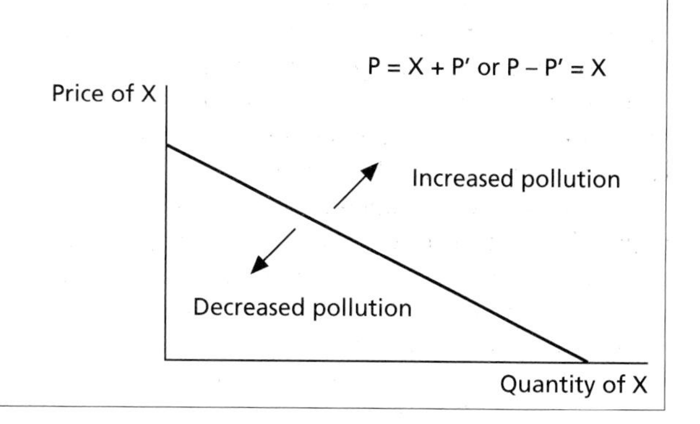

individuals to protect themselves adequately against adverse impacts.
In all cases, the expenditure is generally clearly defined (Box 6.1).

This technique has been applied in a variety of situations, such as studying the costs of noise control in domestic dwellings near London's Heathrow airport (Starkie & Johnson 1975) and in evaluating soil-management techniques designed to stabilize upland soils and to enhance agricultural production (Kim & Dixon 1986). In the latter case, lowland farmers were prepared to incur costs for the construction of dikes to divert water and thus prevent siltation of their fields. The subjective valuation by farmers of the benefit of soil erosion prevention measures was thus at least as great as the cost they incurred to construct the dikes.

Some limitations to preventative valuations should be noted. The fact that such valuations are likely to be minimum estimates has already been touched upon. There is also the possibility that consumers simply spend too much on prevention, so that they end up regretting their level of expenditure which over-compensated for the initial damages. Perhaps more seriously, preventative expenditures are clearly related to ability to pay; expenditures that might over the long term yield high benefits will not be undertaken if people simply cannot afford to do so in the short term. As in all market-based approaches, income distributions therefore play an important part in the final value estimates. Finally, although expenditures are undertaken to mitigate losses in environmental quality, they may provide secondary benefits. Thus double-glazing may prevent noise disturbance, but it also improves insulation and possibly appearance (Winpenny 1991). Such factors will lead to over-estimation of environmental damage if the total expenditure is assumed to go purely to mitigate damages.

REPLACEMENT OR RESTORATION COST

Replacement costs, sometimes considered a special category of preventive expenditures, estimate the value of environmental damage according to the amount that has (or would have) to be spent to restore the environment to its previous undamaged state. Thus, in the case of pollution incidents, they calculate the actual or potential clean-up costs, which may be a good indicator of whether it is worth investing in preventative measures. Winpenny (1991) notes that moving house to avoid an environmental nuisance is itself a form of replacement cost from the perspective of the individual, who incurs a cost in order to 'replace' the old house location with a new one that is free of the nuisance. Such decisions are, however, clearly constrained by a great many factors.

Replacement costs have been the subject of criticism by environmentalists regarding whether damaged habitats can ever meaningfully be restored to their pristine state. The UK Environment Agency uses replacement cost as a guide to the fines levied on water polluters. The fines are set to recover from polluters the costs of clean-up following pollution incidents. However, the rationale here is that the fines are intended to act as a deterrent. They do not assume that polluters are free to damage watercourses as long as they subsequently pay for their restoration.

SHADOW PROJECTS

Another similar technique to replacement cost is that of valuing a shadow project, which is a project that provides an equal, alternative environmental good or service elsewhere in the area that suffers an environmental loss. The cost of the shadow project is calculated and used as an estimate of the value of the original good.

The possible alternatives for recreating an environmental good are asset reconstruction (providing an alternative habitat site for a threatened wildlife habitat); asset transplantation (moving the existing habitat to a new site); and asset restoration (enhancing an existing degraded habitat). The cost of the chosen option is added to the basic resource cost of the proposed development project in order to estimate its full cost. Inclusion of shadow-project costs gives an indication of how great the benefits of the new project must be in order to outweigh the losses it causes.

The problems noted above for replacement costs are clearly applicable to shadow project proposals too. The proposed alternatives, by definition, will have differing qualities to the original site. It may therefore be problematic to determine how successfully any shadow scheme can provide the benefits of the original. Rather than focusing on directly equivalent sites, planners may prefer to consider instead *planning gains*, where developers guarantee protection (perhaps through purchase) of other non-related environmental sites to compensate for damaging the site under development. Such gains are not strictly speaking replacements, and are not evaluated economically, but simply ensure overall improvements in the level of environmental quality as judged by the relevant authority.

SUBSTITUTE COSTS

The substitute or alternative cost approach values a particular environmental service or good according to the cost of available substitutes. For example, private swimming pools may be regarded as substitutes for clean lakes or streams, or private parks may be considered substitutes for national parks. If the two alternatives provide an identical service, the value of the environmental good is the *saved cost of using the substitute*. An example is provided by Misomali (1987), reported by Price (1989). In a study on fuelwood plantations in Malawi, the author priced fuelwood on the basis of the saved kerosene imports. Similarly, Newcomb (1984) looked at fuelwood as a substitute for dung in domestic heating. Dung was thus made available as a fertilizer, and the costs of chemical fertilizer imports (in fact, imports plus internal marketing costs) were saved. Therefore, the resulting value for fuelwood was the saved costs of imports of chemical fertilizers.

The validity of this approach depends upon three main conditions being respected:

1 that substitutes can provide exactly the same function of the good or service substituted for, which is seldom true especially in the case of environmental goods;

2 that the substitute is actually the least-cost alternative; and

3 that willingness to pay (WTP) evidence indicates that per capita demand for the substitute service would be the same as for the original good.

COST-EFFECTIVENESS APPROACH

Strictly speaking, cost-effectiveness analysis (CEA) is an entire decision-making methodology in its own right, comparable in scale to cost–benefit analysis. Quite simply, cost-effectiveness measures costs without attempting to value the benefits of any particular plan. The approach is usually involved in the following circumstances:

1 *insufficient knowledge to establish dose–response relationships*: in this case, the decision-makers usually set a goal and analysts try to find out the least-cost means of achieving it;

2 *availability of a given amount of money to be spent*: the objective here is to spend the money in the most effective way; and

3 *several possible goals to be achieved*: the analyst examines which of them is preferable on the basis of the cost of each.

The most frequent case is the first one, in which the analyst is asked to identify the least-cost means to achieve a pre-established goal, e.g. a pollution emission standard. The CEA does not tell us whether the decision to spend money in meeting the emission standard is worthwhile or not, but if a political decision has already been taken it is an important procedure for ensuring a rational use of financial resources.

6.4.2 Implicit market techniques

These approaches assume that the behaviour of individuals reveals implicit valuations of features of the environment. This may be through wages accepted to work in locations with different levels of environmental quality, prices or rents paid for properties that have particular levels of environmental amenities, or costs associated with specific activities, such as recreational outdoor trips.

HEDONIC PRICING METHODS

Hedonic pricing (HP) valuation methods are based on the theory of commodity characteristics pioneered by Lancaster (1971), and later by Griliches (1971) and Rosen (1974). They seek to isolate the contribution that environmental quality makes to the total market value of an asset. In essence, Lancaster's theory assumes that the total value of any good is a function of the set of characteristics associated with that type of good. Thus, any particular make of car has a set of characteristics, such as speed, fuel-efficiency, colour, etc., and the final value that consumers are prepared to pay for a car is related to how it is evaluated across these characteristics. If the level of any one characteristic, say fuel efficiency, is increased while holding all other characteristics constant, it should be possible to observe how the price of the good changes. This change will give an estimate of the value of that particular characteristic to the purchaser.

Some experiments in HP of environmental goods have focused on property prices (considering either rentals or purchases) to see how they vary with associated environmental attributes. The technique requires analysts to collect a large quantity of data relating to all the factors that influence the value of a property, such as number of rooms, central heating, garage space, proximity to parkland, etc. These data are collected for a large number of houses, and statistical techniques are then used to isolate the variations in purchase or rental price that come about as a result of changes in environmental factors while holding standardized values for all other characteristics constant.

The proportion of the price differential between two otherwise identical houses accounted for by the change in the environmental quality characteristic reveals an individual purchaser's valuation of the importance of environmental quality. This, as for any other characteristic, is likely to vary between individuals according to their personal preferences. Analysis needs therefore to determine how quality valuations differ according to social factors, such as income, age and so on. Once average implicit values for these segments of the population have been found, a total value for a change in environmental quality can be derived by summing the resultant demand curves according to the composition of the affected population.

Brookshire *et al.* (1982) undertook this kind of analysis for 634 households in Los Angeles. They divided house characteristics into four types: housing structure (rooms, etc.); neighbourhood qualities (crime rates, etc.); accessibility (distance to beaches, work centres); and air quality. They then separated out the contribution of individual variations in all the variables thought to influence these types. They were able to account for 90% of the variation in house prices through their model, and found that all the variables correlated as expected with variations in quality and were significant at the 1% level, with the one exception of crime rates. Overall, they found an implicit willingness to pay for air quality improvements ranging from $15.44 to $45.92 for a move from areas of 'poor' to 'fair' quality, and from $33.17 to $128.46 for a move from 'fair' to 'good'.

Hedonic pricing has several limits in its application. The most important of which may be the quantity of variables required, which are seldom recorded in the official statistics even in developed countries. Brookshire *et al.* (1982) identified no less than 18 variables necessary in the analysis of the housing market, most of which must be estimated. Another disadvantage of HP methods is the huge amount of data required in order to undertake the required statistical analysis. The quantity of data required increases when the demand function must be estimated on the basis of income and other socioeconomic data as well as the supply of houses on the market. Reliability of data is also considered a shortcoming of this method. House prices, for example, are often distorted and owners of houses frequently sell or rent at lower prices than the maximum offer received, therefore the observed price may not correspond to a genuine valuation. Finally, this method does not capture non-use values and does not take into account the effect on prices of individuals/households' expectations on the future quality of landscape (Abelson & Markandya 1985).

TRAVEL COST METHOD

Travel cost method (TCM) was first proposed in 1947 by Harold Hotelling and has since become one of the most widely used techniques in environmental valuation. It centres on the expenditures incurred by households or individuals in order to reach recreational sites, and uses these expenditures as a means of measuring

Box 6.2 Theoretical concepts of zonal travel cost method (ZTCM)

In mathematical terms the trip demand curve is defined as:

$$Vhj/Nh = f(Ch,Xh),$$

where Vhj as the total number of trips by individuals of zone i per unit of time; Nh as population of zone i; Ch as travel cost from zone i;[4] Xh are socioeconomic explanatory variables.[5]

The visitor rate Vhj/Nh is calculated as visits per 1000 population in zone h. In each zone the household consumer surplus for all visits to the site is calculated by integrating the equation of the type:

$$Vhj/Nh = a + b\,Ch$$

between the price (cost) of visits actually made from each zone and the price at which the visitor rate would fall to 0 (that is, the vertical intercept of the demand curve at point P in the figure)

$$C.S. = \int_{Ch=B}^{P} (a + b\,Ch)dCh$$

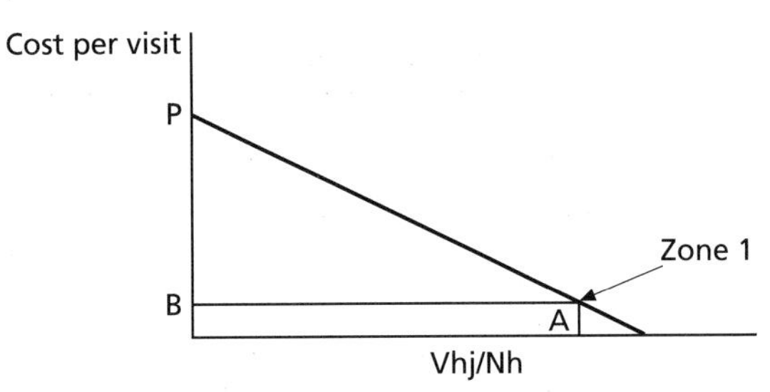

Annual total consumer surplus for the whole recreation experience can be estimated in each zone by first dividing total household consumer surplus (BAP in zone 1) by the zonal average number of visits made by each household to obtain the zonal average consumer surplus per household visit. Then the result can be multiplied by the zonal average number of visits per annum to obtain annual zone consumer surplus. Finally, aggregating zonal consumer surplus across all zones gives the estimate of total consumer surplus per annum for the whole recreational experience of visiting the site.

willingness to pay for the recreational activity (Trice & Wood 1958; Clawson 1959). The sum of the cost of travelling, including the opportunity cost of time, and any entrance fee gives a proxy for market prices in estimating demand for the recreational opportunity provided by the site under investigation. By observing these costs and the number of trips that take place at each of the range of prices, it is possible to derive a demand curve for the particular site being visited.

Two main variants of TCM exist: the zonal travel cost model (ZTCM) and the individual travel cost model (ITCM). The ZTCM divides the entire area from which visitors originate into a set of visitor zones, and then defines the dependent variable as the visitor rate (that is, the number of visits made from a particular zone in a particular period divided by the population of that zone). The alternative ITCM defines the dependent variable as the number of site visits made by each visitor over a specified period. Box 6.2 provides a simple mathematical formalization of the ZTCM method and Box 6.3 for ITCM.

There is a general agreement in considering TCM as one of the most effective approaches in valuing recreation services (Ward & Loomis 1986; Smith 1989; Bockstael *et al.* 1991). Nevertheless, as Smith (1993) points out, this model has been used so far to define 'the demand for and value of services provided by specific types of recreation sites and not to estimating the value people place on changes in the sites' quality feature'. Furthermore, the decision to use either zonal or individual TCM approaches is likely to have a significant impact on the results obtained. Similarly, as with the other techniques addressed above, TCM can only measure the use value of recreation sites.

Other potential problems encountered with this method include the following.

1 *Determination of the opportunity cost of on-site and travel time.* It is not at all clear how time spent in recreation should be valued. Should this be valued at full salary equivalents for those on holiday?

2 *Treatment of substitute sites.* If alternative targets for visits exist, how do these affect the values that are derived for any particular site?

4 Travel cost is the sum of expenditures incurred for petrol, opportunity cost of time for travelling and for visit on-site.
5 These include factors such as income levels, spending on other goods, the existence of substitute sites, entrance fees, quality indices of *n* substitute sites, etc.

Box 6.3 Example of a zonal travel cost method application (adapted from: Bateman 1993)

Assume the following situation:

Zones	Zonal Population (Nh)	Household visits (Vhj)	Average number visits household	Average travel cost per household (Ch)	Consumer surplus per household all visits p.a. ($)	Consumer surplus per household per visit ($)	Total consumer surplus p.a. ($)
1	10 000	12 500	1.25	0.16	2.60	2.08	26 040
2	30 000	30 000	1.00	1.00	1.67	1.67	50 100
3	10 000	7 500	0.75	1.83	0.94	1.25	9 400
4	5 000	2 500	0.50	2.66	0.42	0.84	2 100
5	10 000	2 500	0.25	3.50	0.10	0.40	1 000
Total consumer surplus of the whole experience							88 000

These data are available from questionnaires and from census records.

Column 1 Identification of zones of increasing travel cost.
Column 2 Total population (number of households) in each zone.
Column 3 Households visits per zone and per annum calculated by allocating sampled household visits to their relevant zone or origin.
Column 4 Household visit rate calculated by dividing column 3 by column 2.
Column 5 Zonal average cost of a visit calculated with reference to the distance from the trip origin to the site.
Column 6 Demand and consumer surplus estimates using the hypothetical linear demand function,

$Vhj/Nh = 1.3 - 0.3\ Ch.$[6]

Consumer surplus for each zone is obtained by integrating the demand curve between the actual cost of visits and that price at which the visitor rate would fall to 0.

$$\text{C.S. Zone } 3 = \int_{Ch\,=\,1.83\,=\,B}^{P} (1.3 - 0.3Ch)\ dCh.$$

Column 7 Consumer surplus is divided by the zonal average number of visits made by each household to obtain the zonal aggregate consumer surplus per household visit.
Column 8 The zonal average consumer surplus per household visit is multiplied by the zonal average number of visits per annum to obtain annual zonal consumer surplus. Finally, annual consumer surpluses are cumulated across all zones to obtain the total consumer surplus per annum for the whole recreational experience.

3 *Multipurpose visits*. What component of overall cost should be allocated to one particular site on trips that visit several sites?
4 *Travel motivations*. Do visitors travel specifically to see one site, and visit others incidentally along the way?

Although many applications of this method are available in industrialized countries, only a few have been undertaken in developing countries. Some examples of these are Grandstaff and Dixon (1986) who used the TCM to value the benefits associated with a city park in Bangkok, Thailand and Amacher *et al.* (1991) who assessed willingness to pay for collected fuelwood in Nepal.

6.4.3 Constructed market

CONTINGENT VALUATION METHOD

The contingent valuation method (CVM) estimates environmental values simply by asking people, usually via some form of questionnaire, to state what they are willing to pay (WTP) for an environmental benefit or what they are willing to accept (WTA) in compensation

[6] This function assumes that demand is explained only by visit cost and that this relationship has an (un)likely linear form. It is also assumed that households in all zones react in a similar manner to visit costs.

Table 6.4 Empirical differences between willingness to pay (WTP) and willingness to accept (WTA) bids for the same good.

Study		WTA/WTP
Knetsch & Sinden (1984)		4.0
Coursey *et al.* (1983)	a	3.8
	b	1.6
Brookshire *et al.* (1980)	a	1.6
	b	2.6
	c	6.5
Bishop & Heberlein (1979)		4.8
Banford *et al.* (1977)	a	2.8
	b	4.2
Hammack & Brown (1974)		4.2

a, b and c refer to different experiments reported in the same paper.

for a loss. The questionnaire format thus simulates a hypothetical (contingent) market of a particular good, for example landscape quality. In effect, individuals are asked to reveal their indifference between sums of money (known as 'bids' or 'responses') and the supply of the good in question. Average bids can then be summed for the relevant affected population to give a final value for the good under study.

Following its initial development in the early 1970s and 1980s, CVM has developed into something of a major industry in the environmental valuation field. Its popularity can be attributed to two features. First, although it requires a certain amount of survey work, it does not require the huge amount of data necessary for many of the other techniques and it relies on relatively simple estimation techniques. Secondly, and perhaps more significantly, it may, in theory, be applicable to all goods and services, including use and non-use values, and, in particular, it is the only developed technique available for the evaluation of non-use values. This makes it particularly significant both economically and philosophically.

Much of the work on CVM has been in response to the number of problems which surround the basic technique. Many of these are recognized as systematic biases which can lead to unreliable or inaccurate bids or responses, or which lead to varying final estimates of value depending on the particular experimental design used to undertake the CVM survey. This clearly raises problems of defining the validity of final CVM value estimates and establishing appropriate questionnaire design. A further set of issues surround the underlying philosophy of the technique and respondents' reactions to it. An extensive examination of the overall methodology is given by Mitchell and Carson (1989) and more recently the technique has been reviewed by a so-called 'blue ribbon' economic panel (Arrow *et al.* 1993) which made a number of recommendations on design and applicability.

Below, we briefly review a few of the principal problems of bias that have been encountered in applying the technique.

Willingness to pay or willingness to accept?

One of the first points in questionnaire design is to decide whether or not to utilize willingness to pay (WTP) or willingness to accept (WTA). Willingness to accept questions may be phrased as 'What would you be willing to accept in compensation for good X?', whereas WTP questions may be phrased as 'What would you be willing to pay in order to maintain/achieve good Y?' Willingness to pay would seem to be the most appropriate measure for gainers from a resource allocation decision, and WTA the proper measure for losers from that same reallocation. Problems arise when it is not easy to identify gainers and losers, as this judgement is itself influenced by the individual's own perspective. It has been suggested that WTP and WTA should produce estimates of monetary value that are fairly close (within 5%) (Willig 1976). However, empirical studies have shown that WTA is usually significantly greater than WTP for the same good (Table 6.4). Also the variance of WTA is greater than WTP, and they are less accurate predictors of actual buying/selling decisions.

Strategic bias and the free-rider problem

Strategic bias arises when respondents deliberately misrepresent their true WTP or WTA in order to manipulate the results of the study: for example, to generate very high values. This is a well-recognized problem and some analysts have stated that it is so serious that consumer preferences cannot be obtained from direct questioning. However, several studies conclude that strategic bias is not a significant problem and can be ameliorated by good survey design (Schulze *et al.* 1981; Thayer 1981). The free-rider effect occurs when respondents decline to pay for goods, such as local amenities, because they anticipate being able to enjoy them without payment, thereby free-riding on the payments made by everyone else. Such attitudes would lead to systematic undervaluation of such amenities;

good survey design may again alleviate some of these problems.

Form of bidding

A related problem in formulating the questionnaire concerns how the WTP/WTA question is posed. It could be in a single closed-format question: 'Would you be willing to pay £X to preserve Y?'; an iterative bidding process, starting at some bid and increasing the value until the respondent finally declines to pay; or a continuous open-ended bid, with or without a payment card with possible values displayed on it, where the question is posed in the form 'What would you be willing to pay to preserve Y?' The choice of the form of questioning may influence the final value obtained.

A case in point is that of starting point bias. The bias arises if respondents interpret the initial bid suggested by the questionnaire as being indicative of market information, or as reflecting a guide to the sum they should be willing to pay. Responses may then tend to gravitate towards this point. Interestingly, the empirical evidence for starting point bias is not good. Of six studies reported in Garrod and Willis (1990), only one study found significant starting point bias (Rowe *et al.* 1980). In this work, which sought to value visibility over long distances in Arizona and New Mexico, three starting points were used: $1, $5 and $10. These were found to be positively correlated with the final WTP bid, such that the effect of increasing the starting bid by $1 increased the final bid by approximately $0.60.

Payment vehicle

The payment vehicle is the form or method of payment by which the hypothetical bids given by respondents will be collected; for example, through income tax, entrance fees, higher utility charges or charitable trusts. This recognizes that respondents may vary their bids dependent on the acceptability of the valuation scenario; payments made through taxation increases may be evaluated differently from specific entry charges, for example.

Studies on this phenomenon have reported higher WTP with a payroll tax, compared with increased entrance fees (Schulze *et al.* 1981). Residents' WTP for the option value of water quality using an increased water-sewer fee was only 25% of that estimated through a sales tax. Similarly, in Colorado sales tax, values always exceeded entrance fee values (Daubert & Young 1981). Many recent studies have adopted a neutral charitable trust as the payment vehicle, as this is perceived to have few political undertones and therefore should minimize bias.

Mental account bias (or part–whole bias) and embedding

Mental account bias relates to the inability of some individuals to isolate a specific case from an overall consideration (Hoevenagel 1990). Thus, when respondents are asked to value an improvement in air quality of a specific area (Berlin) they actually value an improvement over a larger area (all of Germany). A second type of mental account bias arises when respondents allocate a greater proportion of their available resources to one scenario than seems rational; for example, pledging more money to environmental concerns than their total income and savings. It may be possible to introduce some element of rationality into the WTP bid through the questionnaire design, by reminding respondents of their budget constraints, about their total annual WTP for all environmental issues, and noting the payments/subscriptions that they have already committed.

A related phenomenon to mental account bias is that of embedding. Embedding occurs when the WTP for a given good does not vary much from the WTP given for a much larger good which actually includes the first good within it. For example, Kahneman and Knetsch (1992) asked respondents their WTP to maintain the quality of fishing in lakes in Ontario. The authors reported no significant difference between mean WTP for a small number of lakes (about 1% of the total number of lakes in the region) and the mean WTP for all lakes in Ontario. A similar result was obtained by Desvousges *et al.* (1992) who considered the WTP to prevent birds being killed by oil-related accidents in the USA, and reported no statistical difference in the WTP to save 2000, 10 000 or 200 000 individual birds. Kahneman and Knetsch (1992) suggest that such bids could be more accurately interpreted as 'warm glows' or 'the purchase of moral satisfaction' deriving from giving to perceived good causes, such as environmental protection, rather than well-considered evaluations of environmental goods.

Information bias

An important element in most CVM studies is the level

of information about the environmental good under evaluation which is given to the respondent. A number of studies have suggested that the type and amount of information provided to respondents may influence the WTP bid. Such variations pose problems as to what constitutes the 'correct' level of information and how this can be determined.

In a classic experiment on attitudes to wildlife preservation, Samples *et al.* (1986) asked respondents to allocate $30 among preservation funds for three animals. Four subgroups of respondents were given different amounts of information about the animals, and the results were as follows:

1 When no information was provided, WTP was the same across all three species.

2 When information about physical appearance was provided through pictures, average bids were: rat = $7.17; rabbit = $10.21; monkey = $12.58.

3 When species were unknown, but endangered status was provided, valuations became: endangered but saveable (rat) = $23.09; no salvation possible (monkey) = $4.05; non-endangered species (rabbit) = $2.68.

4 When endangered and physical information was provided the distribution was almost the same as for endangered only.

In conclusion, information clearly influences bids and represents a significant challenge to good questionnaire design. Ineffective description of the good under evaluation has been blamed for a number of detected biases including embedding, and the difficulty of ensuring a proper understanding of the evaluation scenario was highlighted by Arrow *et al.* (1993) as a problem area.

CHOICE EXPERIMENTS

A final technique considered here is that of the choice experiment (CE). Choice experiments are still in an early stage of development but they offer some potential advantages over CVM. Recent work in the UK has been undertaken on CE by Hanley *et al.* (1998).

The theory of the CE approach is similar to that underlying conjoint analysis, on which it is modelled. Conjoint analysis seeks to discover, in similar fashion to hedonic pricing, the contribution of different characteristics of a good towards its overall value. Unlike hedonic pricing, however, conjoint analyses are carried out through evaluation of *hypothetical* scenarios in which different combinations of characteristics of a 'composite good' (a good composed of a number of valuable characteristics) are compared. Relative rankings of the characteristics or attributes of such a composite good can then be drawn up depending on the scores assigned to different scenarios. It is a technique widely applied in consumer marketing studies, where analysts try to identify the importance of the various characteristics of new products for consumers, and their willingness to pay for them.

In a CE, a respondent would be presented with several short descriptions of a composite good, each description being treated as a complete package and differing from the other packages in respect to one or more of the good's characteristics. Such characteristics in relation to a nature reserve might be overall wildlife diversity, accessibility, numbers of a particular rare species, total area and, most importantly, a price term, such as WTP for entry to or maintenance of the particular package in question. Respondents then make pairwise comparisons between scenarios, and accept or reject a scenario on the basis of their personal preferences. After building up a series of such responses, it is possible to isolate the effects that variations in individual characteristics have on the movement of the price term.

The number of characteristics that can be investigated within scenarios is limited by the ability of respondents to cope with the detailed descriptions involved. Generally, seven or eight characteristics is the upper limit. The calculation of the price–quality characteristics relationship itself is quite complex, but despite these and other problems, the CE approach may be particularly valuable in addressing the problem of 'benefits transfer' in relation to environmental valuation (Luken *et al.* 1992). A major limitation of CVM studies is that they are generally site-specific. Respondents state WTP for a particular good, such as maintenance of a local park, but there is no acceptable way to use this stated WTP to value other parks in other areas, unless park and relevant valuing population are close to identical. Obviously this will rarely, if ever, be the case. In fact, the wide divergence of quality and characteristics between different environmental goods makes even the use of some kind of average CVM bid highly problematic (Bergland *et al.* 1995).

The CE methodology aims to determine the effects of changes in the *quality* of certain characteristics on the value of a site. If this can be done effectively then, in theory, valuations could be derived for widely differing environmental sites by appraising their quality levels across the relevant characteristics. This would require data on site quality characteristics, and demand curves for those characteristics developed through a number of CE experiments. In effect, the benefits or values attributed to separate quality characteristics would be transferable to the valuation of new sites.

As an example, recall the case of car characteristics in relation to hedonic pricing above. If the part-worth of each characteristic, like fuel efficiency and acceleration, in determining the overall value of a car is known, it should then be possible to assess the value of any new car by assessing its individual quality characteristics and summing up the relevant part-worths. The analyst needs to establish the price–quality relationship in general, and a valuation for any particular case can then be estimated.

Seen in this light, the CE methodology is very 'reductionist', assuming that goods can be decomposed into a 'set of characteristics'. As Hanley *et al.* (1998) acknowledge, this can be a contentious issue. Particularly in cases where aesthetic judgements are made, for example, the decomposition of value into different component strands is highly problematic: what part-worth might result from the colour of Rembrandt's paintings, as opposed to their structure, light and shade or characterization? In these extreme examples, value is a holistic concept, not meaningfully reducible to a set of constituent parts. In fact such criticisms are not limited to CE-style evaluation, but often extend to the 'reductionist scientific' approach to evaluation in general, for example that underlying the rationale for cost–benefit analysis (Chapter 8).

6.5 Summary

Environmental goods and services are generally not traded in a market and can be termed non-market goods. Assigning a monetary value to non-market goods is intended to assist in environmental management decisions, based on a utilitarian view that changes in money values can reflect changes in human welfare. Valuation aims to calculate a total economic value for environmental goods, composed of use, non-use and option values.

Techniques for such valuation are of three kinds. Conventional market approaches utilize information from traded goods markets to place values on non-traded goods and services; these techniques include productivity change, dose–response, opportunity cost and defensive expenditures. Implicit market approaches use observed behaviour to infer values for non-market goods, by identifying the implicit willingness to pay for these goods evident in consumer behaviour; hedonic pricing and travel cost are the main techniques. In constructed market approaches, a market is hypothetically 'constructed' and consumers are directly asked their willingness to pay for goods; these methods include contingent valuation and choice experiments. These constructed market techniques are the most controversial valuation methodologies but are the only methods currently available for estimating non-use values.

Further reading

Dixon, J.A., Scura, L.F., Carpenter, R.A. & Sherman, P.B. (1998) *Economic Analysis of Environmental Impacts*. Earthscan Publications Ltd, London.

Hufschmidt, M.M., James, D.E., Meister, A.D., Bower, B.T. & Dixon, J.A. (1983) *Environment, Natural Systems and Development: An Economic Valuation Guide*. John Hopkins University Press, Baltimore and London.

Mitchell, R.C. & Carson, R.T. (1989) *Using Surveys to Value Public Goods: The Contingent Valuation Method*. Resources for the Future, Washington DC.

Winpenny, J.T. (1991) *Values for the Environment*, HMSO, London.

7 The Ecological Approach to Environmental Evaluation

'A tree's a tree. How many more do you need to look at?' [Ronald Reagan]

7.1 Introduction

Ecologists are often asked to help make a range of decisions concerned with the assessment of species and areas of land. Most obviously these are required for selecting and designating land as nature reserves, and in the UK alone there are 5600 Sites of Special Scientific Interest (SSSIs), 240 national nature reserves (NNRs), 43 special protection areas (for birds), and 49 Ramsar sites (for wetland birds), each of which has been carefully selected from all of the land in the UK (Spellerberg 1992). In addition to strictly designated reserves, much effort is being made in the European Union towards providing some level of protection for much larger areas of agricultural land. In the UK, several agri-environment schemes aim to encourage farmers to manage their land in a certain way which will bring landscape and conservation benefits. None of these schemes could go ahead unless somebody actively designated certain areas for protection. Ecologists fulfil this role.

A parallel set of decisions has to be made about selecting certain species for protection under the law. In the UK, 92 plants and 329 animals are protected under the Wildlife and Countryside Act. Somebody has to decide which species are eligible for such protection. Ecologists also fulfil this role. Finally, ecologists may also be asked to contribute to an environmental impact assessment (see Chapter 9), and here they will be asked to evaluate a site's importance and assess the magnitude and significance of any environmental impacts which may arise from the project.

This chapter will examine some of the techniques used by ecologists to help value and assess sites and species for designation and protection. First, the basic process of assessing an ecological good will be discussed. This will be followed by the examination of methods used to evaluate species and habitats, and a discussion of some of the problems arising from these evaluation methods. In the second part of the chapter, we consider the relatively recent investigation of the ecological services provided by ecosystems, an increasingly important area of study within ecological economics research.

7.2 Making decisions about the quality of ecological goods

Many of the decisions with which ecologists are involved are in some way related to nature conservation, and it has been suggested by Usher (1989) that such conservation assessment decisions proceed in three steps. First, attributes are identified which could be used to reflect the conservation interest of a site. Examples of commonly used attributes include species diversity, species rarity and naturalness. Secondly, criteria are developed for the expression of the attributes in a form which allows evaluation; for example, a species list (the criterion) may be used to assess species richness (the attribute). Finally, values are attached to particular states and levels of the criteria. This final step is not a scientific task, and Usher suggests that the value attached to any given criterion should reflect the values of the society which owns the site.

In essence such evaluations seem to be relatively straightforward. However, several problems exist which render conservation assessment quite difficult in practice. One of the major problems is associated with the identification of the relevant criteria for use in the conservation decision. Although some criteria are more frequently used than others (Table 7.1) there is no clear guidance on the correct criteria to use in any given situation and many different criteria have been developed over the years (Box 7.1).

The large number of criteria available for evaluating ecological goods in some ways represents the difficulties that exist in conceptualizing their value. The value of a site cannot always be measured in terms of species diversity or aesthetic beauty. If one or two criteria were universally adopted for use in conservation evaluation, then many areas of land would not be protected. For example, heather moorlands are naturally species-poor

Table 7.1 Frequency of use of different criteria in conservation evaluation. Data derived from 17 studies. (Source: Usher 1986.)

Criteria	Frequency of use
Richness (of habitats and/or species)	16
Naturalness, rarity (of habitats and/or species)	13
Area (or size, extent)	11
Threat of human interference	8
Amenity value, education value, representativeness	7
Scientific value	6
Recorded history	4

but are felt to be important, and coastal mud-flats are visually unattractive but are rich in invertebrate species which provide valuable food for wading birds. Despite the problems inherent in the use of these criteria for assessing ecological goods, they are frequently used to help establish conservation priorities. One of the best documented examples of their use is that of the Red Data Books of the International Union for Conservation of Nature and Natural Resources (IUCN).

7.2.1 Assessing the conservation priorities for species

RED DATA BOOKS

The Red Data Books were originally conceived in the 1960s as an internal publication of the IUCN, and their primary purpose was to identify threats or causes of decline for different species around the world. They aim to place species in categories related to their threat of extinction in order to:

1 provide synoptic information on which to base conservation programmes;
2 help with the drafting of legislation; and
3 to convey information to the non-specialist.

The IUCN classification is primarily concerned with threats and causes of decline in species and not with rarity or abundance. They place species in one of eight different categories: extinct, extinct in the wild, critically endangered, endangered, vulnerable, lower risk, data deficient and not evaluated. The definition of these categories is given below, and the criteria A–E to which they refer are given in Box 7.2.

Extinct

A taxon is extinct when there is no reasonable doubt that the last individual has died.

Extinct in the wild

A taxon is extinct in the wild when it is known only to survive in cultivation, in captivity or as a naturalized population (or populations) well outside the past range. A taxon is presumed extinct in the wild when exhaustive surveys in known and/or expected habitat, at appropriate times (diurnal, seasonal, annual), throughout its historic range have failed to record an individual.

Critically endangered

A taxon is critically endangered when it is facing an extremely high risk of extinction in the wild in the immediate future (as defined by any of the criteria A–E).

Endangered

A taxon is endangered when it is not critically endangered but is facing a very high risk of extinction in the wild in the near future (as defined by any of the criteria A–E).

Vulnerable

A taxon is vulnerable when it is not critically endangered or endangered but is facing a high risk of extinction in the wild in the medium-term future (as defined by any of the criteria A–D).

Lower risk

A taxon is lower risk when it has been evaluated, does not satisfy the criteria for any of the categories critically endangered, endangered or vulnerable.

Data deficient

A taxon is data deficient when there is inadequate information to make a direct, or indirect, assessment of its risk of extinction based on its distribution and/or population status. Listing of taxa in this category indicates that more information is required and acknowledges the possibility that future research will show that threatened classification is appropriate.

Box 7.1 A range of criteria which have been used for nature conservation evaluation

Criteria	Definition and rationale for use in conservation evaluation
Species richness	The number of different species which exist in a given area. It is generally true that the higher the species richness the better is the site.
Diversity	A measure of the number of different types of species, or sometimes habitats which exist in a given area and the equity with which individuals are distributed between species (Hurlbert 1971). According to this measure the more equitable the distribution of individuals between the species the more diverse is the site. High diversity is generally felt to be better than low diversity.
Area	Generally the larger the site the better as, all other things being equal, larger sites support more species and greater densities of species than smaller sites.
Rarity	A measure of how frequently certain species, or habitats, are encountered.
Naturalness	An assessment of how disturbed, or undisturbed, by humans a given habitat is. The more natural (less disturbed) the better. The use of this attribute implies there is some natural condition for all habitats, this is conceptually difficult to define as most, if not all, habitats have been influenced by humans in some way.
Representativeness	A measure of how well any site, or collection of sites, reflects all of the communities which may be expected to occur in that geographical area. The more representative a site is of an area, the better.
Typicalness	A measure of how typical the species assemblages in a certain site are for that type of site. Usually, sites are chosen for conservation because they are the best examples of a particular habitat. Unusual sites are not typical.
Fragility	Fragility reflects the 'degree of sensitivity of habitats, communities and species to environmental change' (Ratcliffe 1977). It is the inverse of resilience. Fragility is related to a complex of other attributes, and generally habitats are categorized as fragile on the basis of experience with that habitat elsewhere. For example coral reefs in Sri Lanka are known to be fragile to the sediments produced from mining operations, so all coral reefs can be assumed to be susceptible to that source of pollution too (UNEP 1992). The more fragile a community, the greater is the requirement for its protection. Fragility applies easily to climax communities, as you would not expect them to change over time. It is more difficult to apply to several communities which change naturally. Measuring fragility requires data over time, i.e. a measurement of change.
Threat	The probability that a species or habitat will be damaged or destroyed within a given time-scale. Generally, the greater the threat the more immediate is the need to preserve. Rates of loss are usually a good indicator of threat.
Amenity value	The value a habitat or site provides to the general public as a place to visit and/or see.
Educational value	The demand for educational use of reserves is high and increasing, and in some ways the assessment of educational value is assessed in similar manner to amenity value as ease of access and proximity to demand (schools, colleges, universities) are both important attributes.
Recorded history	The extent to which a site has been used for education, natural history and research. Sites with good histories enhance our understanding of ecology and as such are valuable to science. For example, if the moths in a certain woodland have been recorded for 100 years, then that long run of data provides good baseline data against which the impacts of environmental change may be felt, e.g. changes in species composition and abundance with increasing pollution or climate change.
Position in ecological or geographical unit	Where practical it may be desirable to include within a single geographical location as many as possible of the important characteristic formations, communities and species of an area, e.g. a woodland area adjoining a riparian area.
Potential value	A measure of how under given management/natural change the site will develop features of particular value for conservation. It is probably a function of other criteria.
Endemicity	Endemic species are those species which only occur within the defined area, usually a country.
Type locality	When a new species is described the locality in which the specimen (holotype) was collected is designated the 'type locality'. Many species are known only from their type locality. Hence it is related to endemism.
Intrinsic appeal	Although science may be objective, humans tend to give more weight to certain taxonomic groups/habitats. For example, species such as mammals (cuddly), birds (colourful and/or tuneful), wild flowers (not weeds) and rivers have high public appeal, while beetles, spiders, snakes, limestone pavements and mud-flats have lower appeal.

Box 7.2 The criteria for critically endangered, endangered and vulnerable. Prepared by the IUCN Species Survival Commission as approved by the 40th Meeting of the IUCN Council, Gland, Switzerland, 30 November 1994. (Source: IUCN web site: HTTP://iucn.org/themes/ssc/redlist/categor.htm)

Critically endangered (CR)

A Population reduction in the form of either of the following.

1 An observed, estimated, inferred or suspected reduction of at least 80% over the last 10 years or three generations, whichever is the longer, based on (and specifying) any of the following: (a) direct observation; (b) an index of abundance appropriate for the taxon; (c) a decline in area of occupancy, extent of occurrence and/or quality of habitat; (d) actual or potential levels of exploitation; and (e) the effects of introduced taxa, hybridization, pathogens, pollutants, competitors or parasites.

2 A reduction of at least 80%, projected or suspected, to be met within the next 10 years or three generations, whichever is the longer, based on (and specifying) any of (b–e) above.

B Extent of occurrence estimated to be less than 100 km^2 or area of occupancy estimated to be less than 10 km^2, and estimates indicating any two of the following.

1 Severely fragmented or known to exist only at a single location.

2 Continuing decline, observed, inferred or projected, in any of the following: (a) extent of occurrence; (b) area of occupancy; (c) area, extent and/or quality of habitat; (d) number of locations or subpopulations; and (e) number of mature individuals.

3 Extreme fluctuations in any of the following:
(a) extent of occurrence; (b) area of occupancy; (c) number of locations or subpopulations; and (d) number of mature individuals.

C Population estimated to number less than 250 mature individuals and either:

1 An estimated continuing decline of at least 25% within 3 years or one generation, whichever is longer, or

2 A continuing decline, observed, projected, or inferred, in numbers of mature individuals and population structure in the form of either: (a) severely fragmented (i.e. no subpopulation estimated to contain more than 50 mature individuals); or (b) all individuals are in a single subpopulation.

D Population estimated to number less than 50 mature individuals.

E Quantitative analysis showing the probability of extinction in the wild is at least 50% within 10 years or three generations, whichever is the longer.

Not evaluated

A taxon is not evaluated when it has not yet been assessed against the criteria.

Since their introduction, the Red Data Books and Red Lists categories have become widely recognized internationally, and they are now used by numerous governmental and non-governmental organizations, as well as in a wide range of publications and listings produced by the IUCN. The Red Data Book categories provide a method for highlighting those species under higher extinction risk, so as to focus attention on conservation measures designed to protect them. The allocation of a species to one of the Red List categories requires only that the species meets any one of the criteria listed above.

As the relevance of the criteria to species will differ from case to case it will never be clear in advance which criteria are appropriate for a particular species. 'For this reason each species should be evaluated against all the criteria, and any criterion met should be listed' (IUCN 1994). Unfortunately, placing a species in a category of threat is not sufficient to guarantee that conservation action will be started. Initiating such action is a long and largely unscientific process. The IUCN recognize this.

> 'The category of threat simply provides an assessment of the likelihood of extinction under current circumstances, whereas a system for assessing priorities for action will include numerous other factors concerning conservation action such as costs, logistics, chances of success, and even perhaps the taxonomic distinctiveness of the subject.' (IUCN, 1994)

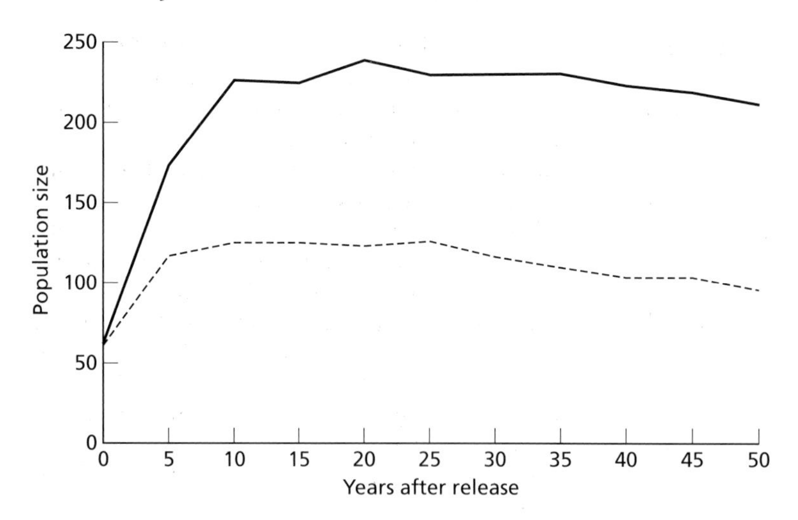

Fig. 7.1 Part of a population viability analysis undertaken in order to assess the feasibility of reintroducing the capercaillie (*Tetrao urogallus*) to forests in southern Scotland. The solid line assumes the new habitat has a carrying capacity of 300 individuals, while the dotted line assumes a carrying capacity of 150. (Source: Marshall & Edwards-Jones 1998.)

POPULATION VIABILITY ANALYSIS

The quantitative analysis referred to in criterion E of the IUCN categories in Box 7.2 refers to population viability analysis (PVA). This uses modelling techniques in order to explore the population dynamics of rare or declining species, and to estimate the probability of their extinction (Soulé 1987). Population viability analysis can provide predictions, as well as highlight areas for further research, in order to improve management of a given species and its habitat for conservation purposes. Almost by definition, PVAs require some computer model of the species of interest be developed and used. Several generic models have been developed for this task, e.g. VORTEX (Lacy 1993) and METAPOP (Akcakaya 1994), and these can be used to consider the complex interactions between demographic, environmental and genetic influences on a population.

The VORTEX model simulates the dynamics of populations based on user-specified data on mortality and fecundity rates for each sex and age class. The outputs from VORTEX include graphs of numbers over time (Fig. 7.1) and also changes in the genetic parameters of the population over time, such as heterozygosity levels and degree of inbreeding. These demographic and genetic variables are then combined in order to calculate a probability of extinction for the species in question over a prespecified time period. The information on changes in numbers, genetics and probability of extinction can then inform the conservation decision to a greater degree than any one set of data alone, and manipulation of the model inputs can serve to highlight aspects of habitat or behaviour which are most pertinent to the survival of a population.

One specific output of a PVA can be the calculation of a minimum viable population (MVP), which is the smallest population which has an acceptable probability of persisting over a given period of time. This can then be used to find the minimum dynamic area (MDA) that will represent the area of suitable habitat required in order to support the MVP (Soulé 1986; Fiedler & Jain 1992; Caughley & Gunn 1996). Identifying the MDA can give some guidance as to the appropriate size of habitat that may need to be protected in order to conserve the species. While VORTEX can be used to simulate the population dynamics of threatened or endangered species in the wild, it may also be used to predict the dynamics of a population following the reintroduction of a species back into the wild. In such cases the results from a PVA can be used to help increase the probability of a project's success. There is increasing interest in the use of PVAs to aid conservation decisions and studies have been published on Leadbeater's possum (*Gymnobelideus leadbeateri*) by Lindenmayer *et al.* (1993), the eastern barred bandicoot (*Perameles gunnii*) (Lacy & Clark 1990), the Puerto Rican parrot (*Amazona vittata*) (Lacy *et al.* 1989), wild boar (*Sus scrofa*) (Howells & Edwards-Jones 1997), and the giant panda (*Ailuropoda melanoleuca*) (Zhou & Pan 1997).

7.2.2 Conclusion on evaluating species

Making decisions about species conservation does have some scientific basis, namely degree of extinction threat, which in turn is based on observations of changes in natural populations of the species of interest. Further, given our understanding of population dynamics and population genetics, we are able to develop computer models which can simulate the changes which may occur in species of interest. We can also use these models to explore the implications of various management scenarios. So at least when considering species there is an objective and transparent mechanism for categorizing threat—the probability of extinction—thus any decision about allocating resources to species could be done on the basis of this criterion alone. In practice, the probability of extinction is not the only criterion used in conservation decisions, and even if it were, resources would be allocated according to some combination of ecological data and sociopolitical pressure.

Indeed, even though one of the aims of the IUCN's Red List categories is to improve the objectivity of evaluation of the different factors which affect risk of extinction, it is interesting to note that when considering the quantitative values presented in the various criteria associated with threatened categories, the IUCN carried out a wide consultation exercise with relevant experts and set quantitative criteria at what were felt to be appropriate levels 'even if no formal justification for these values exists' (IUCN web page: http://iucn.org). So, while the classification of a species into a threat category appears to be highly objective in nature, the actual definition of these categories largely relied on a subjective analysis. As we shall see in the next section, the reliance on subjective opinion as the basis for making decisions about habitats is, if anything, even more prevalent than in the evaluation of species.

7.3 Assessing the conservation priorities for habitats

7.3.1 Sites of special scientific interest in the UK

One of the tasks of the original Nature Conservancy Council (NCC) in the UK was to identify areas of land which, because of their 'value', were worthy of legal protection. Indeed, the existing country-specific agencies within the UK continue to have a responsibility for the management and designation of such areas. Perhaps the first attempt at identifying such areas was made by Charles Rothschild in 1915. He compiled a list of desirable nature reserves for the Society for the Promotion of Nature Reserves, which was based largely on his own, and other leading figures', perceptions (NCC 1989). This largely subjective process set the standards for much conservation decision-making, as summarized in the following quote from the NCC (1989): 'The standards of nature conservation value thus became established through the practice and precedents of a collective wisdom.' Indeed, while the concept of the 'site of special scientific interest' first appeared in a report to the Parliament of the Wild Life Conservation Special Committee (England and Wales), *Conservation of nature in England and Wales* in 1947, and has been enshrined in Acts of Parliament in 1949 and 1981, the government has given no guidance on how these sites should be selected. Rather, the NCC was required simply to exercise its 'opinion' in the selection of sites for notification as SSSIs.

The 1949 and 1981 Acts define SSSIs as 'any land of special interest by reason of any of its flora, fauna or geological or physiographic features'. Initially it was anticipated that about 7% of land in the UK, including geological and physiographical sites, would be designated as SSSIs, and that when taken as a whole 'the series of sites should contain adequate representation, in the form of best examples, of the total countrywide range of variation in natural and seminatural ecosystem types, with their associated assemblages of plants and animals, considered both as communities and species' (NCC 1989).

Much discussion ensued about the relative merits of the methods for selecting SSSIs and, as time passed, it became clear that some form of analysis was needed to help with the designation process. This requirement was partially met in 1977 when *A Nature Conservation Review* was published by Ratcliffe (1977). This important document identified 10 criteria which could be used for identifying areas of land for protection in the UK. These 10 criteria were size, richness (of habitats and species), naturalness, rarity, typicalness, fragility, recorded history, ecological fragility, position in ecological or geographical unit, potential value and intrinsic appeal (see Box 7.1).

These criteria were central to the development of an important document entitled *Guidelines for the Selection of Biological SSSIs* (NCC 1989) which details the framework and methods for selecting SSSIs. This manual specifically discusses 19 types of habitat and considers what renders certain patches of these habitats worthy of designation as SSSIs. For example, when considering upland habitats it is suggested that 'the overall requirement is that sites be selected which

Box 7.3 Recommendations on the designation of the four types of salt-marsh which exist within Wales as Sites of Special Scientific Interest (source: NCC 1989)

- Large areas of grazed marsh with extensive communities dominated by a few species, especially *Puccinella maritima*; the upper marsh often includes *Juncu maritimus*. Examples will normally fall within the sites selected primarily for their ornithological interest, though the upper marsh may include species such as *Althea officinalis*
- Spartina marshes, mainly with *Spartina anglica* dominant. This marsh type should not be included on botanical grounds in the site selection process
- Ungrazed salt-marshes with *Limonium* prominent; upper transition communities include *Juncus maritimus/Oenanthe lachenaii* marsh. All areas above 50 ha should be selected
- In a limited number of areas on lightly grazed or ungrazed sites sand-dune transitions similar to those of the north Norfolk coast occur which may include *Frankenia laevis* and *Limonium binervosum*. All examples should be selected

represent adequately the variety and extent of upland habitats, flora and fauna present'. The guidelines also note regional variations in habitats and make relevant recommendations. For example, when considering salt-marshes, six different regions are noted in the UK. Details are given of what type of salt-marsh is important, and thereby worthy of protection in each region. Box 7.3 presents this information for the four main types of salt-marsh which occur in Wales, and this is obviously very detailed in nature.

Despite the presentation of these detailed habitat analyses, no objective framework is presented which would enable the selection of sites for designation as SSSIs. Indeed, it is clearly stated that 'It is not possible to provide rigid rules for SSSI selection which require only the measurement of attributes to determine whether sites pass a critical total "score" of value and thereby absolve those concerned from exercising any personal judgement . . . [I]n the last analysis, each case (of selection of SSSI) rests on matters of opinion.' (NCC, 1989)

7.3.2 The search for an evaluation framework

EARLY DEVELOPMENTS

> 'When it comes to an assessment of an actual site or tract of land, or a consideration of possible methods of management, the matter may often be decided by a single over-riding factor or by simple rules of thumb. Unless, however, there is some rational basis on which such rules of thumb are based, decisions can easily be swayed by passing fashion or by personal prejudice. In addition to this, it will also be difficult to explain the assessment to people who are not particularly interested in nature conservation, or to incorporate the assessment into a planning model to examine the outcome of different courses of action.' (Helliwell 1985)

The requirement for an objective framework assessing conservation importance is well stated by Helliwell above, and while statutory agencies in the UK have so far ignored the need for using an objective framework to aid site evaluations and conservation designation, many such frameworks have been developed by other scientists and organizations. However, it should be recognized that many of these have their drawbacks, and in some ways these drawbacks may partly justify the decision of statutory agencies to avoid such frameworks.

An early example of a multiattribute method for evaluating conservation value was proposed in 1969 to evaluate coastal habitats (Ranwell 1969) and it provides a useful example from which we can examine some of the basic strengths and weaknesses associated with the 'objective' approaches ecologists have utilized in making decisions. This method sought to combine nine criteria into a single score, the comparative biological value index (CBVI). The criteria used were:

1. size (S);
2. physicochemical features (Ph);
3. optimum populations (O);
4. diversity (D);
5. geographical limits (G);
6. purity (P);
7. education and research use (E);
8. combinatory value (C); and
9. unknown factors (X).

Table 7.2 Rating of the criteria utilized in Ranwell's system for evaluating coastal habitats (excluding unknown factors).

Criteria	Type		Rating
Physicochemical features	High speciality		3
	Some special features		2
	Type example		1
Optimum populations	Best populations of one or more local species		4
	Large populations of local species		3
	Large populations of common species and/or small populations of local species		2
	Representative populations		1
Diversity	Outstanding diversity		3
	High diversity		2
	Species range small		1
Geographical limits	Many species at limit		3
	Some species at limit		2
	Few or no species at limit		1
Size	Mud-flats (ha)	Cliffs (km)	
	> 4000	> 80	5
	1600–3999	40–79	4
	800–1599	24–39	3
	400–799	8–23	2
	< 400	< 8	1
Purity	Little disturbance		3
	Moderate disturbance		2
	Much ground disturbed or polluted		1
Education and research use	Much used		3
	Some use		2
	Potential use		1
Combinatory value	Adjacent to another habitat of likely national value		4
	Adjacent to another habitat of likely regional value		3
	Adjacent to another coast habitat site not spoilt by development		1
	Surrounded by developed coastline		0

Each of these criteria were scored on a simple rating scale (Table 7.2) and the final score is obtained by summing the scores for all nine criteria:

$$\mathrm{CBVI} = \mathrm{S} + \mathrm{Ph} + \mathrm{O} + \mathrm{D} + \mathrm{G} + \mathrm{P} + \mathrm{E} + \mathrm{C} + \mathrm{X}.$$

Indices of this type are often used by non-specialists to help them make decisions about nature conservation. As in environmental impact assessment, there is a need to consider a number of different criteria, and one frequently used way to simplify this process is to combine all the attributes into a single index, here the CBVI. Such indices seek to be intuitive, consistent and easy to interpret. However, a closer look at this apparently simple and robust assessment technique reveals several anomalies which are common in other indices of this type, and which may serve to confuse their users.

Excluding unknown factors, we can work out that the maximum potential value for CBVI is 28 and the minimum value is 7. The minimum value is not 0, as seven of the eight criteria have 1 as the lowest possible score, while one has 0 as its lowest score. This sort of counterintuitive scale can easily confuse non-specialists who have to act on the outputs of the index. Furthermore, the simple addition of the attribute scores may lead to the assumption that all attributes are of equal importance in determining the overall value of the site. In fact, because of the uneven scales used to measure attributes (with maximum top scores of 3, 4 and 5), the overall importance of some criteria is implicitly weighted within the system.

A further point of interest concerns the constancy of type definition between attributes and the mapping of these types into scores. When compiling scores for geographical limits and size, quantitative data are converted into a score. So for size a mud-flat of 1600–3999 ha scores 4 for size as does a salt-marsh of 400–799 ha, and a cliff with 40–79 km of undisturbed length. Are these categories really equivalent? While there may be some means of estimating the relative importance of physical attributes—for example, by considering the size distribution of all habitats of that type in the country/region—it is very difficult to decide whether or not the type category for physicochemical of 'some special features' deserves the same score as the type category 'large populations of common species and/or small populations of local species' in 'Optimum populations'. Assigning a score of 2 to these two type categories is really only saying that sites with this type category are not the worst examples for this type, and similarly are not the best.

A final problem with this, and other indices, is that the output serves to hide information and that several of the inputs require some subjective evaluation (Goldsmith 1991). As such they may not be as simple to use, or as helpful to the decision-making process as originally intended. Despite specific problems of this nature being relevant to many situations, the allocation of quantitative scores has remained a popular practice in environmental management; for example, assessment of woodland amenity, landscape and urban habitats (Spellerberg 1992). In addition, other quantitative methods for evaluating conservation status have been devised.

For example, when considering birds it has been suggested that any site supporting more than 1% of the biogeographical population of one species of waterfowl should be regarded as internationally important (Stroud *et al.* 1990). Whereas 1% of a national population can be considered nationally important (unless population < 100 breeding pairs). A more complex system for evaluating wetlands of ornithological interest was devised by Nilsson and Nilsson (1976). They considered the species breeding at each site, each year and the number of breeding pairs of each species in western Europe. They then used the following index in order to rank the sites in descending order of importance:

$$CV = \sum n_i / N_i$$

where CV is the index of conservation value, n_i is the number of pairs of the ith species at a site and N_i is the number of pairs in western Europe.

COMPUTER-BASED SYSTEMS

While the entirely subjective nature of decision-making for designating SSSIs may appear to have many weaknesses, the above examination of the supposedly more objective means of making conservation decisions has also highlighted some of the problems inherent in these methods. Perhaps one of the main reasons why an acceptable objective framework has remained elusive is because of the simple fact that assessment and evaluation cannot be undertaken on a purely quantitative basis. It remains undeniable that many of the attributes on which assessments are based are largely qualitative, and as Usher (1989) states the whole evaluation system is dominated by the personal value system of the assessor.

The influence of personal value systems on assessment decisions may not necessarily be bad, but it does seem important that such values are applied transparently and consistently. For example, it would seem desirable that value systems used by conservation agencies when designating sites are open to scrutiny, both by the general public and by others in their own and other organizations. Similarly, when a habitat is under threat from a development, it would seem fair to all parties that the value system used to assess that habitat was also open to scrutiny. In addition to these individual uses, there may also be merits in applying the same value system across a country or region, and thus preventing similar developments being treated differently in different areas because of variation in the value systems of local decision-makers.

While many benefits may accrue from the adoption of a formal transparent assessment and evaluation framework, those developed to date (e.g. Helliwell 1967, 1985) are largely quantitative and do not place adequate emphasis on the subjective elements of the assessment process. This problem is confounded as accessing and representing personal values in a formal sense is not a task usually undertaken in standard mathematics or computing. Recently, however, several computer-based systems have been developed which seek to include elements of subjectivity within a transparent and repeatable framework. The rationale which has enabled the development of these systems has come

Table 7.3 Objectives of the SERCON computer program (System for Evaluating Rivers for Conservation). (Source: Boon *et al.* 1997.)

1 To encourage greater rigour and repeatability in data collection and evaluation
2 To identify gaps in the scientific knowledge for specific rivers
3 To enable the assessment of rivers within a wide range of environmental quality (not just those of the highest quality)
4 To provide a simpler way of communicating technical information to planners, developers and policy-makers
5 To aid in the assessment of the rehabilitation potential of degraded rivers
6 To construct a tool to assist in predicting the impact of different development options on river conservation value
7 To establish a framework for extending the principles of SERCON to the evaluation of other attributes of rivers (e.g. geomorphological, landscape or recreational features) as well as those relating to fauna, flora and habitats

from the branch of computer science known as artificial intelligence. As such, these systems take some of the heuristics utilized by human assessors and utilize them within their algorithms to reach decisions.

SYSTEM FOR EVALUATING RIVERS FOR CONSERVATION (SERCON)

One such system is SERCON (System for Evaluating Rivers for Conservation (Boon *et al.* 1997)). The rationale for developing SERCON was the perceived need for a more rigorous method which could help support conservation decisions. The development of SERCON was undertaken by a team from Scottish Natural Heritage, the statutory conservation agency in Scotland, and had the objectives presented in Table 7.3.

The system was developed through consultation with 161 different individuals representing 24 organizations and 19 areas of specialism (e.g. mammals, invertebrates, landscape, public perception). Through consultation with this group a list of 35 attributes related to conservation were identified for inclusion in SERCON and these were grouped in order to contribute to one of six criteria (physical diversity, naturalness, representativeness, rarity, species richness and special features) (Table 7.4). After agreeing the list of attributes the members of the specialist group ranked each attribute in order of importance where 1 was the most important. They then assigned weights to each attribute and criterion, with a value of 1 given to the attribute ranked lowest and with the other attributes being weighted in accordance to their perceived importance, i.e. in line with the rank order. The highest weight permitted was 5.0. These results were then utilized to form the weights for each attribute as shown in Table 7.5.

SERCON utilizes these weights to express a percentage of a maximum score for each criterion. For example, consider the criterion physical diversity (PDY) comprising three attributes (1, substrates; 2, fluvial features; 3, structure of aquatic vegetation). When utilizing SERCON to evaluate this criterion, users refer to a table of nine substrate types which presents scores for each type as shown below:

0 Only 1 natural substrate type present;
1 2–3 natural substrate types present;
2 4–5 natural substrate types present;
3 6–7 natural substrate types present; and
4 8–9 natural substrate types present.

The score assigned by the user is then manipulated as shown in Table 7.5 in order to produce a physical diversity criterion index. The final index is expressed as a percentage of the maximum score possible for that criterion given the number of attributes for which data were available. This has two advantages. First, it allows a fair direct comparison of criteria which are composed of different numbers of attributes (for example, it permits equal consideration of a criterion with 10 attributes which could score a maximum of 50, against one with only two attributes could only score a maximum of 10, and thereby does not bias the evaluation process). Secondly, it permits some evaluation to be made even when some of the data are missing, which may be a common occurrence.

WOODLAND EVALUATION SYSTEM (WES)

A similar system, also developed in Scotland, sought to evaluate lowland woodlands. In essence the Woodland Evaluation System (WES) is very similar to SERCON, however, in addition to including ecological data within its evaluation, WES also seeks to include information on the amenity and landscape value of a woodland. A summary of the results from applying WES to three different

Table 7.4 The full list of attributes used by SERCON (System for Evaluating Rivers for Conservation). These attributes are grouped under 'impacts'. The weights allocated to criteria and attributes are given in the right-hand column. (Source: Boon *et al.* 1997.)

Conservation criteria		Weight
PDY	Physical diversity	3
NA	Naturalness (NA 1–4: NA 'A')	5
	Naturalness (NA 5–8: NA 'B')	1.5
RE	Representativeness	3
RA	Rarity	2
SR	Species richness	2.5
SF	Special features	1
Physical diversity		
PDY 1	Substrates	4
PDY 2	Fluvial features	5
PDY 3	Structure of aquatic vegetation	1
Naturalness		
NA 1	Channel naturalness	5
NA 2	Physical features of the bank	3.5
NA 3	Plant assemblages on the bank	2
NA 4	Riparian zone	2
NA 5	Aquatic and marginal macrophytes	4
NA 6	Aquatic invertebrates	3.5
NA 7	Fish	2.5
NA 8	Breeding birds	1
Representativeness		
RE 1	Substrate diversity	3
RE 2	Fluvial features	5
RE 3	Aquatic macrophytes	4
RE 4	Aquatic invertebrates	4
RE 5	Fish	2.5
RE 6	Breeding birds	1
Rarity		
RA 1	EC Habitats Directive/Bern Convention species (+ rare in UK)	4.5
RA 2	Scheduled species	4
RA 3	EC Habitats Directive Species (but not rare in UK)	3.5
RA 4	Red Data Book/nationally scarce macrophyte species	2
RA 5	Red Data Book/nationally scarce invertebrate species	2
RA 6	Regionally rare macrophyte species	1
Species richness		
SR 1	Aquatic and marginal macrophytes	4.5
SR 2	Aquatic invertebrates	4.5*, 3†
SR 3	Fish	2
SR 4	Breeding birds	1
Special features		
SF 1	Influence of natural on-line lakes	3.5
SF 2	Extent and character of riparian zone	4.5
SF 3	Floodplain: recreatable water-dependent habitats	2
SF 4	Floodplain: unrecreatable water-dependent habitats	4.5
SF 5	Invertebrates of river margins and banks	2.5
SF 6	Amphibians	1.5
SF 7	Wintering birds on floodplain of ECS	1.5
SF 8	Mammals	2.5

Continued

Table 7.4 (*cont'd*)

Conservation criteria		Weight
Impacts		
IM 1	Acidification	3
IM 2	Toxic/industrial/agricultural effluent	4
IM 3	Sewage effluent	4.5
IM 4	Groundwater abstraction	3
IM 5	Surface water abstraction	2
IM 6	Inter-river transfers	1
IM 7	Channelization	4.5
IM 8	Management for flood defence	2.5
IM 9	Man-made structures	2.5
IM 10	Recreational pressures	2
IM 11	Introduced species	2

Abbreviations: BMWP, Biological Monitoring Working Party; ECS, Evaluated Catchment Sections; NA, Naturalness of physical features. * Weighting used for species data. † Weighting used for BMWP family data.

Table 7.5 An example of the mathematical manipulation of scores and weights of three criteria used to evaluate the attribute physical diversity (PDY) in SERCON. (Source: Boon *et al.* 1997.)

Attribute	Score	Weight	Weighted score	Maximum possible score	Maximum possible weighted score
PDY 1	5	4	20	5	20
PDY 2	3	5	15	5	25
PDY 3	4	1	4	5	5
Total			39		50

woodlands is given in Table 7.6, and it can be seen that while Drummond Reach scores highest on the ecological assessment, it scores lowest for amenity, and is intermediate for landscape.

7.3.3 Conclusions on valuing ecological goods

While many people would argue that holistic assessment of sites is necessary, the results of holistic assessments often seem to confuse conservation decisions. For example, consider the example above (see Table 7.6) and try and answer the question: 'If all of these three woodlands were under threat and you only had the resources to save one, which would you conserve?' If you feel that ecological attributes are more important than landscape and amenity features, then you may decide to conserve Drummond Reach; conversely, if you feel landscape and amenity to be most important, then you may conserve Roslin Glen. Alternatively, you may choose to argue that you cannot afford to lose either of the woodlands as one of them is important for ecology and the other for landscape and amenity.

In the WES example, the woodlands are providing the composite goods: ecological value, landscape value and amenity value. But if you look at Table 7.6 a little more closely you will realize that even within the category 'ecology' there are multiple values. For example, while Garrion Gill and Drummond Reach both achieve the same score for plant species richness, Drummond Reach has a much greater score for rare plants. In the final analysis, we can only conclude that the choice between 'apples and pears' is an inherently subjective one, to which there is no 'objectively correct' answer.

7.4 Ecosystem services

Previous sections of this chapter have considered how and why species and habitats should be conserved

Table 7.6 Output from the Woodland Evaluation System for three woodlands in southern Scotland. Only the summary scores of major attributes are presented, some of which are aggregates of the scores of many constituent attributes, e.g. the overall vegetation cover score is a function of the score for canopy, shrub layer, field layer and ground layer. In this situation the attribute weight is given as variable. (Source: Edwards-Jones *et al.* 1996.)

			Potential score		Actual score for test woodlands		
Assessment categories	Attribute	Att. wt	Minimum	Maximum	Roslin Glen	Garrion Gill	Drummond Reach
Ecology	Tree species richness	3	0	1	0.50	0.48	0.50
	Plant species richness	3	0	1	0.46	0.59	0.59
	Rare species	5	0	1	0.00	0.00	0.38
	Invasive species	var.	−1	1	−0.20	−0.20	−0.20
	Vegetation cover	var.	0	1	0.36	0.36	0.29
	Habitats	var.	0	9	4.33	4.07	5.10
	Age distribution	3	0	1	0.60	0.60	0.60
	Area	5	0	1	1.00	0.40	0.60
	Adjacent habitats	var.	−0.5	1	0.26	0.27	0.00
	Total ecology	—	−1.5	17	7.31	6.57	7.86
	Percentage of maximum score	—	—	—	43.00%	38.64%	46.23%
Recreation	Access	var.	0.1	1	0.60	0.60	0.27
	Infrastructure	var.	0	6	1.53	1.00	0.00
	Activities	var.	−1.87	6	1.42	1.00	1.00
	Total recreation	—	−1.77	13	3.55	2.60	1.27
	Percentage of maximum score	—	—	—	27.31%	20.00%	9.77%
Landscape	Compatibility	var.	−1	1.0	0.00	0.00	0.30
	Screening	3	0	1.0	0.60	0.00	0.00
	Composition	4	−0.8	1.0	0.80	0.80	0.80
	Views from wood	3	0	1.0	0.60	0.00	0.00
	Total landscape	—	−1.8	4.0	2.00	0.80	1.10
	Percentage of maximum score	—	—	—	50.00%	20.00%	27.50%
	Overall total		−5.07	34.00	12.86	9.97	10.23
	Percentage of maximum score		—	—	37.82	29.32	30.09

Abbreviations: Att. wt, attribute importance weight is the mean of the importance weights given to each attribute by the experts, these can vary between 1 (least important) to 5 (most important); var., variable.

from a perception of intrinsic value, which is not related to any direct tangible benefits flowing to humankind. However, much recent work in ecological economics has focused attention on more tangible conservation benefits linked to important ecosystem services. The ecologist Paul Erlich suggests:

> 'many of the less cuddly, less spectacular organisms that *Homo sapiens* are wiping out are more important to the human future than are most of the publicized endangered species . . . the most anthropocentric reason for preserving diversity is the role that microorganisms, plants and animals play in providing free ecosystem services, without which society in its present form could not persist.' (Ehrlich 1988)

An ecosystem service is some attribute of ecosystems which provides value to humankind. These services are usually derived from some attribute of the ecosystem's structure or function, but there is not necessarily a direct one-to-one mapping of function to service. For example, the natural decomposition of the large quantity of organic wastes produced by humans is an ecosystem function which actually provides more than one service to humankind. It removes the vast volume of waste, it ensures that important nutrients re-enter ecosystems and thereby enable continued plant growth, and it reduces the risk of such wastes causing environmental pollution and disease (Pimental 1998). So here one function provides many services. The opposite can also be true, when several functions combine to provide one service. Thus, successful crop growth depends upon adequate soil structure (partially maintained by earthworms and other soil-dwelling invertebrates), soil fertility (provided by the microorganisms and invertebrates responsible for breaking down and recycling nutrients), maintenance of hydrological cycles

Table 7.7 The 17 types of ecosystem services and functions defined by Costanza *et al.* (1998) when estimating the value of the planet's ecosystem services. (Source: Costanza *et al.* 1998.)

Number	Ecosystem service*	Ecosystem functions	Examples
1	Gas regulation	Regulation of atmospheric chemical composition	CO_2/O_2 balance, O_3 for UVB protection, and SO_x levels
2	Climate regulation	Regulation of global temperature, precipitation and other biological mediated climatic processes at global or local levels	Greenhouse gas regulation, dimethyl sulphide production affecting cloud formation
3	Disturbance regulation	Capacitance, damping and integrity of ecosystem response to environmental fluctuations	Storm protection, flood control, drought recovery and other aspects of habitat response to environmental variability mainly controlled by vegetation structure
4	Water regulation	Regulation of hydrological flows	Provisioning of water for agricultural (such as irrigation) or industrial (such as milling) processes or transportation
5	Water supply	Storage and retention of water	Provisioning of water by watersheds, reservoirs and aquifers
6	Erosion control and sediment retention	Retention of soil within an ecosystem	Prevention of loss of soil by wind, run-off, or other removal processes, storage of silt in lakes and wetlands
7	Soil formation	Soil formation processes	Weathering of rock and the accumulation of organic material
8	Nutrient cycling	Storage, internal cycling, processing and acquisition of nutrients	Nitrogen fixation, nitrogen, phosphorus and other elemental or nutrient cycles
9	Waste treatment	Recovery of mobile nutrients and removal or breakdown of excess of xenic nutrients and compounds	Waste treatment, pollution control, detoxification
10	Pollination	Movement of floral gametes	Provisioning of pollinators for the reproduction of plant populations
11	Biological control	Trophic–dynamic regulations of populations	Keystone predator control of prey species, reduction of herbivory by top predators
12	Refugia	Habitat for resident and transient populations	Nurseries, habitat for migratory species, regional habitats for locally harvested species, or overwintering grounds
13	Food production	That portion of gross primary production extractable as food	Production of fish, game, crops, nuts, fruits by hunting, gathering, subsistence farming or fishing
14	Raw materials	That portion of gross primary production extractable as raw materials	The production of lumber, fuel or fodder
15	Genetic resources	Sources of unique biological materials and products	Medicine, products for materials science, genes for resistance to plant pathogens and crop pests, ornamental species (pets and horticultural varieties of plants)
16	Recreation	Providing opportunities for recreational activities	Ecotourism, sport fishing, and other outdoor recreational activities
17	Cultural	Providing opportunities for non-commercial uses	Aesthetic, artistic, educational, spiritual and/or scientific values of ecosystems

* We include ecosystem 'goods' along with ecosystem services.

and avoidance of damaging ultraviolet light from the sun (provided by the ozone layer, greenhouse gas production and di-methyl sulphide (DMS) production affecting cloud formation) and successful pollination (which, depending on the crop, is provided by bees and other invertebrates). Although there remains some debate about the exact details of the functions and the services provided by ecosystems, there is some agreement as to the general classes of service provided (Table 7.7).

Increasing amounts of attention have been given to analysing and understanding the values provided by ecosystem services by academics and policy-makers. An

Table 7.8 Summary of the average global value in $US per hectare per year of the 17 ecosystem services listed in Table 7.7 across a range of biomes. (Source: Costanza *et al.* 1998.)

		Ecosystem services (1994 $US per hectare per year)								
Biome	Area (ha × 10^6)	1 (Gas regulation)	2 (Climate regulation)	3 (Disturbance regulation)	4 (Water regulation)	5 (Water supply)	6 (Erosion control)	7 (Soil formation)	8 (Nutrient cycling)	9 (Waste treatment)
Marine	36 302									
Open ocean	3 322 038				–	–	–	–	118	
Coastal	3 102			88	–	–	–	–	3 677	
Estuaries	180			567	–	–	–	–	21 100	
Seagrass/ algae beds	200				–	–	–	–	19 002	
Coral reefs	62			2 750	–	–	–	–		58
Shelf	2 660				–	–	–	–	1 431	
Terrestrial	15 323									
Forest	4 855		141	2	2	3	96	10	361	87
Tropical	1 900		223	5	6	8	245	10	922	87
Temperate/ boreal	2 955		88		0			10		87
Grass/ rangelands	3 898	7	0		3		29	1		87
Wetlands	330	133		4 539	15	3 800				4 177
Tidal marsh/ mangroves	165			1 839						6 696
Swamps/ floodplains	165	265		7 240	30	7 600				1 659
Lakes/rivers	200				5 445	2 117				665
Desert	1 925									
Tundra	743									
Ice/rock	1 640									
Cropland	1 400									
Urban	332									
Total	51 625	1 341	684	1 779	1 115	1 692	576	53	17 075	2 277

Number in the body of the table in $US per hectare per year. Row and column totals are in $US per year × 10^9; column totals are the sum of the products of the per hectare services in the table and the area of each biome, not the sum of the per hectare services themselves. Dashes indicate services that do not occur or are known to be negligible. Open cells indicate the lack of available information.

example of the latter is a report by the Environmental Protection Agency (EPA) Science Advisory Board in the USA.

> 'The value of natural systems was inadequately considered in setting priorities and valuation is generally insufficient for decisions facing EPA in two important ways: (1) current economic models focus on structural parts of ecosystems, not functions or relationships and (2) current ecological models do not adequately describe the "services" of ecosystems.' (Brody & Kealy 1995)

Indeed, very few countries pay much attention to the value of ecosystem services, and examples of conserving areas of land because of the important ecosystem services they provide are rare. One example of such a conservation effort relates to the protection of water through designating areas of land around aquifers and water bodies as buffer zones. These areas of land then serve to 'buffer' the water from potential pollutants and eroded soil, which could enter the water body. Thus, these areas of land are supplying a purifying or filtering service.

While it is easy to state that these areas of land provide a filtering service, it is more difficult to specify the exact attributes (plants, microorganisms, chemical processes) of the ecosystem which undertake the filtering. Many processes occur in ecosystems and not all these processes provide direct value to humankind, but given the interconnections within ecosystems it is generally impossible to isolate and preserve only the important services.

The problems of making decisions about preserving ecosystems in the light of such uncertainty are real and

10 (Pollination)	11 (Biological control)	12 (Habitat/ refugia)	13 (Food production)	14 (Raw materials)	15 (Genetic resources)	16 (Recreation)	17 (Cultural)	Total value per hectare ($US per hectare per year)	Total global flow value ($US per year × 10^9)
–								577	20 949
	5		15	0			76	252	8 381
–	38	8	93	4		82	62	4 052	12 568
–	78	131	521	25		381	29	22 832	4 110
–				2				19 004	3 801
–	5	7	220	27		3 008	1	6 075	375
–	39		68	2			70	1 610	4 283
							804	12 319	
	2		43	138	16	66	2	969	4 706
			32	315	41	112	2	2 007	3 813
	4		50	25		36	2	302	894
25	23		67		0	2		232	906
		304	256	106		574	881	14 785	4 879
		169	466	162		658		9 990	1 648
		439	47	49		491	1 761	19 580	3 231
			41			230		8 498	1 700
14	24	–	54					92	128
–	–	–	–	–	–				
117	417	124	1 386	721	79	815	3 015		33 268

difficult, and recent ecological economic analyses have sought to develop methodologies to help make these difficult decisions. One of these analyses sought to undertake a meta-analysis of all available data on ecosystem functions and services and estimate their current value to humankind (Costanza *et al.* 1997). The first stage in this analysis necessitated the identification of the different functions and services provided by the different ecosystems on the planet, ranging from oceans to tropical forests (see Table 7.7). After identifying these services the study sought to attach monetary values to them. The values used were derived from the many valuation studies which had been conducted over recent years, and as such were undoubtedly subject to some errors. These and other errors were acknowledged by the authors, who state their work 'merely opens the door for more and better research and valuations that go further to address the problems and limitations we identify and clearly acknowledge' (Costanza *et al.* 1998).

When the values for all the services provided by all the ecosystems of the world are combined, their summed value to humankind lies between $US16–54 trillion ($10^{12}$) per year, with an average of $US33 trillion per year. A further breakdown of the contribution of different ecosystems to each of the 17 classes of service is provided in Table 7.8 and, as can be seen from this table, the most valuable services provided by ecosystems are gas regulation and nutrient recycling. The ecosystems which provide the greatest value per hectare are estuaries, wetlands and swamps/floodplains. The value of these ecosystems is largely related to their role in nutrient recycling and waste treatment. In a further

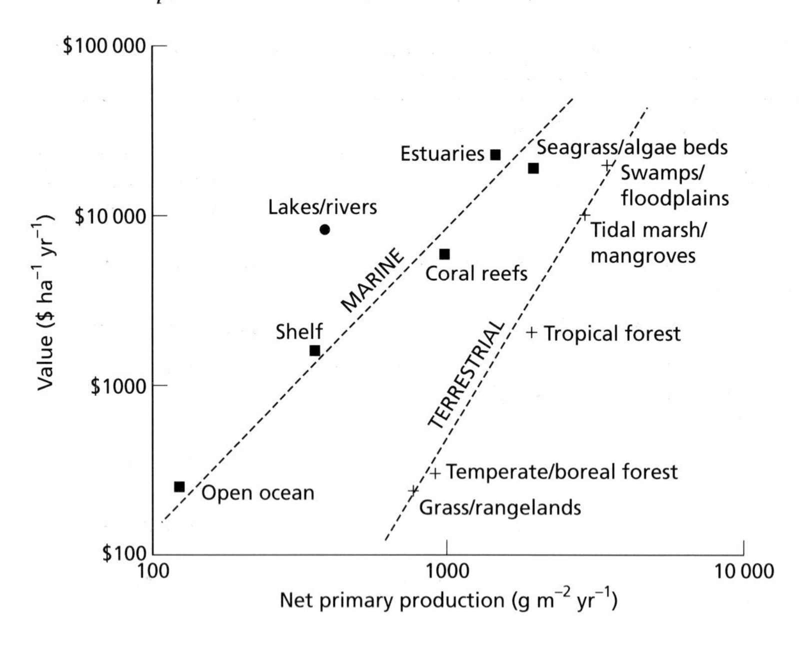

Fig. 7.2 Relationship between net primary product of ecosystems and the value of the services they supply. + terrestrial ecosystems; ● freshwater ecosystems; ■ marine ecosystems. (Source: Costanza *et al.* 1998.)

analysis of these data Costanza *et al.* (1998) suggest that the value of ecosystem services increases with the net primary production of the different ecosystems (Fig. 7.2). This analysis also suggests that marine ecosystems seem to provide higher value services to humankind per hectare per year than do terrestrial ecosystems of similar net primary productivity. It is also noticeable that even amongst terrestrial ecosystems it is the aquatic ecosystems (lakes and rivers) that provide the highest value services.

There has understandably been considerable discussion of this analysis (see Costanza *et al.* 1998), and some typical reactions are reflected in those of Norgaard *et al.* (1998):

> 'Will ecological economists bring us the value of God next? . . . will this be the end of history for economic valuation? . . . and now that we know the exchange value of the earth we wondered with whom we might exchange it and what we might be able to do with the money, sans earth.'

These rhetorical questions may sum up many people's feelings about this analysis. However, documenting estimated values for ecosystem services does seem to be a useful exercise if only to highlight the general importance of ecosystem functioning to humankind. Furthermore, it serves to highlight the specific importance of what may otherwise be rather unattractive and unspectacular ecosystems, such as wetlands and swamps, and in this way counterbalances more traditional nature conservation values and methods.

At a more local level, the identification of ecosystem services can also highlight the importance of ecosystem services and act as an aid to making decisions about the future of ecosystems and habitats. For example, in many tropical areas of the world there is considerable pressure to develop areas of coast currently covered by mangroves for alternative uses, such as intensive shrimp production. Mangroves are salt-tolerant trees found around sheltered tidal shorelines in tropical and subtropical areas and the function and services provided by the mangrove ecosystem to local economies are listed in Table 7.9.

This information can be used to help make decisions about the future management of mangrove ecosystems. For example, Gilbert and Janssen (1998) use this information to help develop a series of system diagrams which represent the production of environmental goods and ecosystem services by their study mangrove area, Pagbilao in the Philippines (Fig. 7.3). These system diagrams represent the inherent interconnection of different elements and functions of ecosystems and can be used to trace the impacts of certain changes to the mangrove system. For example, in Fig. 7.3 the basic productivity of mangrove ecosystems, which itself is dependent on nutrient recycling and climate regulation, can lead to the existence of fish, timber and other biota.

Table 7.9 A summary of the environmental functions and services provided by the Pagbilao mangroves in the Philippines, and the ecological processes which generate these services. (Adapted from: Gilbert & Janssen 1998.)

Ecological processes	Environmental function	Good/service	User
Hydrological cycle	Production of water	Water	Aquaculture adjacent to mangrove forest
Fixation of solar energy and biomass production	Production of food and nutritious drink	Offshore fish and shellfish	Artisanal fisheries
	Production of fuel and energy	Wood, charcoal	Local communities
	Production of raw materials for building, construction and industry	Wood, leaves, tannins, *Nipa* shingles	Local communities
	Production of other biotic resources	Medicinal resources	Local communities
		Uptake of carbon dioxide	Global population
Maintenance of nursery and migration habits	Production of food and nutritious drink	Offshore fish and shellfish	Artisanal fisheries
		On-site crabs	
	Production of juveniles for cultivation	Mangrove propagules	Government (afforestation and reafforestation programmes)
	Maintenance of biodiversity	Biodiversity	Global population
			On-site ecotourism
			Off-site ecotourism
Storage/recycling of organic matter and nutrients	Regulation of environmental quality	Disposal of wastes and surplus	Aquaculture adjacent to mangrove forest
Sediment control	Prevention of soil erosion	Shoreline protection	Local communities Aquaculture adjacent to mangrove forest
Buffering of storms	Flood mitigation	Flood mitigation	Local communities Aquaculture adjacent to mangrove forest
All ecological processes	Scientific and educational information	Knowledge	Scientific and educational community

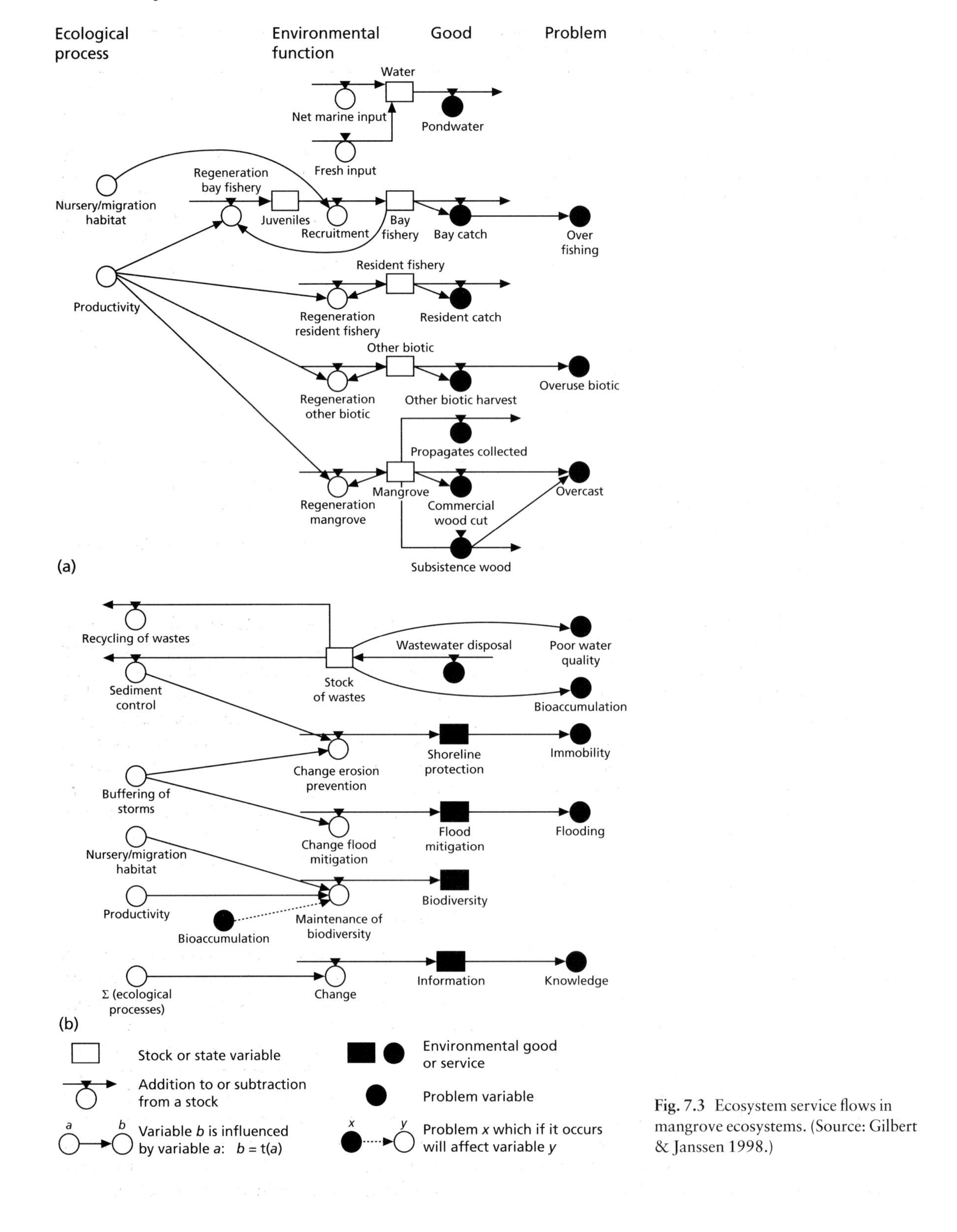

Fig. 7.3 Ecosystem service flows in mangrove ecosystems. (Source: Gilbert & Janssen 1998.)

Table 7.10 The net value in Philippino pesos of the goods and services provided by the Pagbilao mangroves under eight different management regimes (A–H). (Source: Gilbert & Janssen 1998.)

	Unit	A	B	C	D	E	F	G	H
Goods									
Fisheries	'000 pesos	165	161	161	124	8	8	40	40
Subsistence forestry	'000 pesos	0	349	0	0	0	0	0	189
Commercial forestry	'000 pesos	0	0	416	218	0	0	229	0
Aquaculture: fish	'000 pesos	0	0	0	5 648	18 801	13 577	4 992	4 992
Mangrove nursery	0/+++	+	+	+	0	0	0	0	0
Total goods	'000 pesos	165	510	577	5 990	18 809	13 585	5 261	5 221
Services									
Aquaculture: waste	0/+++	0	0	0	+	++	+++	++	++
Damage control	0/+++	+++	+++	+++	++	+	+	++	++
Ecotourism	0/+++	+++	++	0	0	0	0	0	+
Existence value	0/+++	+++	++	++	+	0	0	+	+
Information value	0/+++	+++	+++	++	+	0	0	+	+
Total services	0/+++	+++	+++	++	+	+	+	+	+

Key to management regimes: A, preservation; B, subsistence forestry; C, commercial forestry; D, aqua-silviculture; E, semi-intensive aquaculture; F, intensive aquaculture; G, commercial forestry/intensive aquaculture; and H, subsistence forestry/intensive aquaculture.
Key to non-monetary values: +++, large contribution to value; ++, moderate contribution to value; +, small contribution to value; 0, no contribution to value.

However, the over-harvest of any of these goods will lead to problems for their users.

When considering the services provided by mangroves, we can see from Fig. 7.3 that ecological processes in mangrove systems may affect the structure of waterways, but apart from this have little influence on water volumes and flows. Water is used to flush fish-ponds, releasing contaminated waste to the environment. Some proportion of these wastes enter the mangrove ecosystem where they can be taken up by aquatic organisms. If the waste load exceeds the system's capacity for removal then this will lead to a problem of poor water quality. This may then have numerous impacts on the mangrove's species and functioning. Similar analyses can be undertaken for the services related to flood mitigation, shoreline protection and provision of biodiversity. Subsequent to completing these analyses for all goods and services provided by the Pagbilao mangroves, Gilbert and Janssen (1998) estimated the impact of eight different management regimes on the mangrove system. The net annual flow of goods and services provided under each of these management regimes is shown in Table 7.10.

This sort of integrated analysis offers much potential to aid decisions about the balance of ecosystem conservation and development. From a biological point of view it is important that such analyses explicitly recognize the structure and function of ecosystems, and the complex and interactive nature of many of their processes. From an economic viewpoint, they can provide opportunities for extending monetary estimates of value to non-market goods and ecosystem services, as well as integrating the effects of ecosystem disruptions into economic appraisals. This kind of integrated ecological–economic modelling is a key topic within advanced ecological economics research.

7.5 Summary

In this chapter we have examined the criteria which have been used by ecologists to assess conservation value of sites. A wide range of these criteria exist ranging from species diversity to educational value and, despite much work in this area, there is no one criterion or set of criteria which are applicable in all situations. Assessing the conservation threat to species is a little easier and population viability analysis provides a useful tool for use in this context. However, it is clear that despite the existence of much biological knowledge, the essence of conservation decisions remains subjective.

Despite this subjectivity it is important that decision-making frameworks remain transparent and consistent. Computer-based systems which combine quantitative and qualitative data have been developed which offer some help towards improving decision-making. However, while these systems may render the decision-making process more open, they should be considered as assisting rather than replacing the essential human input to the process.

Recent economic analysis has focused on the value of the services provided by ecosystems to humankind: elements such as soil stabilization from forest cover, and water purification through wetlands. Although contentious, these analyses have shown that even if a site possesses little of wildlife or botanical value it may provide very important services to human society. Combined ecological–economic modelling demonstrating the extent of these relationships is an important direction for future research.

Further reading

Helliwell, D.R. (1985) *Planning for Nature Conservation*. Packard Publishing Ltd, Chichester.

Spellerberg, I.F. (1992) *Evaluation and Assessment for Conservation*, Chapman and Hall, London.

Ratcliffe, D.A. (ed.) (1977) *A Nature Conservation Review* (2 vols), Cambridge University Press, Cambridge.

Part 3
Frameworks for Decision-Making

8 Cost–Benefit Analysis

'Counting is the religion of this generation: it is its hope and its salvation.' [Gertrude Stein]

8.1 Introduction

Cost–benefit analysis (CBA) is a methodology which aims to select projects and policies which are efficient in terms of resource use. As the name suggests it is, in principle, extremely simple. All the positive and negative effects of a proposed project or policy are valued in monetary terms, providing a list of benefits and costs. If benefits are seen to outweigh costs, then the proposed plan represents a potential gain in terms of social welfare.

The underlying justification for CBA lies within the theory of welfare economics, introduced in Section 2.6. It is founded on a directly utilitarian approach to decision-making which should, under certain circumstances, enable government to select those projects which maximize social welfare. According to Leonard and Zeckhauser (1986), the practice of CBA is therefore often considered to have the 'hypothetical consent of the citizenry', because the citizens of any country elect governments precisely on the basis that those governments will work to raise the people's welfare. In fact, although the citizenry may 'consent' to the general principle of CBA in decision-making, this does not mean that they will endorse its application in every particular case. For example, in certain cases the US Endangered Species Act is in conflict with policies based on CBA outcomes.

In Section 2.6 we noted that welfare economists try to improve the Pareto efficiency of the economy, which is to find ways to make some (or all) people better off but *without* making others worse off. The Pareto criterion is a welfare-based rule for government decision-making which states that proposed projects bringing about Pareto gains are justifiable in terms of the public interest. Unfortunately, the problem with this decision rule is that finding projects that benefit everyone and most crucially *make no one worse off* is extremely difficult, if not impossible. Almost all projects have direct winners and losers.

A very widely accepted variation to the Pareto criterion is what is known as the Kaldor–Hicks 'potential compensation' principle, which asks whether the winners from a development project could *in theory* 'compensate' the losers and still remain better off after the project than before. In other words, the key test is whether the net gains from the project are positive. If so, society as a whole gains, and it is at least possible in theory to make the final outcome a genuine Pareto gain simply by transferring some of the winners' gains to the losers, for example through taxation. Cost–benefit analysis itself is intended to help decision-makers to identify projects with potential Pareto gains by evaluating all the relevant costs and benefits. It is worth noting that the 'potential compensation' is, in fact, very rarely if ever paid; however, given the large number of government projects undertaken, it could be argued that overall the gains and losses will even out.

Cost–benefit analysis has been widely applied and endorsed in the context of both public and private decision-making. Gramlich (1990) cites Benjamin Franklin and Abraham Lincoln as early proponents of the CBA. The US Presidential Executive Order 12291 (1981) made the application of CBA a necessary requirement for all new policies. In the UK, the government endorsed the inclusion of environmental appraisal methodologies into formal procedures governing both policy and projects in a White Paper entitled *This Common Inheritance* (HMSO 1990).

8.2 Objections to cost–benefit analysis

Although CBA is widely practised and accepted, it has also been strongly criticized from perspectives outside economics, particularly the environmental and social spheres. Problems with the technique are noted throughout this chapter but it is useful to summarize the main objections here. Criticisms generally revolve around seven issues.

1 The *uncertainty* with which the physical costs and

benefits associated with a project can be estimated. This is particularly relevant where complex ecosystems are affected by development projects, such as a water irrigation scheme affecting downstream wetlands.
2 The *accuracy and acceptability* of monetary valuations of impacts. In the case of rare habitats, for example, controversy may surround both the precise figures estimated for damages, and the philosophical implications of these monetary estimates (as reviewed in Chapters 5 and 6).
3 The *distribution* of costs and benefits amongst the population. Standard CBA practice does not distinguish between the sections of society that receive benefits or endure costs as a result of a project, though these can be included as shown later in this chapter.
4 The practice of *discounting the future* in estimating total costs and benefits. Discounting reduces the significance of future costs and benefits, and its rationale is explored in detail below.
5 The treatment of *irreversibility* in development decisions. This is particularly pertinent to preservation/development decisions concerning the environment because a number of such decisions result in irreversible changes.
6 The *institutional impartiality* of the CBA process. Critics point out that the CBA process can be manipulated to serve particular interests by varying the assumptions used, at several stages of the process.
7 The lack of a *sustainability criterion* in cost–benefit decision-making. As CBA is generally applied to individual projects, the methodology is not designed to ensure overall sustainability of the impacts of a series of projects or developments, or of the size of the economy as a whole, although this criticism is certainly not exclusive to CBA.

For the time being, as a response, but not a direct answer, to these concerns, it is helpful to note more carefully the conditions and assumptions under which CBA is undertaken.

- Cost–benefit analysis summarizes the values of economically relevant costs and benefits over the lifespan of a project in a figure termed *net present value* (NPV). This figure should allow comparisons to be made between projects over their lifetimes on a like-for-like basis.
- The various costs and benefits over time are made commensurate through a process known as *discounting*, which converts them into what they would be worth today, i.e. their respective present values.
- Cost–benefit analysis is based on the utilitarian ethic (see Section 4.3.3). Thus, in order that a cost or benefit be economically significant, it must affect an individual's utility.
- Changes in utility arise out of changes in the quantity and/or quality of both marketed and non-marketed commodities that an individual consumes.
- Even though there may be no market for environmental commodities, such as clean air, if a project leads to a change in air quality *and* this affects utility, then a CBA study should value this change in monetary terms.

The following sections present an overview of the important elements of the CBA methodology, followed by a more detailed discussion of discounting. At the end of this chapter we present a brief case study of an appraisal carried out on a reforestation project in Tunisia.

8.3 An outline of cost–benefit analysis methodology

The following sections outline the various stages of the methodology, and therein explain some of the concepts sketched out above. These stages are the following:

1 Project definition.
2 Classification of impacts.
3 Conversion into monetary terms.
4 Discounting.
5 Project assessment under the net present value (NPV) and internal rate of return (IRR) tests.
6 Sensitivity analysis.

8.3.1 Project definition

It is necessary in the first instance to define precisely what the proposed project implies in terms of resource allocation. This process is also critical in that the *boundary of inquiry* of the CBA is thereby established. Consider a proposed project to construct an oil rig; the boundary might feasibly extend to an appraisal of UK national energy policy, or be confined to the local impacts alone of the construction. A wider boundary generally implies a more expensive CBA study.

A related task is to define the *population* of affected people that are to be considered by the study. Oil extraction ultimately could lead to global warming through the burning of fossil fuel. This process affects both current and future generations. The defined population for the oil rig construction study could theoretically be enormous. However, the scope of analysis might be restricted to only those currently working in the close vicinity of the construction site.

8.3.2 Classification of impacts

After project definition, the impacts arising from the project must be identified. For the oil rig project the impacts would include the following.

- Consumption of materials, such as steel and concrete.
- Effects on local employment levels. This should include *direct* employment, initially by the construction company and then by the rig's operatives, and *indirect* employment generated by increased spending in the local economy. Labour hours are subdivided into the various skilled and unskilled input categories.
- Impacts on the environment, from the construction process itself, through the operational life of the rig and finally to the disposal of the rig.

TIMING OF IMPACTS

All the impacts to be included in the CBA must be quantified and then tabulated in terms of *when* they arise. If the construction project is estimated to take 18 months for completion, then there will be an associated estimate for the input of, for instance, engineers' labour hours for each of these months. The same procedure is applied to each impact for each discrete time interval. This is important for the process of discounting, explained in Section 8.3.4. Determining such timings will generally involve some element of *estimation* (considered below).

IRRELEVANT IMPACTS

Some impacts are irrelevant in economic terms. The decision criterion for including an impact in the CBA is whether either utility or production levels are affected by the impact. If the oil rig releases petrochemical residues then these are included if they change production by depleting the fish stocks available to fishing trawlers, or change utility by contaminating a beach which is used for recreation.

An important class of irrelevant impacts is *transfer payments*. Examples of transfer payments which might arise from the oil rig construction include an increase in revenues from local sales taxes as a result of increased consumption in the local economy, and a decrease in unemployment benefit payments resulting from the job creation of the construction project. Cost–benefit analysis was defined above as a performance yardstick to optimize resource allocation in society. Such taxes and social security payments do not use up resources in society but merely *redistribute* them; whereas the steel and concrete used up in construction are then not available for an alternative project. More taxation revenue coming into the state's coffers implies more being paid by wage-earners: the state's gains are directly counterbalanced by citizens' losses.

ESTIMATION OF IMPACTS

Estimation across a wide range of impacts is likely to be necessary in any CBA. Some impacts will, of course, be more uncertain than others. For instance, the expected useful lifespan of the rig is known with more certainty than the effects of petrochemical residue releases on the fish populations.

This varying degree of certainty can be captured, to a degree, by using probabilities. Assume, somewhat arbitrarily, that the consultant engineers estimate that the rig's lifespan has a 5% chance of being 15 years, a 55% chance of being 20 years and a 40% chance of being 25 years. (Note that the total probabilities must add up to 1, or 100%.) The expected lifespan can then be calculated as follows:

$(0.05 \times 15) + (0.55 \times 20) + (0.4 \times 25)$, or
$0.75 + 11 + 10 = 21.75$ years.

This does not imply that the rig will last exactly 21.75 years; it implies that this is its expected lifespan, given the engineering estimates. For the lifespan estimation, there is a 95% probability that the rig lasts at least 20 years. This is a relatively low level of uncertainty concerning lifespan. Relatively higher uncertainty would be represented by a greater spread of values for a given probability.

ADDITIONALITY OR WITH-MINUS-WITHOUT

An important principle to be applied at the classification stage is *with-minus-without* (also termed *additionality*), which involves isolating project-related impacts from trends in background conditions. For example, some changes in, say, employment would arise even without the project. The correct methodology is to calculate the project impact on employment *net* of these changes. Thus it considers only those changes directly attributable to the project.

Consider Fig. 8.1. Each of the three projects commences at time t_0 and finishes at time t_F. The employment rate at t_0 and t_F are the same for each project, that is E_0 and E_F, respectively. However, between t_0 and t_F

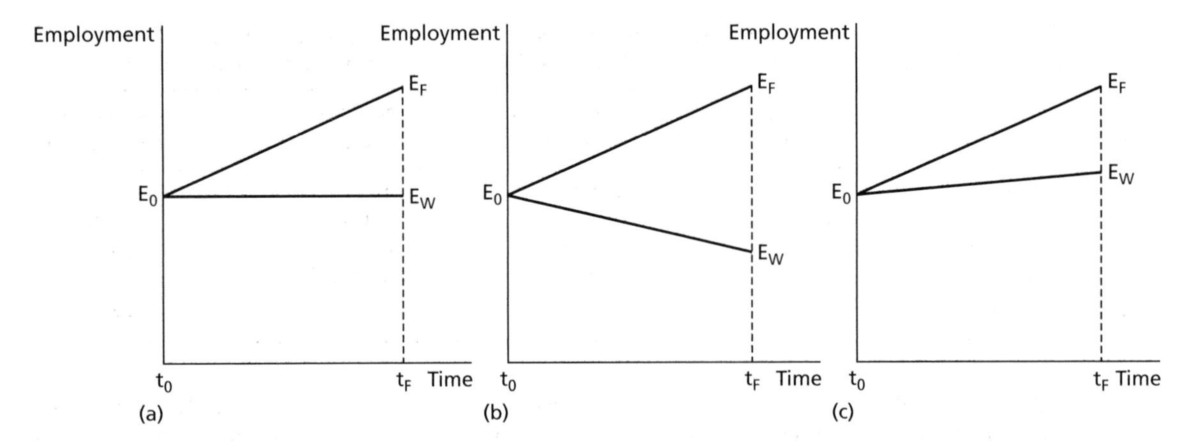

Fig. 8.1 With-minus-without scenarios for three trends in employment on hypothetical projects. The lower line represents the 'without project' employment level and the upper line the 'with project' level.

the employment rate without the project (E_W) would have stayed constant in the region in Fig. 8.1(a), decreased in the region in Fig. 8.1(b) and increased in the region in Fig. 8.1(c). The with-minus-without procedure thus provides a true picture of the impacts attributable to each project in each respective region. Omitting this procedural stage would imply that each of the three projects performed equally well in terms of employment generation.

The with-minus-without procedure should, where appropriate, be applied inter-regionally. Consider Fig. 8.2. There is a proposal from a multinational microprocessor manufacturer to set up a production plant in East Lothian in Scotland. This proposed investment is conditional on the construction project being subsidized by the UK government. The government might then commission a CBA to assess the trade-off between the cost to the taxpayer of this subsidy and the expected generation of income and employment in East Lothian. The CBA might produce results as given in Fig. 8.2(a).

However, a CBA exclusively considering East Lothian might not provide the full picture; any changes in income or employment outside East Lothian, but within the UK, which result from this proposed inward investment should be included in decision-making. Say that a competing microprocessor manufacturing plant already exists in Glasgow, and that the effect of the proposed inward investment on Strathclyde region, which contains Glasgow, is given by Fig. 8.2(b). The net effect of the proposal on employment is given by:

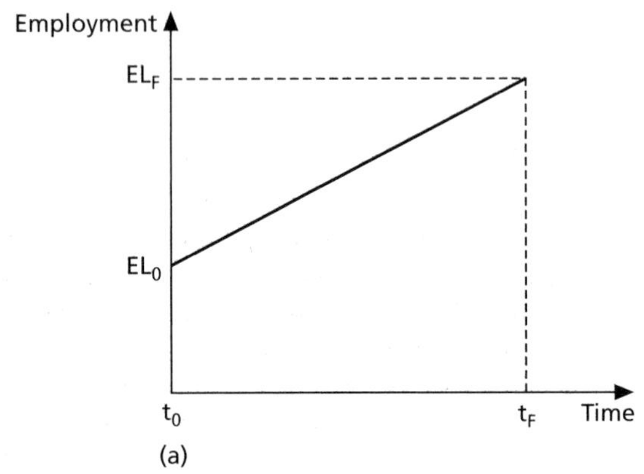

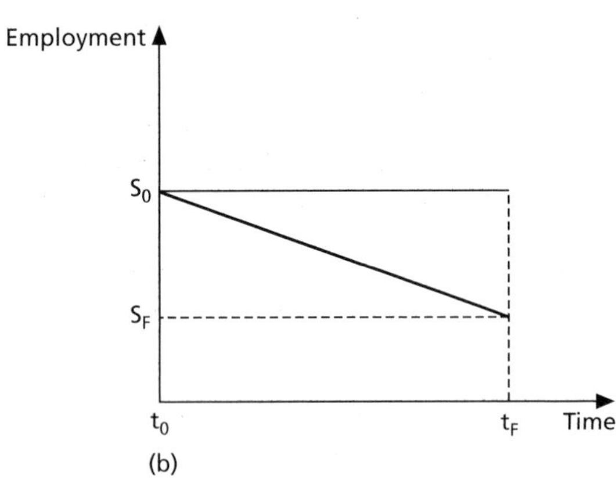

Fig. 8.2 The effects of displacement.

$$(EL_F - EL_0) + (S_F - S_0).$$

In this case, the proposed East Lothian plant 'crowds out' some jobs from Strathclyde through competition, and $(S_F - S_0)$ is negative.

Whether conducting such a displacement analysis

Table 8.1 The conversion from nominal to real prices. The nominal labour cost is found by multiplying the labour hours by the wage rate in each year. The real aggregate labour cost is calculated by multiplying this figure by the inflation adjustment factor.

Year	Labour hours	Wage rate (£) (nominal)	Nominal aggregate labour cost (£)	Inflation adjustment factor	Real aggregate labour cost (£)
0	265 000	5.5	1 457 500	1	1 457 500
1	320 000	5.75	1 840 000	0.943396	1 735 848
2	285 000	6.05	1 724 250	0.889996	1 534 575
3	125 000	6.4	800 000	0.839619	671 695
Total	995 000		5 821 750		5 399 618

within a CBA is appropriate or not is dependent on the domain of the funding agency. In this case, as the funding agency is the UK government, the CBA study should ideally account for all significant changes across the whole of the UK.

8.3.3 Conversion into monetary terms

Money is used as the value scale in CBA. The fact that money has been adopted as an impact-valuation measure is not intended to signal that those impacts that cannot feasibly be monetarized are insignificant, but it is undeniable that there is a tendency for this to happen in practice. Cost–benefit analysis practitioners generally state that money has been chosen through convenience. Money is indeed convenient in that prices do signal scarcity, at least when markets are functioning correctly, and steel and concrete and labour are all priced in monetary terms. Further, it is difficult to conceive of an alternative measure which is any less problematic.

But these arguments do not detract from the fact that using money to value impacts which are conventionally non-monetarized raises a host of problems. Mark Twain's famous phrase is sometimes quoted by critics in exasperation at CBA: 'when the only tool in your toolkit is a hammer, then it's no wonder that everything begins to resemble a nail'. The only tool that CBA uses is a conversion of impacts to a single point on a monetary yardstick. If the systems under study are, by their very nature, complex and multidimensional, then conversion to a single dimension offers gains in terms of simplicity but only at the expense of an associated loss in relevance. Given this caveat, we now turn to the conversion process itself.

ADJUSTMENTS FOR INFLATION

Some impacts will have a market price which is only revealed at some time in the future. Part of the valuation stage then is to estimate the monetary values of impacts arising in the future. The monetary values might differ from current prices in two different respects. First, a change in the price of a particular cost or benefit can occur relative to other prices, in which case it is a genuine shift in value as a result of increasing scarcity or other factors. Secondly, there may be a universal rise in all prices, called *inflation*. Rises in prices of goods caused by inflation do not signal 'real' changes in their value, they are simply a result of the imperfect functioning of the monetary system as a result of factors such as rising wages and shifts in money supplies. Cost–benefit analysis studies are therefore carried out in inflation-adjusted real terms. Pre-adjustment prices are termed *nominal*, and post-adjustment prices are termed *real*. In the UK, the retail price index (RPI) is used as a measure of general inflation.

An example of such an inflation-adjustment is shown in Table 8.1. The figures represent labour costs for the rig construction. The labour-hours per annum would have been classified at the previous methodological stage. This classification is shown in column 2. An estimate for the wage rate in nominal terms for each year is given in column 3. Multiplying the figures in columns 2 and 3 gives the nominal aggregate labour cost for each year, given in column 4.

Column 5 shows the inflation-adjustment factor for each respective year, assuming in this case that inflation stays constant throughout the construction period at a rate of 6% per annum. The formula used to convert from nominal to real figures for each year is given by

$$(1 + p)^{-t},$$

where p is the inflation rate per year, and t is the year under consideration. We discuss this formula in relation to discounting in the next section. If the inflation rate is positive then

$(1 + p)^{-t} < 1,$

and $(1 + p)^{-t}$ decreases as t increases. In this example, $p = 0.06$. Column 6 gives the real costs of labour for each year, in year 0 prices: the figure in column 6 is obtained by multiplying the respective figures in columns 4 and 5.

ADJUSTMENTS FOR SHADOW PRICES

As mentioned above, prices do signal scarcity, but only when markets for goods are functioning perfectly. Therefore, part of the conversion process involves correcting market prices, if necessary, so that they represent measures of true scarcity. The corrected price is then termed the *shadow price*. The need to use shadow prices can arise as a result of several factors.

Government intervention

One significant category of divergence between market and shadow prices is caused by government intervention. In the discussion on economically significant impacts above, it was noted that transfer payments, such as social security payments, are excluded from the CBA. However, it is possible for transfer payments to be built into the market price for goods, and they must also be excluded from the CBA. Perhaps the most obvious form of such government intervention is in markets for agricultural produce. Under the European Union's Common Agricultural Policy (CAP), prices for some agricultural commodities inside European Union countries are held artificially high (that is, above the going price on world markets). This is to assist the farming industry for historical and strategic reasons. In these cases the price—say for wheat—in a particular European Union country does not reflect genuine scarcity but is affected by this government price intervention. In this case an adjustment using the shadow price of the affected goods, which is basically their price on world markets, should be made within the CBA.

Imperfect competition

If markets are imperfect—in particular, if the suppliers of some goods have market power and can set prices as they wish—then the market price does not accurately represent resource scarcity; it is rather a reflection of the power of the dominant firm. Such imperfect prices should be adjusted within the CBA to reflect the genuine scarcity value of the good in question.

Externalities

Perhaps the key problem for CBA to contend with is that many environmental and social impacts arising from a project are not traded and therefore have no explicit market price. These problems were noted as externalities in Section 3.6.3, and specific techniques used to attempt to include these kinds of effects in economic evaluation was the subject of Chapter 6. The problems with these techniques are not reviewed again here. Such externalities have tended to dominate the concern over CBA in application to environmental impacts, hence the ongoing research effort to develop better methods for valuing environmental externalities. Many of the concerns with CBA noted at the start of this chapter relate to this 'internalizing the externality' issue, and specifically how far CBA can be extended to encompass all relevant concerns regarding projects with significant environmental impacts.

8.3.4 Discounting

The fundamental supposition of discounting is that in the estimation of the value of a project, future costs and benefits count for less than present ones. Two immediate qualifications must be made here.

1 Discounting is entirely separate from the process of adjusting for inflation. All calculations of discounting take place in real terms; that is, keeping the 'purchasing power' of money values constant across time by deflating by a chosen rate. Even if the rate of inflation was zero, the arguments for discounting still stand.

2 Discounting applies to the values attributed to future consumption, not to future utility or welfare. This is an extremely important distinction. The utility, or welfare, of all individuals in all time periods is considered as equally valuable; what is discounted is the value of consumption in different time periods, because the relationship between consumption and utility is generally thought to be time-dependent, for reasons given below.

The practice of discounting is at the hub of much controversy between economists and environmentalists. Some of this relates to the validity of discounting, some to the overall validity of CBA as a decision-making tool, and some to limitations within the actual practice of discounting itself. Ramsey (1928) suggests that discounting is 'ethically indefensible and arises merely from a weakness of the imagination', and Harrod (1948) rather more colourfully describes it as 'a polite expression for rapacity and the conquest of reason by passion'.

Here we first explain the rationale for applying some form of discounting to future values, and its effects. After continuing with the survey of CBA methodology, we then return to discuss various criticisms of discounting, both technical and philosophical.

With the above qualifications in mind, there are two basic reasons for discounting: time preference and the productivity of capital.

TIME PREFERENCE

Time preference describes the preference of an individual for when he or she receives a benefit, or has to bear a cost. Is £1000 received today considered as valuable as £1000 received this time next year, or in 10 years' time? If the individual's assessment of the value of the benefit does change depending on when it is received, that individual is said to have a time preference.

What justifies having a positive (sooner rather than later) time preference? The justification is a result of two factors: pure time preference and opportunity costs.

Pure time preference

Pure time preference relates essentially to the factors of *impatience* and *uncertainty* when considering any future good. As Ramsey's observation bears out, individuals are often short-sighted in their behaviour, acting for the short term because the future somehow simply seems less important than the here and now. This is basically impatience; it may not have any very rational justifications, but nevertheless it is a preference that many consumers do have. The desire for goods now is, however, also related to uncertainty about the future. Goods may prove to be less valuable in the future than now because of a change in circumstances; of course, the reverse might also be true. This generally leads the future to be discounted because of uncertainty about future benefits, whereas we are much more certain of how a good will impact on our immediate welfare.

Opportunity costs

The second reason for a positive time preference rate is opportunity costs: that is, the opportunities for generating further benefits that come from having a good now rather than in the future. In the simple case of £1000 now or in a year's time, and with a prevailing bank interest rate of 10%, you can choose either to spend immediately or deposit the money in the bank. If you choose the latter option then you will have £1100 at the end of the first year. If you keep the money on deposit for a second year you will have £1210, the year after £1331, and so on. The computation of these figures is:

$$1100 = (1000)(1 + 0.1) \qquad 1210 = (1100)(1 + 0.1)$$
$$1331 = (1210)(1 + 0.1)$$

$$1100 = (1000)(1.1)^1 \qquad 1210 = (1000)(1.1)^2$$
$$1331 = (1000)(1.1)^3.$$

The generalization of the above calculation gives the following expression for the total value of a deposit under compound interest:

$$V_T = V_0(1 + r)^T,$$

where V_T is the amount in the account after T years; V_0 is the initial deposit and r is the annual rate of compound interest.

Assuming a fixed interest rate, £1000 deposited now is worth £1331 in three years' time. Looking backwards then, £1331 in three years' time is worth as much to you as £1000 today. The *present value* of this £1331 is thus £1000. Generalizing this result, using the same notation as above, gives

$$PV = V_T/(1 + d)^T,$$

where d is discount rate.

During the discounting process in CBA, this formula is applied to all costs and benefits for the relevant time periods and at a specified discount rate, to provide present value figures. (Note in passing that the same formula is applied to convert inflation-affected figures back to real ones, because the same discounting logic applies. Under inflation, all prices rise by a given rate each year, and the conversion formula is used to deflate these prices, to give their real value relative to the present year.) Additional justification for why future values should be discounted is provided by the productivity of capital.

PRODUCTIVITY OF CAPITAL

The productivity of capital argument for discounting is basically the flip side of the opportunity cost argument

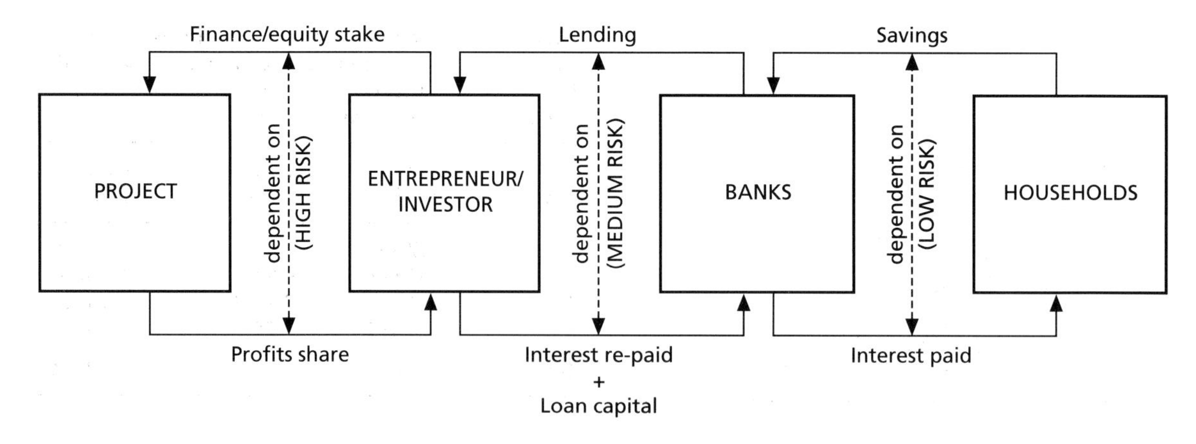

Fig. 8.3 A simplified model of the circulation of money from banks to investors to projects.

for having a positive time preference. That is, the productivity of capital accounts for the fact that there are opportunities within the economy to earn interest on savings. This is because the process of saving, with that money reinvested by banks in productive projects, leads to increased overall output in the future. If the economy was completely static, that is, not growing at all, there would essentially be no rationale for discounting on the basis of future opportunity costs, because money invested now would yield no surplus returns in the future.

Given that most economies do offer investment opportunities, the extent of those opportunities will determine the demand, coming from entrepreneurs, for capital or finance for investment. In fast-growing economies, everyone wants to invest in projects because potential future profits are high. This demand for investment capital has to be met in part from savings held by banks, which are strictly limited. If demand is strong for investment capital, banks can correspondingly set higher interest rates on the loans they make to entrepreneurs. Such entrepreneurs are willing to borrow at high rates because the returns from investment opportunities still look good; good enough to repay those high loan interest rates and still make a profit. These factors also interrelate with consumers' decisions to deposit money in banks, because banks offer interest on consumers' deposits in order to attract savings which they can then lend to investors. This relationship is shown schematically in Fig. 8.3.

The essence of the productivity of capital rationale for discounting is that if present consumption is deferred into the future by savings, and corresponding investment, future consumption will be greater as a result. The consumption made possible in the future per unit of investment (or per unit of saving) in the current time period is known as the *gross productivity of capital*. The *net productivity of capital* is the increase in consumption made possible in the future per unit of investment, but excluding the cost of the productive capital itself. The net productivity of capital is also known as the *marginal efficiency of capital* (in other words, what that capital contributes to society as a productive asset beyond its own value as an asset). It is also called the *internal rate of return* (IRR), meaning the return in value to investors per unit of their investment.

The IRR, or net productivity of capital, is thus the justification for paying interest on savings. Interest paid is itself one reason for a time preference of individuals with respect to consumption now or later; interest on savings is the 'added value' that consumers get from deferring consumption until later. The social rate of time preference (SRTP) for society as a whole is the rate at which future consumption will be traded off against current consumption. It reflects a balance between the desires of consumers for immediate wants satisfaction (current consumption) and the desire to be able to enjoy more satisfaction but at a later date.

If the STRP is high, desires for immediate consumption are strong, and vice versa. The rate itself reveals the additional future consumption required per unit of current consumption to remain indifferent between consuming in the current and in the future time periods. Thus a STRP of 5% per annum means that consumers are indifferent between consuming £100 of goods today and consuming £105 of goods in a year's time. If bank interest rates are lower than the SRTP, then people in

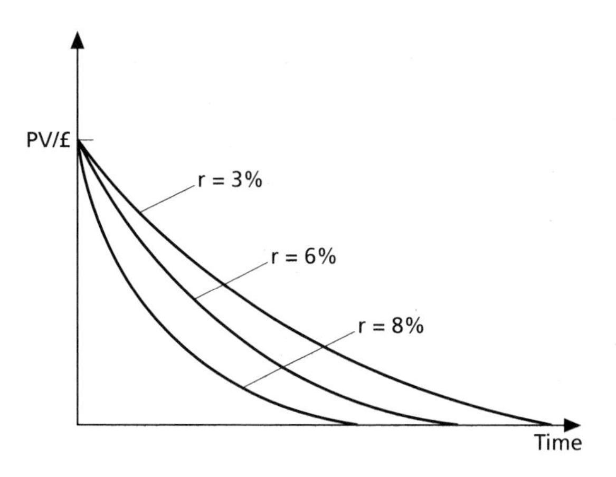

Fig. 8.4 Present value equivalents of £1 for three discount rates over time.

general will spend today and not save. If bank rates are above SRTP, people will forgo current consumption by saving and gain greater utility from higher levels of spending in the future.

Effects of varying the discount rate

Having considered the rationale for discounting, we now consider briefly its effects. The effect of discounting values is that costs and benefits that occur in the future are worth less in financial terms than identical costs and benefits arising in the present. This tends to support decisions in favour of those projects that have greater short-term benefits, and those projects in which costs arise in the medium to long term. Figure 8.4 shows the effect that discounting has on the present value of a £1 cost or benefit occurring in the future. Three discount rates (DRs) have been chosen for illustration: 8%, 6% and 3%. In the UK, an 8% DR is applied to publicly financed roads, a 6% DR to management decisions, and a 3% DR to land acquisitions for new tree planting carried out by the Forestry Commission (HMSO 1991).

The figure indicates that the net present value (NPV) of any cost or benefit under any positive discount rate asymptotically approaches zero in the distant future. Any project that provides a continuous but fixed-value stream of benefits or costs (i.e. £100 per annum, in perpetuity) can therefore be represented from today's perspective by a single NPV: the NPV is the area under the curve. It is this aspect of assigning a fixed value to assets that may provide a constant benefit stream

Table 8.2 Calculating the cost of labour in present value terms. For each year in the study, the real (inflation-adjusted) labour cost is multiplied by the discount factor to give each year's labour cost in present value terms.

Year	Labour costs (£)	Discount factor	Present value
0	1 457 500	1	1 457 500
1	1 735 848	0.90909	1 578 042
2	1 534 575	0.82644	1 268 234
3	671 695	0.75131	504 651
Total	5 399 618		4 808 427

indefinitely into the future that makes many people intuitively suspicious of discounting. Some of these concerns are taken up later in this chapter.

8.3.5 Project assessment

Having followed the steps of analysis above, analysing a wide range of possible impacts, several options are then available to assist judgements about the overall value of the project under investigation. Most important amongst these is the calculation of project NPV. Some form of social weighting can also be considered at this stage.

NET PRESENT VALUE

This methodological stage of the CBA requires that all monetary values be discounted so that they are all expressed in present value (PV) terms. For illustration, consider the labour input to the construction project, shown in Table 8.2. The figures for labour costs in column 2 are estimated real shadow prices. Column 3 has the respective discount factors, given an assumed annual discount rate of 10%. Column 4 then shows the PVs for labour costs in each year. Summing these PVs gives the aggregate cost of the labour input to the project.

This process is then repeated for all costs and all benefits arising from the project. The sum of the discounted benefits minus the sum of the discounted costs is equal to the NPV of the project. The NPV is the keystone of the CBA appraisal methodology. If the NPV is positive, then the project represents an efficient shift in resource allocation according to the estimations and calculations of the CBA. It therefore represents a project that, on balance, increases social welfare.

INTERNAL RATE OF RETURN

An alternative appraisal criterion to the NPV is the internal rate of return (IRR), often used in commercial assessments. The objective of evaluating the IRR is to discover what discount rate, when applied to the project, would yield a zero NPV figure. The reason for this calculation is that it allows the project's net value to be compared to the opportunity cost of capital. If a project is found to have an IRR of 5%, and the prevailing bank interest rate is 7%, then the project is not worthwhile. In this case the IRR indicates that the project expenditure is generating added value of 5% over the lifetime of the project, whereas simply putting the money in the bank would yield 7%. Investors will therefore reject investments in projects that have an IRR at or below the prevailing interest rate. The IRR has become increasingly criticized as a less reliable appraisal criterion than NPV (Pearce & Turner 1990; Hanley & Spash 1993) and is becoming less popular in practice.

DISTRIBUTIONAL ASSESSMENTS

Another option in project assessment is to apply distributional weights to the cost and benefits that have been calculated. Such a procedure can be applied in order to address inequalities in the distribution of income. It can be termed *social CBA*. Generally, a higher weight is put on the costs or benefits affecting lower income groups than those affecting higher income groups. A CBA without distributional weightings implicitly assumes that society values a £1 change equally regardless of the socioeconomic condition of the individual enjoying or suffering that change.

With regards the methodology of weighting, consider Table 8.3. There are two alternative projects, *A* and *B*, with associated figures for effects upon the 'rich' and 'poor' arising from each. All affected agents in society are categorized as either 'rich' or 'poor' in accordance with respective incomes: half of the population is 'rich' and the other half 'poor'. All of the figures are in £ millions.

The standard NPV test, with the distributional weightings d being 1 for each category, implies the selection of project B $[(160 - 80) > (150 - 100)]$. However, if a social CBA, with a weighting $d = 0.5$ for the 'rich' and $d = 2$ for the 'poor', then project A is chosen as the social NPV of A is 250, whereas the social NPV for B is -80 $[(150 \times 2) - (100 \times 0.5) > (160 \times 0.5) - (80 \times 2)]$.

There are some problems associated with the application of social NPV. By applying weights, the role of CBA as a method of allocating resources to their most efficient uses is undermined. If income inequalities are to be reduced then it is preferable from an economic point of view to do this through transfer payments, using a progressive income tax[1] to correct for income disparities, and using CBA simply to pick projects that maximize society's wealth overall. However, if well-functioning tax systems are not operational, as can be the case in developing nations, the case for social CBA may be stronger. A more direct problem is that of tracing net income changes accruing from a project to different groups in society. Even if this can be estimated, the CBA practitioner must then determine the appropriate weights. This is an inherently political issue.

Table 8.3 Conventional and social net present value calculations. A conventional cost–benefit analysis would pick project *B* over project *A*, whereas the application of weightings in the social net present value calculations means, in this case, that *A* is preferred to *B*.

	Project *A*		Project *B*	
	Poor	Rich	Poor	Rich
Costs	0	100	80	0
Benefits	150	0	0	160
$d =$	1	1	1	1
NPV	50		80	
$d =$	2	0.5	2	0.5
Social net present value	250		−80	

8.3.6 Sensitivity analysis

Uncertainties and estimations pervade CBA. By reconsidering the CBA methodology in chronological order, the following major classes of uncertainty become apparent:

- the quantity and quality of physical inputs and outputs to and from the project;
- the respective shadow prices for these inputs and outputs;
- the value of changes in environmental quality;
- the appropriate discount rate; and
- the correct distributional weightings (if applied).

During sensitivity analysis, the influence upon NPV of changing each of these estimations is calculated. From this calculation, it should be possible to define

1 A progressive income tax charges increasing rates of tax as incomes increase.

which estimations NPV is relatively sensitive to. Thus, in the oil rig construction case, it may be that a doubling in the real price of skilled labour would not significantly alter NPV, but that a small error in estimation of environmental costs would.

Such analysis allows the targeting of research expenditure into the better estimation of certain critical costs and benefits. It can also provide a spread of estimated NPVs for a project. Thus, although the estimated NPVs of two projects might be identical at, say, £10 million, one might be significantly more sensitive to errors in estimation. This provides important additional information for decision-makers.

8.4 Problems and arguments surrounding discounting

Before introducing some problems relating to discounting, it is helpful to summarize the arguments in its favour. A discount rate arises in a society in order to reflect the following:

1 Individuals do simply care more for immediate rather than deferred consumption.

2 Future consumption should count for less than current consumption if there is any positive rate of growth in the economy. This assumes that future consumption is greater than current consumption, and that there is *declining marginal utility* of consumption in the future. This means that an extra unit of consumption in the future is considered less valuable than an extra unit of consumption now because more units in total are being consumed in the future.

3 Future consumption, and its effects on welfare, is relatively uncertain, and this uncertainty leads to a reduced value of possible benefits and costs, related to the level of uncertainty or risk that pertains to them.

Having analysed the theory underlying discount rates, we now look at some of the more problematic elements of discounting.

8.4.1 Divergence of social and private rates of discount

We should note initially that it is important to distinguish between the social rates of time preference determined in the market-place through individual consumption and saving decisions, and a societal rate of time preference applied by government to public projects. That is, governments may be expected to have a different view from individuals on the optimal allocation of savings and consumption through time. The reasons for this are principally because of the following:

1 There may be positive externalities deriving from investment for the future but these will not be realized if individual consumers act alone because, as individuals, they have no effective incentives to generate these positive externalities. Governments, however, can realize these effects for society as a whole.

2 Society is effectively immortal, barring the extremely long term or extremely bad decision-making, and so future consumption will always have positive benefits. However, individuals can be expected to have a finite time horizon and this will cause them to operate with a more short-term outlook. There are, however, instances where individuals do demonstrate 'negative time preference', for example, in bequests to their descendants.

3 People may themselves make different judgements about the optimal rate of savings and investment if they adopt some sort of 'socially responsible' outlook than if they act as independent individuals without any particular future responsibilities. The 'market' for savings and consumption is, however, determined by individual consumer choices, and there is no direct mechanism to take account of any alternative 'citizenship' perspective that individuals may have on how goods should be allocated between generations.

8.4.2 Criticisms of discounting

We noted above that the societal discount rate may diverge from the free-market or socially determined rate. This indicates reasons for using a different discount rate for government projects. Below, we briefly note five sets of objections to the rationale for discounting, the first related to use of the market rate itself, the rest to the wider issue of discounting itself.

PRESENT GENERATION BIAS

Prevailing interest and discount rates are determined by the behaviour of the current generation, and the current generation may act myopically. For example, 'irresponsible' individuals, and indeed governments, might be happy to build a cheap nuclear plant with a high risk of collapse in 100 years' time because they are simply not going to have to worry about that collapse. It seems fair to argue that a responsible government should take responsibility for future generations alongside the current one. One school of thought argues, however, that the current generation is the only one which can possibly express its preferences, and so there is no other way of making consumption decisions other than in accordance with its preferences. In a democratic society, this does present real problems, both of political responsibility and political feasibility.

UNCERTAINTY AND RISK

The issue of risky future costs and benefits is separate from that of deliberately ignored future costs; the latter being caused by what we might call 'irresponsible' pure time preference on the part of the current generation. If outcomes are genuinely uncertain, there is clearly a rationale for their being treated differently to costs or benefits which are certain. The issue remains whether the discount rate is the correct vehicle through which to deal with this kind of uncertainty; for example, the possibility of a catastrophic loss caused by the breakdown of a hydrological cycle under pollution loads. It appears that it may be better to deal with this kind of potential problem through other non-financial mechanisms (see below). In effect, in the face of uncertainty there is a need to define the appropriate range of effects to which discounting should be applied in assessing the overall benefits of a project.

FALLACY OF INCREASING CONSUMPTION

Two objections can be raised regarding the assumption that future benefits should be discounted because future consumption will be greater. The first is the so-called Easterlin paradox (Easterlin 1997), which relates the utility of consumption not to absolute levels but to relative, and intragenerational, levels. Easterlin suggests that individual happiness relates more to relative consumption with regard to other individuals than to absolute consumption levels. An overall increase in consumption which maintains or, in particular, increases the levels of *inequality* in society—as may well be the case with CBA under a 'potential compensation' criterion—may then act actually to lower welfare.

Secondly, and following the arguments surrounding uncertainty above, if current consumption decisions act so as to damage important ecological services then future welfare may actually decline as a result. The assumption that the future will be richer than the present relies on the basic productivity of capital (natural and man-made) being able to compensate for any declines in ecological services. It is not immediately clear that all such damage can be compensated in this way, or that valuation techniques are particularly good at attempting to include such valuations in CBA.

UNCOMPENSATED DAMAGES

One feature of discounting addressed by Pearce and Turner (1990) is that of intergenerational compensation. They suggest that if future generations bear costs and the present receives benefits as a result of some development decision, then the present generation should compensate the future for bearing those costs. This would mean specifically putting aside enough in investments today to pay for the damage when it accrues in the future (as a kind of super-fund to pay for decommissioning nuclear power plants, for example). It is therefore not unfair to 'discount' future costs in financial terms, because they will be paid for out of the increased wealth generated by the compensatory investments. Problems remain concerning estimating damages, of course. Pearce and Turner's argument is that it is unacceptable to apply an intergenerational Kaldor–Hicks potential compensation test in deciding to proceed with a project without actually setting up a mechanism for making the compensating transfers.

INHERENT RIGHTS AND OPPORTUNITIES

A final set of arguments, in many ways the strongest and circumscribing all those above, are those that see intergenerational responsibilities in terms of 'respecting the rights' of future generations rather than 'managing the welfare' of those generations. The important distinction is essentially the same as that made in relation to issues of distributional justice in Section 5.3.2. If future generations are considered as having a moral right to a healthy and diverse environment, then it is not admissible to leave them an unhealthy or impoverished one, regardless of what else we leave them by way of compensation in the form of other forms of wealth. In a similar vein, it is not morally acceptable to deprive a free person of their liberty, regardless of how many luxuries are provided to try to make that loss of liberty enjoyable. On this view, liberty is a precondition for attaining higher welfare levels, not an alternative to it. Utilitarian reasoning—applied through discounting and CBA—can therefore only be applied once basic rights to a wide set of alternative development options have been recognized and secured.

Arguments about intergenerational equity thus rotate around leaving future generations with an undiminished set of opportunities compared to the current generation. The issue is, of course, particularly pertinent to the natural environment as it can be argued that many aspects of the natural environment are inherently unique and converting them to man-made capital is an irreversible process. In this sense, the critical aspect in intergenerational transfers is the level of substitutability between these goods and others, a theme developed in relation to sustainable development in Section 2.10, and in relation to value in Section 5.4.

Right-based theories of intergenerational justice claim, in effect, that development options are non-substitutable with welfare levels. On this view, leaving each generation with an undiminished set of options for development—including preservation of species and non-reproducible natural habitats—is better than confining them to a narrower set of options (predominantly man-made capital) at a higher welfare level. In terms of planning for an unknown future, this position equates with a risk-averse strategy. It is very difficult to know exactly how a future generation will value different goods and services; leaving them more options—and, in particular, avoiding irreversible decisions as far as possible—limits the risk of their welfare declining as a result of their being unable to take up certain options. This theme is developed in more detail in Chapter 13.

8.4.3 Concluding comment on discounting

At the heart of most of these criticisms is the issue of whether adapting the discount rate in some way can bring about a social improvement when applied in CBA, or whether discounting should operate on market or near market determined rates and other ethical or political criteria should be adopted separately as constraints on development. In conclusion, it should be noted that changing the discount rate—in particular, using a lower rate as many environmentalists advocate—may in itself have very adverse environmental consequences. This is for three reasons:

1 Low rates encourage wide investment in projects because of the cheaply available capital; this accelerates the rates of depletion of resources overall.

2 High rates concomitantly reduce this rate of growth, tending towards less development overall and a greater stock of resources inherited by the next generation.

3 Primary industries (agriculture, timber, fishing) tend to have relatively low rates of return compared to more high-tech industries. A high rate of discount tends to discourage investment in these kinds of primary industry projects on an NPV basis, which is beneficial because they are often the ones most likely to damage environmentally sensitive areas.

Although these are merely ‘accidental’ properties of discount rates as regards their effect on the environment, they suggest caution in trying to achieve environmental objectives by tampering with a mechanism that has only an indirect effect on natural resource use.

Instead of adjusting discount rates themselves, Hanley and Spash (1993), and Pearce (1989) advocate adopting parallel criteria on environmental protection alongside CBA. One such criterion is the safe minimum standard (see Section 13.2.3); a similar kind of idea is expressed in the notion of protecting critical natural capital (Section 2.10.1); and the end result of an environmental impact assessment (EIA) (Chapter 9) often assists in this kind of judgement. Such criteria limit the range of developments for which CBA is considered an appropriate decision-making framework. They act to filter out projects according to important environmental criteria, principally because no effective methods exist whereby these criteria can be included within CBA itself.

We now turn to an example of a CBA applied to a multipurpose forestry project in Tunisia (Cistulli 1996). The framework developed by the project team gives a good outline of the levels of complexity that can be attempted within the CBA methodology.

8.5 An example of cost–benefit analysis: forestry development and soil moisture conservation project in Tunisia

8.5.1 Project background

The project reported here formed part of a 10-year programme of forestry development involving the government of Tunisia, the World Bank and the UN Food and Agriculture Organization (FAO). Among the major objectives of the forestry development programme was the development and application of a CBA model suitable for incorporating environmental costs and benefits of the project. The model developed was applied to two watersheds in Tunisia: Marguellil and Bou Hertma, the first being characterized by prevailing soil/moisture conservation actions and the second consisting mainly of forestry actions. Only the Bou Hertma project is considered here.

Table 8.4 Impacts of project actions. (Source: Cistulli 1996.)

Actions	Wood and forest products	Agriculture and Livestock	Hunting	On-site soil conservation	Off-site soil conservation	Community income	Recreation	Carbon store	Biodiversity
Reforestation of pines	X	X	X	X	X	X	X	X	
Plantations around water basins	X		X	X				X	
Regeneration of cork-oak	X	X						X	
Fodder plantations		X	X	X					
Creation of prairies		X		X					
Rangeland improvement	X	X	X	X					
Correction of 'ravines' and construction of cordons	X			X	X	X	X		
Thinning of pines	X		X	X		X	X		X
Infrastructure	X		X			X	X		
Fruits, nuts, herbs									X

The total area affected by the Bou Hertma watershed project extends to 33 636 ha, with 27 858 ha within the watershed itself. Land use within the watershed comprises 61.1% (20 565 ha) arable land, 33.9% (11 399 ha) forests, of which 8877 ha are natural forests, 2522 ha are plantations and 5% (1672 ha) are non-agricultural lands. The dam of Bou Hertma has a capacity of 117.5 million cubic metres and was constructed in 1976 at a cost of 7.8 million Tunisian dinar (1976 prices). Total siltation in the period 1976–90 was 3.99 million m^3, or 285 000 m^3 per year on average. It follows that erosion in the watershed can be estimated roughly as 10.2 m^3 per hectare per year. The volume of irrigation water supplied by the dam is 58–60 million m^3 per year on average, which allows irrigation of an area of about 10 000 ha.

8.5.2 Project actions

The project is multiobjective and includes the following activities which are to be executed within a period of 5 years:

1. Reforestation of pines.
2. Establishment of plantations around water basins.
3. Regeneration of cork-oak.
4. Establishment of fodder plantations.
5. Creation of prairies.
6. Rangeland improvement.
7. Correction of 'ravines' and construction of cordons.
8. Thinning of pines.
9. Creation of infrastructure.
10. Planting of fruits, nuts and herbs.

8.5.3 Project impacts

The extent of the major impacts occurring as a result of the project are given in Table 8.4. Their quantification in physical terms was undertaken from a variety of sources, including national statistics and official documents, the general scientific literature, and project-specific data available from similar projects undertaken in other Mediterranean countries. In the absence of relevant data, proxies or qualitative evaluations were undertaken. The impacts taken into account in this application are not comprehensive, but given the constraints of time and resources, analysis focused on the most important impacts as identified by local experts.

PRIMARY IMPACTS

Positive impacts arising from the project include increased availability of timber and other forest products, such as fuelwood, cork, fruits, nuts and herbs. Agriculture and livestock will be affected by changes in the production of fodder and forage. The project will also have a positive impact on developed hunting because of increased forested area.

SECONDARY IMPACTS: INTERNAL TO THE MARKET

This category comprises all the costs and benefits of a project occurring inside and/or outside the project area whose economic values can be estimated on the basis of

Table 8.5 Summary of valuation techniques used in Bou Hertma cost–benefit analysis. Note that community income impacts from the improved infrastructure were not estimated, an increase in per capita income of the people living in the area was assumed (data were provided by Tunisian experts). Similarly, no estimate for biodiversity was made directly, rather data from other studies in the Mediterranean region were used. (Source: Cistulli 1996.)

	Demand curve approach			Market priced goods					
Actions	Hedonic price	Conting. Valuation	Travel cost method	Market prices	Wage differential	Productivity change	Substitute costs	Shadow project	Dose–response
1 Timber				*					
2 Fuelwood				*					
3 Non-wood products				*					
4 Fodder				*					
5 Agricultural products						*			
6 Hunting						*			
7 On-site soil conservation						*			
8 Off-site soil conservation						*			
9 Community integrity					*				
10 Recreation		*	*						
11 Carbon store									*
12 Biodiversity									

the observed market prices. Among these impacts are on-site and off-site soil protection, which in turn have a positive impact on water supply and crop productivity. Other secondary impacts include the positive effects of road construction (infrastructure) on community incomes (resulting from market access and shifts from self-consumption to marketed agricultural production).

SECONDARY IMPACTS: EXTERNAL TO THE MARKET

These impacts include those for which no market price exists, meaning essentially public goods and services. Among these are informal recreation, carbon storage and biodiversity.

8.5.4 Valuation techniques used

The range of valuation techniques used is shown in Table 8.5. In outline, these involved the following elements.

1 For the value of timber, fuelwood, cork, fodder and the other marketed non-wood products of the forest, direct market prices were available.

2 The development of hunting benefits was assessed by estimating the saved expenses incurred by government in the area for management and repopulating of hunting areas. This was a productivity change in the ability of the local land to support a wild harvest.

3 The economic value of on-site and off-site soil erosion was also assessed through the productivity change on crops and water supply. As an example, the estimation of off-site impacts of soil erosion prevention involved the following steps:

(a) assessment of the annual loss of soil in the without and with project situation using the modified universal soil loss equation (MUSLE) method;

(b) assessment of dam siltation in the with and without project situation;

(c) assumption of a linear relation between water supplied by the dam and irrigated area downstream; that is, a percentage increase of water availability in the dam will allow an increase in the irrigation area by the same percentage;

(d) changes in irrigated areas were valued by estimating the incremental benefits of agricultural production (net benefits with the project, less net benefit without the project); and

(e) the value of the off-site soil conservation service of the project is the incremental value of agricultural production in the irrigation scheme.

4 Community income effects resulting from the opening of roads were estimated assuming an increase in per capita income of the people living in the area. This was assessed through expert consultation, given the low sensitivity of the project to these benefits and the lack of readily available data.

5 Non-developed recreation benefits were estimated using proxy average willingness to pay (WTP) and travel cost estimates for recreation, obtained from other Mediterranean areas, again resulting from resource constraints and the relatively low importance of these impacts.
6 Carbon storage value was estimated with a dose–response approach. An average net growth of forests of 3.5 million cubic metres (CM)/ha, a coefficient for carbon/CM of 0.26 and a value of $US4 per ton of carbon stored (this value is among the lowest price obtained in different applications). It was assumed that carbon storage service of forests is temporary, and lasts 20 years on average.
7 Biodiversity impacts were only considered in qualitative terms because the present state of the area and foreseeable developments do not appear to involve irreversible impacts on the ecosystem. In fact, the overall reforestation activities might be expected to have a beneficial effect on biodiversity.

8.5.5 A five-step cost–benefit analysis

In order to take account of the various types of environmental impacts analysed in the previous sections, as well as of the different levels of complexity concerning their economic evaluation, a five-step framework for project analysis was devised. This provides the decision-maker with increasingly detailed information on the overall impact of the project at each step of the analysis. It is evident that an extended and deeper analysis will involve increasing margins of uncertainty. As uncertainty increases, so therefore does the need to carry out sensitivity analysis on the project's impacts.

The five analytical steps were as follows.
1 *Financial analysis (FA)*. In this module, market prices alone are directly taken into account. Costs are defined according to different actions undertaken in a geographical unit. Their items, units (cm, litres, tons, etc.), prices, quantities and values for the total duration of the project must be identified. Similar specifications apply to benefits. Cost and revenue items may be more or less detailed according to the availability of information and data.
2 *Conventional economic analysis (CEA)*. Prices (costs and benefits) are modified to take account of the real value of resources, thus eliminating transfers, taxes and the various effects of market distortions. For each item a specific conversion factor must be used, whether developed *ad hoc* by the analyst or provided by the public administration or those responsible for financing the project.
3 *Extended economic analysis 1 (EEA1)*. Besides conventional costs and benefits, other effects are taken into account which are external to the actions (off-site) but internal to the markets. Given the interrelationship between actions, both costs and benefits must be considered as a whole by geographical units rather than by individual actions—as was the case with FA and CEA.
4 *Extended economic analysis 2 (EEA2)*. Besides the previous costs and benefits, those external to the market (intangibles) are also taken into account. It is mainly at this stage that the total economic value (TEV) of natural resources enters into the CBA. Non-market values can be estimated using the methods reported in the previous sections. EEA2 is also carried out by geographical units as a whole, because it is almost impossible to evaluate and isolate costs and benefits of individual actions.
5 *Socioeconomic analysis (SEA)*. The previous costs and benefits can be weighted according to the welfare accruing to the various social groups. Three types of basic information are required:
(a) the socioeconomic groups involved with the project;
(b) the weight to assign to each socioeconomic group; and
(c) the distribution of the various types of costs and benefits among the groups.

The SEA can be carried out at each step of analysis: FA, CEA, EEA1 and EEA2. In the present study it is, however, limited to EEA2 level.

MAIN RESULTS

The results of the five-step CBA procedure are given in Table 8.6. It can be seen that in moving from financial (FA) to economic (ECA, EEA1 and EEA2) and social CBA (SEA), the NPV of the project steadily increases. Financial (FA) and conventional economic analysis (CEA) using 10% discount rate show that the project is profitable. EEA1, taking into account off-site market effects, improves the IRR to 17.2%, while EEA2, including intangibles and non-market effects, gives an IRR over 19.1%. Social analysis, taking account of net welfare increase per income groups, produces a 23.1% IRR. It can therefore be argued that the Bou Hertma forest development project represents a potentially sound investment in welfare terms for government and agency funding.

Table 8.6 Main results of the Bou Hertma cost–benefit analysis. (Source: Cistulli 1996.)

Geographical Units	NPV (DT)	IRR (%)
Financial analysis (FA)		
Bou Hertma	1 186 039	12.68
Marguellil	−1 435 617	9.19
Project	−249 578	9.89
Conventional economic analysis (CEA)		
Bou Hertma	4 623 708	19.9
Marguellil	5 784 849	13.91
Project	10 408 557	15.40
Extended economic analysis I (EEA1)		
Bou Hertma	6 126 297	23.08
Marguellil	7 746 723	15.22
Project	13 873 020	17.21
Financial analysis II (EEA2)		
Bou Hertma	6 231 660	23.33
Marguellil	10 534 898	17.57
Project	16 776 558	19.13
Socioeconomic analysis (SEA)		
Bou Hertma	9 071 056	27.84
Marguellil	22 474 503	21.55
Project	31 515 559	23.06

Abbreviations: IRR, internal rate of return; NPV, net present value.

Some impacts of the project, characterized by higher uncertainty, were tested through sensitivity analysis. Table 8.7 presents the results of the simulation of the most relevant variables used in the two case studies. The variables were modified according to a variation of ±20% and ±50% of their quantities. The results of these simulations are shown in terms of NPV and IRR variations. Even large variations in individual variables did not significantly affect the general profitability of the project. In this case, this is because of the large number of actions and variables.

8.6 Summary

Cost–benefit analysis is a methodology that aims to identify projects and plans that will maximize overall benefits to society. This is done by evaluating all costs and benefits in monetary terms.

A set procedure for a CBA progresses through several stages. First, the scope of the analysis is established by defining the project and how broadly its impacts are to be considered. Secondly, all the potential impacts that fall within this boundary of inquiry are classified; an impact is considered important if it has the potential to affect human welfare in some way.

Following impact identification, impacts are valued in monetary terms. The timing of impacts is considered

Table 8.7 Sensitivity analysis of Bou Hertma cost–benefit analysis. (Source: Cistulli 1996.)

	Variables		ΔNPV	ΔIRR	%
NPV (FA) = 1 186 039	Revenues action 4 (forage plantations)	±20%	±248 665	+0.55	−0.56
IRR (FA) = 12.68%		±50%	±621 662	+1.38	−1.40
	Revenues action 2 (protective plantations)	±20%	±36 064	+0.06	−0.06
		±50%	±90 161	+0.15	−0.15
	Revenues action 1 (pine plantations)	±20%	±485 656	+1.01	−1.06
		±50%	±1 214 139	+2.45	−2.75
NPV (EEA1) = 6 126 297	Siltation ratio	±10%	±1 193 748	−3.07	+3.10
IRR (EEA1) = 23.08%		±20%	±1 327 626	−3.40	+3.45
	Income increase caused by new roads	±20%	±243 998	+0.55	−0.55
		±50%	±609 995	+1.39	−1.37
NPV (EEA2) = 6 231 660	Shadow price of 1 carbon ton	±20%	±7 790	+0.02	−0.02
IRR (EEA2) = 23.33%		±50%	±19 475	+0.05	−0.04
	Willingness to pay for 1 visit to new forests	±20%	±13 283	+0.03	−0.03
		±50%	±33 207	+0.09	−0.08
NPV (SEA) = 9 071 056	Weight of social group 1	±20%	±1 587 246	+2.86	−2.88
IRR (SEA) = 27.84%		±50%	±3 968 116	+7.20	−7.27
	Weight of social group 2	±20%	±297 444	−0.77	+0.90
		±50%	±743 604	−1.72	+2.62

Abbreviations: EEA1, extended economic analysis I; EEA2, extended economic analysis II; FA, financial analysis; IRR, internal rate of return; NPV, net present value; and SEA, socioeconomic analysis.

to influence their value, and this leads to the process of discounting. Discounting occurs because of rates of growth in the wider economy, which reduce over time the value of outputs from any one project. Discounted values for all costs and benefits are then combined to calculate an NPV for the project. A positive NPV means a potential improvement in human welfare. Sensitivity analysis should be used to check the validity of final estimates.

Concerns over CBA as a decision-making methodology relate to several factors. For example, valuation methods can be inaccurate or of limited applicability in many cases. The practice of discounting values is also questionable, both in terms of the rate adopted and the practice itself. Overall, there are good arguments for defining carefully the range of projects for which CBA may be an applicable decision-making tool, using other tools (e.g. EIA) to establish this range.

Further reading

Hanley, N. & Spash, C. (1993) *Cost–Benefit Analysis and the Environment*. Edward Elgar, Aldershot.

Laylard, R. & Glaister, S. (eds) (1994) *Cost–Benefit Analysis*. Cambridge University Press, Cambridge.

Pearce, D.W. (1983) *Cost–Benefit Analysis*. Macmillan, London.

9 Environmental Impact Assessment

'Usually, thresholds for sustainability—carrying capacities—are substantially exceeded even before we become aware of the nature of the problem.' [Michael Carley]

9.1 Introduction

Development goes on around us through a series of more or less unconnected and discrete projects. New infrastructure, commercial premises, changes in land use, industrial plants and domestic housing are all elements of this development process. In most countries, some system of national and regional planning authority is required to consider the merits of these and other proposals, and to decide whether or not permission should be granted to any particular project.

A range of economic, environmental and social impacts is likely to be associated with any development project. This diversity of impacts confronts the decision-maker with two problems. The first is to identify the impacts and the second is to assess a best option once they have been identified. Some of the impacts arising from the project may be expressed in financial terms (profit), others may be aesthetic (an enhanced landscape), or social (enhanced health care in the case of a plan to build a new hospital). The purpose of an environmental impact assessment (EIA) is to help with these planning problems by identifying, and in some cases evaluating, the full range of environmental and social impacts which will arise from a project.

In Chapter 1, EIA was stated as being one of the techniques which should be central to an ecological economics approach to problem solving. In fact, the European Community's Fifth Action Programme identifies EIA specifically as an important tool in the search for sustainable development: 'Given the goal of achieving sustainable development it seems only logical, if not essential, to apply an assessment of the environmental implications of all relevant policies, plans and programmes' (Commission of the European Communities 1992). While there is little doubt that this should be true, it is important to recognize that EIA is also subject to a wide array of problems in application. These problems are noted throughout the following sections.

9.2 The methodology of environmental impact assessment

Environmental impact assessment is the process of predicting the impact of a planned activity, usually a project or policy, on the environment before that project/policy is initiated. It is not an end in itself, rather the purpose of any EIA is to aid the development decision. A good EIA ensures that decision-makers have available as good information as possible when considering projects.

Structurally, EIA can be thought of as being composed of three layers. The first layer is concerned with the formal legal and administrative procedures associated with EIAs. For example, within the European Union, Directive 85/337 lays out the basic requirements for the contents of an EIA, and the projects to which it should be applied. The second layer within the EIA is concerned with the process and structure of the EIA study itself, while the third layer within the EIA is concerned with the actual activities that take place within the EIA structure. The activities which occur within each of these layers are discussed below.

9.3 Layer 1: legal and administrative processes

Environmental impact assessment was first legislated for in the USA by the National Environmental Policy Act 1969 (NEPA), and since then has become law in many countries. This section does not aim to provide a detailed analysis of the legal and administrative structures associated with EIA, but several issues are of interest.

First, although many countries have passed legislation concerned with EIA, the quality of its uptake has been variable. Much discussion has revolved around certain words in the legislation and, as Masera (1991) states, 'Legal norms do not define methodology, and here they leave a broad field for disputes.' For example, in the USA much litigation ensued in order to clarify the wording of the NEPA. The original act required 'adequate' environmental statements be produced for 'major federal actions' likely to induce a 'significant impact'. Much debate surrounded the definition of these words and legal challenges were made about environmental impact statements (EISs) on the basis of them being inadequate and hence unsuitable to aid a development decision. Eventually the courts decided that an adequate EIS is one that contains 'sufficient detail to ensure that the agency has acted in good faith, made a full disclosure, and ensured the integrity of the process'. It remains a point of debate as to whether it is socially optimal for guidelines on environmental management to be drawn up by the judiciary.

Such issues over definitions are also important in the European situation where the Directive on EIA has two Annexes concerned with the types of project for which an EIA is necessary. The first of these, Annex 1, lists nine types of projects for which an EIA is mandatory in all member states (e.g. crude oil refineries, installations for the permanent storage of radioactive waste, motorways), while the second, Annex 2, lists 80 project types for which an EIA may be undertaken at the discretion of the member state (e.g. pig and poultry rearing installations, extraction of coal, dams, waste-water treatments, etc.). In the UK, EIA is required for projects in Annex 2 which 'are likely to give rise to significant environmental effects' (Department of the Environment (DoE) 1989). In order to aid decisions concerning projects in Annex 2, the DoE suggest that there are three main criteria for significance.

1 Whether the project is of more than local importance, principally in terms of physical scale.
2 Whether the project is intended for a particular sensitive location, for example national park or a Site of Special Scientific Interest (SSSI), and for that reason may have significant effects on the area's environment even though the project is not on a major scale.
3 Whether the project is thought likely to give rise to particularly complex or adverse effects, for example, in terms of discharge of pollutants.

While it appears to be sensible in this context to define the criteria for significance, it is evident there still remains a wide area for dispute. Using terms such as 'local importance', 'sensitive location', 'complex or adverse effects' in an operational situation leave a large area for subjective interpretation. The continued dependence of EIA on subjective analyses is one reason why many believe a more objective quantitative approach would be beneficial.

Another interesting point concerning the production of EIAs is that within the European Union it is generally left to the proponent (the person who wishes to undertake the development—the developer) to conduct the study. The proponent must bear all costs related to the EIA and must make available all relevant data on the project. In many cases proponents hire an environmental consultancy to undertake this work on their behalf, while occasionally bigger companies may undertake the EIA themselves. There are few standards related to the quality of environmental consultancies; anybody can register themselves as a consultant and, if a proponent wished to hire them, they could undertake an EIA.

This may present a problem if proponents act 'strategically', which means that within the given legal constraints they chose a strategy that serves their own interests rather than the wider good of society. As such they could vary the quantity, type and quality of information made available about the project, could select the most appropriate consultancy to undertake the work and manipulate the amount of money they wish to pay for the work to be done. Environmental consultancies also have their own agendas within this system. They need to achieve a steady throughput of work to stay solvent. Having obtained one contract from a proponent, they may wish to maximize their chances of getting a further contract. This places them in a difficult situation, as from a business point of view 'the customer is always right', while from a professional standpoint they should do the best for society.

A third set of players within the EIA system is the decision-makers. These individuals are presented with a range of documents concerning the project, one of which is the environmental statement, and from this information they must decide whether or not the project should go ahead. As EIA is a relatively new tool in planning it is not surprising that many planners are unfamiliar with both the basic structure of EIAs and their scientific content. Several surveys revealed that in Europe to date most planning authorities have only received a small number of EIAs to review (Lee & Dancey 1993). Given that the total number of EIAs received by a planning authority is small, and that the range of topics for which EIAs may be submitted—from forestry to power generation—is large, it is unlikely that many planners will review more than one EIA on a similar topic over a period of several years. In this situation planners have the almost impossible task of having to assimilate a large amount of technical data on a topic which may not be familiar to them. Within the UK, planners are aided in this task by the so-called statutory consultees, who are government agencies responsible for nature conservation, pollution control, heritage, etc., and who comment on the EIA and have a technical input to the decision.

A final point about legal and administrative matters concerns the role of the public in the EIA process. Both US and EC legislation encourages the participation of the public. While current best practice suggests that the public can have a part to play in all stages of an EIA, they have a particularly important role in identifying likely impacts (scoping), and in evaluating the importance of these impacts. Rarely, however, are the public fully involved in the EIA process in practice. Nearly all planning procedures allow the public to comment on new developments in some way, but the onus is placed firmly on the public to seek out the information and to make a deliberate effort to become part of the planning process. Logistically, it may be difficult to design planning frameworks which permit a better integration of the public's views into the EIA decision, but public involvement in decision-making is seen as a vital element in sustainable development, and EIA provides a good opportunity to take this ideal further in an important and practical situation.

This brief summary of the legal and procedural elements of EIA has highlighted several important issues. The defining laws have been, and will probably continue to be, subject to heated discussions about meaning, both semantic and philosophical. For the process itself to be of value it requires three parties—proponents, consultants and decision-makers—to act in good faith. This is despite the fact that proponents and consultants may have a financial incentive not to do so. In addition, decision-makers may lack detailed and appropriate knowledge of the EIA process and the likely environmental impacts of projects. Finally, EIA aims to increase social welfare but, to a large extent, society is kept distant from the process and the decision-making.

9.4 Layer 2: the environmental impact assessment process

The second layer within the EIA is concerned with the process and structure of the EIA study itself. It is generally recognized that there are certain steps which must be followed. These are shown in Fig. 9.1 and their content is summarized below.

9.4.1 Description of the project

This is usually provided by the developer and should include a description of the project comprising information on the site, design, size of the project and information on the construction, operational and decommissioning phases of the project.

9.4.2 Preliminary description of the environment

This comprises general information on the site of the proposed project and should be aimed at informing the screening decision. This is often largely derived from secondary sources, such as existing reports.

9.4.3 Screening

At the screening stage, the description of the project and the preliminary description of the environment are considered together and a decision is made as to whether a full EIA should be commissioned or, alternatively, whether the project should proceed through the planning stages without an EIA. Several sets of guidelines have been produced to aid with the screening decision, such as those devised by the UK DoE presented earlier. Generally, these guidelines recommend that if either the project is large or complex, or it is to be implemented in a sensitive environment, then an EIA should be commissioned. The European Union legislation suggests the decision may be made solely on the basis of project size and type, while overseas donor agencies often include the site's sensitivity. For example, the African Development Bank suggest that an EIA is needed for all projects affecting mangrove swamps, small islands, tropical rainforests, areas with erosion-prone soils and other sensitive areas.

9.4.4 Scoping

Scoping is the process of identifying a number of priority issues from a broad range of potential problems. This is necessary for two reasons. First, it would be very expensive and time-consuming to consider in depth all the issues arising from a project. Secondly, it may not be necessary to consider all theoretically possible issues as only some will be relevant. Scoping requires that the

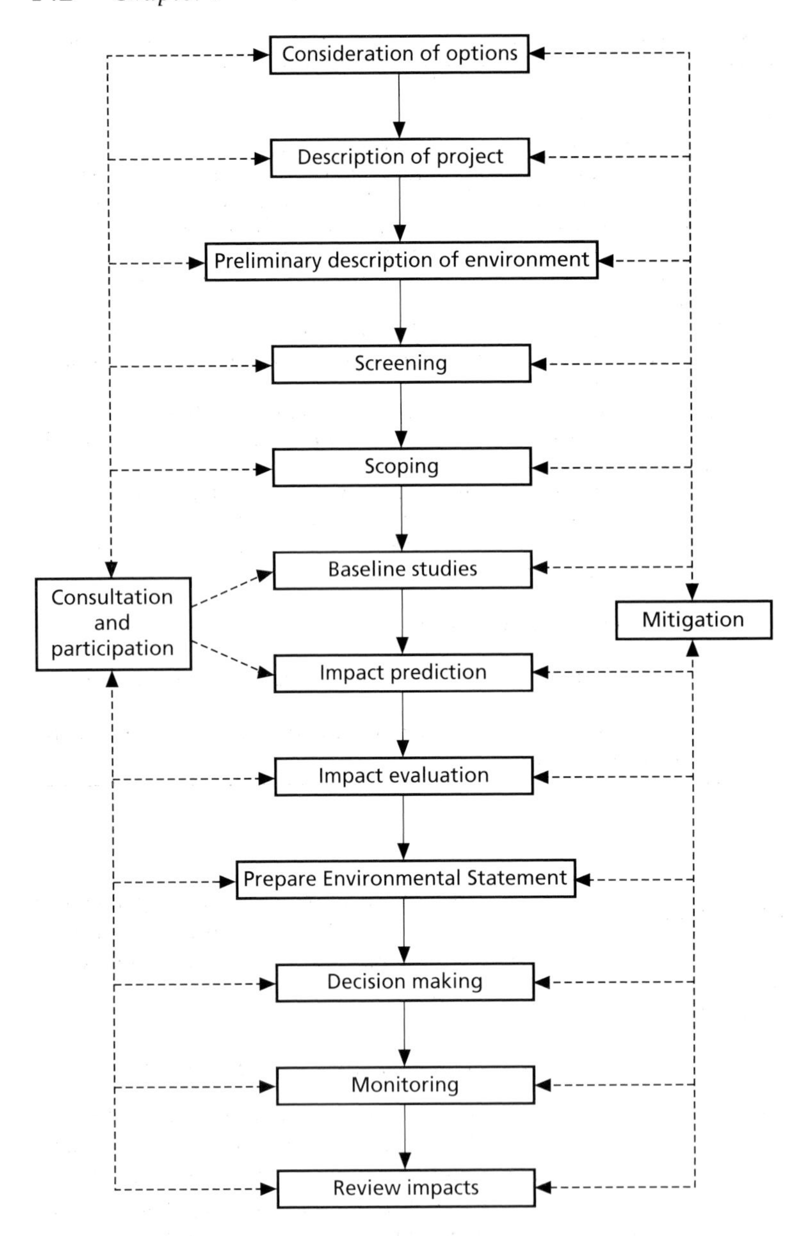

Fig. 9.1 Steps within the EIA process. Solid arrows indicate progression through time. Dotted arrows suggest an iterative process.

issues of primary importance be highlighted. This does not mean that other issues should be ignored, simply that they are not priorities. Public participation should form an important component of the scoping exercise, but this is rarely done well.

9.4.5 Baseline studies

Baseline studies centre on the collection of data on the issues of concern highlighted in the scoping exercise. The collection of these data allow the impacts of the project to be predicted, the management and mitigation of these impacts, and scientifically defensible monitoring to be undertaken subsequent to the project being completed. Data collection during the baseline studies should adhere to the best scientific principles and the quality of the impact prediction depends heavily on the quality of the baseline study.

Table 9.1 Examples of computer models which could be used to predict impacts as part of environmental impact assessments.

Model	Purpose
Crop estimation through resource and environment synthesis (CERES)—maize	To test quickly and easily a variety of different fertilization and irrigation schedules to maximize maize (corn) production from a given piece of land
FIVFIV and SINSIN	To provide policy-making officials with detailed demographic projections
Atmospheric greenhouse model (AGM)	To analyse the consequences for the global climate of various scenarios regarding the production of carbon dioxide from fossil fuel combustion
Range, livestock and wildlife model	To help decision-makers understand and evaluate policy alternatives for rangeland management
Enhanced stream water quality model (QUAL2E)	To provide tools for water quality planning by simulating the behaviour of the hydrologic and water quality components of a branching stream system or lake under the impact of a wide range of pollutants
Waterborne toxic risk assessment model (WTRISK)	To estimate the risks of adverse human health effects from substances emitted into the air, surface water, soil and groundwater from a source such as a coal-fired power plant
Complex terrain dispersion model (CTDM)	To calculate air pollution contractions for various pollutants discharged into the environment by a smokestack near complex terrain
Microcomputer programs for improved planning and design for water supply and waste disposal systems (WSWDS)	To analyse the costs and benefits of alternative designs of water distribution networks and gravity sewer systems
The woodsman's ideal growth projection system (TWIGS)	To project future yields for a small forest
Fishery management model for the Rio Grande basin (RIOFISH)	To explore the effects on angler fishing of alternative management scenarios
Fish population estimate (FPE)	To estimate fish populations from known quantities of fish that are periodically captured and tagged

9.4.6 Identification of the environmental impacts

Subsequent to describing the project and collecting information on the potential sites for its development, it is necessary to identify any environmental impact which may occur as a result of the implementation of that project on that site.

Much research has been directed at developing methods for identifying the potential impacts of development projects but, despite much work in the area, our understanding of natural systems remains poor. We know quite a lot about certain environmental subsystems, e.g. crop growth, soil water movements, pollutant movement through air, and computer models are often available to simulate these processes (Table 9.1). However, while these models may provide extremely precise predictions of changes to attributes, they are often limited in that they may have exacting data requirements and require some specialist knowledge in order to get the best from them. Further, it is unlikely that any single model will adequately be able to simulate all of the potential impacts of any one project in its entirety. Generally, these quantitative techniques are only used in large projects. More often than not, less quantitative methods are utilized for impact prediction.

9.4.7 Evaluation of the environmental impacts

Depending on the institutional context of the project, either the EIA team or some other decision-maker must evaluate the predicted impacts of the project in terms of their importance. This evaluation is a preliminary step in deciding whether or not the damage caused by the impacts is acceptable relative to the benefit provided by the project.

Difficulties in evaluation arise from incomplete data sets, inaccurate or uncertain predictions, the absence of robust techniques for comparing data of different types, e.g. financial, ecological and social, and good working definitions of words such as 'significant' and 'important'. In the absence of strict definitions, criteria for the determination of important impacts have been developed. Commonly used criteria include the following:

1 *Frequency, duration and geographical extent.* An impact which occurs daily, lasts for 12 hours and is felt over a large area is more important than an impact felt only once yearly, for a small time period over a small geographical area.
2 *Reversibility or recoverability.* The faster any environmental attribute (e.g. water chemistry, species populations) returns to its former level after an impact, the better. Any impact which causes an irreversible loss, e.g. species extinction, should be the cause of serious concern. Indeed, in some situations the costs of irreversible loss/damage may be so serious as to outweigh the benefits from a proposed project.
3 *The possibility of mitigation.* If an impact can be mitigated (reduced) then it is less important than one which cannot be mitigated.
4 *Social or political acceptance.* Regardless of the scientific estimation of the magnitude and significance of an impact, sometimes social and political perceptions affect the decision. For example, certain species or areas may have cultural significance, while on the other hand government may feel that a certain development is vital to the national interest regardless of the environmental impacts it may cause.
5 *The pre-established legal limits* (e.g. environmental quality standards). One relatively easy method of evaluating an impact is to consider whether or not any preset standards or laws will be contravened.
6 *Future developments in the same or similar environments.* Some impacts may appear to be relatively harmless in isolation but, if a further development were to occur in the future, the combined impacts may be great.

9.4.8 Management and control of environmental impacts

If, after identifying and evaluating the impacts, the project is permitted to continue then it may be beneficial to manage actively the negative, or unacceptable, impacts caused by the project. It is often better for the environment, and more cost-effective for society, if these impacts are managed as soon as the project is completed and/or during its construction.

9.4.9 Presentation of the findings of the environmental impact assessment: the Environmental Impact Statement

One of the final outcomes of a formal EIA is often the production and presentation of a report, termed the Environmental Impact Statement (EIS), which documents the method and findings of the study. In most circumstances the EIS is presented to some local or national body, who will use it to help decide whether or not the project should proceed. In addition, the study, or part of it, may be available to the public so that they can understand the likely impacts of the project.

9.4.10 Monitoring

Monitoring involves the periodic or continuous assessment of environmental variables. This ensures that the magnitude of the impacts are, and will continue to be, inside the limits of acceptance, and that the conditions that made the project acceptable are not violated. Ideally, monitoring would complement the baseline studies undertaken prior to the project, and would continue during the construction, operation and decommissioning phases as a routine procedure.

9.5 Layer 3: tools and frameworks used in environmental impact assessment

The third layer within the EIA is concerned with the actual activities that take place within the EIA process; for example, the methods of data collection, methods of prediction and evaluation. These compose the deepest layer of knowledge within the EIA process. The actual techniques and analyses undertaken within the EIA are not standardized and respond to the needs of particular project situations. It is at this level that the social impact aspect of assessment can be more easily distinguished from the environmental aspect.

In the following sections some differences in the philosophy and scope of different assessment tools are briefly noted. The philosophy of social impact assessment (SIA) methods are discussed separately from a review of more environmentally focused approaches. This division is for ease of presentation and does not imply that the environmental and social impacts of a project can be analysed separately. Rather, it reflects the different methodological approaches of the sciences from which environmental and social impact assessment practice have evolved.

9.5.1 Philosophies underlying assessment tools

The tools utilized in predicting environmental impacts are highly diverse and differ in philosophy and approach.

Some of the more fundamental ways in which methods differ are listed below.

1 *Identifies first-order impacts and/or higher order impacts.*

A first-order impact is directly caused by the project itself. A higher, or second-order impact is caused by the first-order impact. For example, if pollution from a new factory was released into a river and killed the fish there, this would be a first-order impact of the pollution. If the income of commercial fishermen on the river fell because there were fewer fish to catch, then this would be a second-order impact. Further, if, in an attempt to maintain their incomes, these fishermen began to fish in another river, this would be a third-order impact, and so on. Each level of impact can lead to numerous other impacts. The initial pollution may also discolour the water or cause a bad smell. Similarly, the death of fish from pollution may not only affect fishermen, but other predators of fish may be affected, such as kingfishers and otters.

2 *Treats the environment as probabilistic or a deterministic system.*

In a deterministic system the same inputs always lead to the same outputs. In a probabilistic system there is a chance that similar inputs will lead to different outputs. Some methodologies treat the environment as a deterministic system and assume that when a certain event happens, for example pollution into a river, the same outcome always occurs, all the fish in the river will die. A probabilistic method may assume that the exact number of fish that die depends on many factors, such as state of the river, rainfall, time of year. The output may be in the form 50% chance that all fish die; 25% chance that over half the fish die; and a 25% chance that less than half the fish die. Probabilistic methods recognize the uncertainty in environmental systems, but they may be difficult to develop and use, and their output is inherently more complex than that from deterministic methods.

3 *Provides a static view of impacts or considers how impacts change with time.*

After a pollution event in a river there may be widespread death of fish and other aquatic life, but over time some pollutants become less toxic or get washed out of the system and the plants and animals in the river can start to re-establish. After a pulse of a pollutant with a high biological oxygen demand (BOD), such as sewage, passes through a river, making oxygen unavailable to aquatic life, the river may begin to reoxygenate within days. There is a clear time element in this impact and in many others. Some methods take a snapshot of time and consider the impacts for one time only; others consider how the impact may vary with time.

4 *Utilizes experts exclusively or encourages public participation.*

Some methods depend to a large extent on input from one or more so-called 'experts'. These individuals use their knowledge and experience to determine the magnitude and importance of a project's impacts without recourse to the public. Although in theory EIA should consider the views of the public, in fact very few methods for impact prediction or evaluation allow for formal public input in this way.

5 *Considers single objectives or multiobjectives.*

If a prediction method is only concerned with the impact of a pollutant on the ecology of a river it could be said to consider only a single objective: that of ecology. If, on the other hand, a method is concerned with the impact of pollution on a river's ecology, its use by people for recreation, and the effect of polluted waters on downstream industries, then it is clearly concerned with multiobjectives: ecology, recreation and economic output.

6 *Separates facts and values.*

Some methods are only concerned with predicting the physical level or extent of an impact; for example, the concentration of a pollutant in a river, the amount of fish kill. These are facts. Other methods may be concerned with the degree of importance society places on different levels of impact. For example, society may not believe that pollution that leads to an increase in turbidity is important but they may place great importance on the ecology of a river. In this case, a pollution event that caused significant fish kill would be deemed highly unacceptable. These are value-based judgements, reflecting values that society places on different environmental attriutes. Some methods treat physical levels of an impact (facts) and society's evaluation of these levels (values) separately, others combine them in some way. Ideally, they should be treated separately.

9.5.2 Environmental impacts: approaches and tools

Prior to beginning an EIA, analysts should consider each of the available methods or frameworks for impact prediction and evaluation against the above, and further, criteria before selecting the tool(s) best suited to the particular project. It is important to remember that there is no single 'best' tool or methodology, rather the

Table 9.2 A simplified example of a completed matrix for an irrigation scheme.

	Activities							
	Construction				Operation			
Attributes of the environment	Digging of main irrigation channels	Road construction	Village construction	Immigration of settlers	Drainage of arable area	Use of pesticides	Use of inorganic fertilizers	Sewage disposal
Soil stability	4/8	4/8	6/7					
Soil nutrient status					8/6		9/7	3/6
Flora	1/4	1/4	2/3	2/5		6/8	5/5	1/4
Fauna	1/4	1/4	1/4	3/6		6/8	1/8	
Water quality in lakes					6/8	4/8	8/8	9/7
Employment	5/8	4/8	5/8					
Human health				3/9		2/9	1/9	5/9
Bedouin culture		1/6	3/6	7/6				

Figures in left-hand section of the cells indicate magnitude of the impact on a scale of 1 (low) to 10 (high). Figures in right-hand section of the cells indicate importance of the impact on a scale of 1 (low) to 10 (high).

methods should be chosen after considering the project and the resources available to the analysts. Three contrasting methods for impact prediction and evaluation are described below.

THE LEOPOLD MATRIX

Matrices are based on the idea that, at the simplest level, the impacts of a project are a function of the interaction between specific activities associated with a project (construction, transport, emission of pollutants) and various attributes of the environment (air, soil, water). Many variations of this basic theme have been developed, but one of the earliest and most important was developed by Leopold *et al.* (1971). Although the original Leopold matrix is still occasionally used in EIA, its real importance arises from its position as a baseline from which many other methodologies have been developed.

The horizontal axis in the Leopold matrix lists 100 project activities which may cause environmental impacts. The vertical axis is composed of 88 environmental quality variables grouped under four categories: physical and chemical, biological, cultural and ecological. A subsection of a modified Leopold matrix for an irrigation project is presented in Table 9.2.

The Leopold matrix is usually completed by a group of experts, who follow a set methodology. First, they examine the row of environmental attributes and enter a slash in cells where impacts are possible for each activity. They then evaluate the impacts according to magnitude and importance on a scale of 1–10 (where 10 indicates the largest effect). The score for magnitude is placed in the left-hand section of each cell, and the score for importance in the right-hand section. Magnitude is a compound index and is defined as the degree, extensiveness or scale of the impact. It is a fact. Importance is a value; it relates to how important an analyst feels the interaction is between the project activity and the environmental attribute under consideration. The scores for importance and magnitude should be kept separate and not combined in any way.

One point for consideration prior to its use, is that of numerical ranks. Some analysts criticize their use in the matrix and simpler methods, such as marking with a cross or a dot, have been developed. These sorts of matrices are also extremely useful for communicating the likely environmental impacts to a non-technical audience, i.e. the public and/or government.

THE ENVIRONMENTAL EVALUATION SYSTEM

The environmental evaluation system (EES) was developed with the aim of increasing the consistency and comprehensiveness of EIS in the US Bureau of Reclamation (Dee *et al.* 1973). Originally, it was based on a quantitative aggregated index of 78 environmental attributes falling under 18 subcategories and into four broad

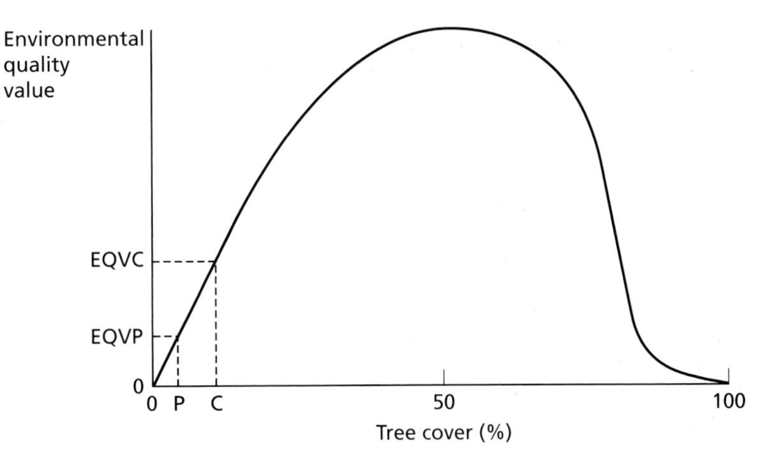

Fig. 9.2 Hypothetical environmental quality values for tree cover in Scotland. C represents the current situation (12%). P represents tree cover after the project (5%). EQVC and EQVP are the quality values in the current and post-project situations respectively.

categories: ecology, environmental pollution, aesthetics and human interest. When using the EES, the first task for the team of experts is to rate physical, chemical and biological measurements in environmental quality values (EQV) on a scale of 0–1, where 0 represents the level of least desirability and 1 the point of maximum desirability for that particular environmental attribute.

In the hypothetical example shown in Fig. 9.2 the environmental attribute is the amount of tree cover in Scotland. Having no tree cover (0%) is not desirable as the presence of some trees are felt to be beneficial for ecological and landscape reasons. Similarly, neither is a tree cover of 100% felt to be desirable as this would not leave any land for agriculture, houses, infrastructure or other habitats. The level of desirability for tree cover increases up to about 30% and remains high to cover values up to 55%, after this point desirability falls to 0 at 100%. When used in an EIA, such curves should be developed for all environmental attributes of interest.

The next stage in EES requires analysts to rate each environmental attribute for its importance. This is done by partitioning 1000 parameter importance units (PIUs) between all the environmental attributes. At the end of this process the EIA team have listed all environmental attributes of interest, they have considered the desirability of different levels of each environmental attribute and they have allocated importance units to each attribute. In the final stage, the evaluation and weightings are combined to obtain environmental impact units (EIUs). This is done by multiplying the scaled values and the associated PIU weights for both with-project and without-project scenarios for each element of the environment separately, and then subtracting the without-project EIUs from the with-project EIUs.

For example, consider the situation for tree cover depicted in Fig. 9.2, and assume that 20 PIUs had been allocated to tree cover. If in the current situation we have 12% tree cover in Scotland, and a project is proposed which will reduce the tree cover to 5%, then the project will cause a decrease in EQV of 0.2 (current EQV is 0.4, post-project EQV is predicted to be 0.2). As 20 PIUs have been allocated to tree cover in the earlier part of the EES process, then the EIUs for this part of the project is −10 (-0.2×20). That is, the project will have a slight negative impact on tree cover. This calculation should be completed for all attributes of the environment and, in a real project, all EIU values are summed for each project alternative to yield a measure of total environmental quality for each alternative. These values can then be compared.

Although the EES has some appealing aspects, it has not been well used in practice. The method is complex to manage and to communicate to the public and/or decision-makers. It also assumes changes in quantity and quality of environmental attributes are substitutable in generating the overall quality score, a problematic assumption in many cases.

ADAPTIVE ENVIRONMENTAL ASSESSMENT AND MANAGEMENT

Adaptive environmental assessment and management (AEAM) is designed to be responsive to changes in the decision-making and environmental domains (Holling 1978). The basis of AEAM is personal contact and communication. It is not a 'cookbook' approach, and although it is based on communication and teamwork, several steps in the process can be identified. After

1. The use of Friesian/Holstein cattle
- **2**, potential for high yields
- **2**, use of cattle poorly adapted to local conditions
  - **3**, high water requirements
  - **3**, poor disease resistance
    - **4**, decreased yields
      - **5**, decreased income
    - **4**, increased use of veterinary medicines
      - **5**, increased cost of production
      - **5**, need access to qualified vets
  - **3**, poor heat tolerance
- **2**, need high-quality feed
  - **3**, use of feed concentrates
    - **4**, decreased fodder production
    - **4**, increased cereal production
    - **4**, increased local traffic
      - **5**, increased human health hazard
      - **5**, increased noise disturbance
      - **5**, increased pressure on roads/infrastructure
    - **4**, increased costs of production

Fig. 9.3 Scoping of environmental impact assessments for agricultural planners with ECOZONE, which is a simple knowledge-based computer system designed to aid planners scope environmental impact assessments in developing countries. (Edwards-Jones & Ibrahim 1997.)

selecting the project manager, all relevant parties associated with the project and the EIA are identified and invited to a workshop. This serves to bring all parties (scientists, planners, decision-makers) together in order to define the problem and identify information requirements during the very early stages of planning.

During the first workshop an attempt should be made to describe the problem (the project–environment interaction). This may be aided through the use of matrices, flow diagrams or modelling techniques. In classic applications of AEAM the outcome of the first meeting should be an outline of a model of the study system. This need not be a computerized or mathematical representation of outcomes, but it should reflect the important attributes of the system.

After agreeing upon the broad outline of the model, the team members should seek to collect information which will help develop the model. This may include site-specific data or information from secondary sources. The idea which permeates AEAM is that through continual negotiated development of the model a good representation of the study system will be attained which will permit environmental impacts to be identified, and possible mitigation plans to be considered. Further meetings are held during which the model is continually refined and, upon completion of the model, the impacts of the project are simulated and management alternatives explored.

REVIEW OF THE EXAMPLES OF TOOLS AND FRAMEWORKS

These three methods are typical of the tools and frameworks that have been utilized in EIA over the last 30 years. None of the three methods explicitly considers higher-order impacts, and none considers the environment to be probabilistic. Matrices are physically constrained to consider first-order impacts only, and while in theory EES could be developed to consider higher-order impacts, it is already a relatively complex method. Through the development of an appropriate model in AEAM it should be possible to consider the higher-order impacts of a development project. This could either be achieved by developing a series of linked quantitative models, or alternatively by devising a qualitative model which explicitly traces an impact tree. One such model has been developed for aiding agricultural planners with scoping of EIAs (Fig. 9.3) and further development of such techniques, which could include assigning probabilities to the likelihood of occurrence of each impact, seems to be a promising avenue for further research (Geraghty 1993).

Both the matrices and EES consider a snapshot in time, which is a severe disadvantage with both of these methods. Another disadvantage of all three methods is that as originally conceived they depend totally on expert input and do not easily allow public participation. While experts may be needed to estimate the magnitude of impacts, there is a clear role for the public in evaluating the importance of any impact. The first two methods would allow such input: the Leopold matrix could be completed by a team which included representatives of local communities, and the general public could help develop EQSs and distribute PIUs in the EES. It is more difficult to engage the public in the process of model development which is central to AEAM, as this is largely a technical process. They could, however, be involved in the evaluation of model output.

The output of each method also represents the typical types of EIA output. Within the Leopold matrix it is not theoretically acceptable to sum or average any of the scores entered into the matrix and hence summarize the results. Facts and values are thus kept apart. Assessment of the results depends solely on viewing the completed matrix, and it is chiefly for this reason that a matrix of 8800 cells, as originally proposed, renders the matrix unwieldy and difficult to use. Conversely, EES provides the decision-maker with a single figure. This is easy to understand, with a positive figure meaning the project has more beneficial impacts than adverse ones. However, this apparently simple output hides the very large number of value judgements and assumptions inherent in this method. The AEAM method does provide a more complex model output, but this too will still hide the assumptions and simplifications inherent in any analytical model.

While each method clearly has its faults, at least two provide useful frameworks for assessment: the Leopold matrix and AEAM. Perhaps the most important advantage of a matrix-based method is its ability to provide a focus for discussion. It is very difficult for a multidisciplinary group to complete a matrix without entering into discussion. During this discussion each member of the team has the opportunity to express their views and identify any potential impacts. For this reason it may be advisable to complete a matrix at an early stage of an EIA, say during scoping, as this serves to bring the team together and to identify information gaps easily and cheaply.

Having utilized a matrix during scoping, the framework of AEAM may prove useful during impact assessment. Despite having advantages and disadvantages, AEAM has proved a robust model for EIA and has been copied and amended in many situations. The dependence on a mathematical model has not been central in all of these situations, rather it is the idea of establishing a multidisciplinary team which discusses the project and meets iteratively and collects data which has proved so useful.

9.5.3 Social impacts: approaches and tools

Social impact assessment (SIA) 'aims to evaluate the effects of environmental and technological change on individuals, communities and societies' (Burningham 1995). As such, SIA is a distinct process which generally runs in parallel with the scientific impacts assessments of an EIA process. Although the basic structure of SIA generally follows that of the overall EIA process outlined previously, it is widely recognized that the social aspects of EISs have played second fiddle to the environmental aspects. There are at least four main reasons for this:

1 Social impacts have often been equated with economic impacts, and thereby SIAs focused on a few relatively easily measured factors, such as employment generation and associated multiplier effects.

2 Social assessment was often mistakenly identified with the process of public consultation, rather than as a separate research process in its own right.

3 A preoccupation amongst planners and decision-makers with quantification has tended to mitigate against reporting of impacts that are difficult to express quantitatively, these characterize many social impacts.

4 Social analysts often found it difficult to adapt to the constraints of the EIA process: social anthropologists, used to spending months or even years studying complex social patterns, did not adjust well to short assessment deadlines. As a result, early assessments often lacked focus and assessors found it difficult to see how they fitted into the overall scheme.

Attitudes to SIA have changed over recent years and the importance accredited to social impacts has increased in project appraisal, with a number of major organizations adopting their own substantial guidelines on assessment methodologies; for example, the World Bank, the US Environmental Protection Agency and the Asian Development Bank. In the following section we review two areas of interest: the scope of social impact assessment and an outline of research methods.

SCOPE OF SOCIAL IMPACT ASSESSMENT

Finsterbusch (1980) advocates considering impacts within an individual 'quality-of-life' framework 'which includes both descriptions of measurable changes in a person's *objective conditions* and *subjective responses* to these changes' (emphasis added). This framework acknowledges that individuals are:

1 organisms with biological needs;
2 personalities with psychological needs;
3 friends and relatives with social needs;
4 workers with employment or production needs;
5 consumers with desires for goods and services;
6 residents desiring attractive and compatible habitats;
7 commuters and travellers with transportation needs;
8 citizens with freedoms, rights and political opportunities;
9 cultural beings with intellectual, cultural and spiritual needs; and
10 pleasure-seekers who enjoy entertainment, recreation and leisure.

Such a list of categories is only a suggestive guide to areas where human well-being may be impacted by disturbances. Clearly, there are also complex synergies between the aspects of the list. However, it is helpful to delineate the number of areas where social impacts may be relevant. As in the assessment of environmental impacts, the scoping exercise in EIA is critical to producing a manageable set of issues to address.

The problems of developing an effective SIA include the following:

1 Difficulty. Many of the variables of importance in social impact analysis may be abstractions of a fairly high level (Freudenberg & Keating 1985). As such, the specification of these social impacts may be difficult and time-consuming.

2 Limited time and resources. Most classic social studies of communities took place over at least a year and, furthermore, these studies produced no more than the equivalent of a baseline description of the project area. (Jacobs 1978). Indeed, the more complex the system, the more time is required to analyse it effectively.

3 Organizational resistance. This may occur through ignorance of the needs of SIA practitioners, direct opposition or deliberate manipulation. Social impact assessment inputs are generally most effective when applied as early as possible in the development cycle (Dietz 1984). The lowly status of SIA can, however, be self-perpetuating in so far as weak early studies provide the opportunity repeatedly to marginalize subsequent studies, generating a self-fulfilling prophecy that SIA work does not provide valuable input.

4 The political aspect. Social impacts are generally highly politicized. The impacts of a project are generally highly distributionally uneven, and SIA needs to recognize that there is no undivided public interest which can be served, but that all projects affect a variety of different social groups differently. This often generates conflict, with discussion forums inclined to be dominated by the most vociferous and powerful vested interests.

Notwithstanding these difficulties, SIA continues to grow in importance. The emphasis of Agenda 21 on more local level activity as a key element in achieving sustainable development has provided a stimulus to increasing participation in planning and assessment, where social impacts are seen as a key component of such assessments.

SOCIAL ASSESSMENT APPROACHES

Probably one of the greatest problems in social assessment is making analyses count when they are not presented in quantitative form. Substantial efforts have gone into developing more quantitative methods for assessing the effects on human welfare of certain direct impacts; for example, human responses to noise pollution. However, the directly measurable aspect of the impact may still fail to take account of its overall effect on the perceptions of the affected community. In this regard, SIA is fundamentally impossible without contact with affected individuals and communities. Participation and cooperation with these communities is therefore the *sine qua non* of SIA. Recently, rapid rural appraisal (RRA) and participatory rural appraisal (PRA) have been the focus of much interest in this area, emphasizing the active engagement of local people in the assessment process through more informal approaches to assessment rather than utilizing formal surveys and rigid consultation structures. These approaches are considered in detail in Chapter 10.

Some of these approaches are evident in an SIA related to a trunk road development on the outskirts of Wellington, New Zealand (Rivers & Buchan 1995). As in environmental methodologies, the range of approaches has to be adapted to individual circumstances, and the elements reported here give a good example of a triangulated approach which addresses the issues through several sources. The research team made use of the following elements:

1 Population statistics. These included the social composition of affected communities and variables, such as car ownership and use, location of schools and travel routes, ethnicity and other variables.
2 The effects of road development recorded in previous social, psychological and transport studies. These covered issues such as effects of noise, pollution, isolation of communities as a result of new road construction ('severance') and the knock-on effects on public transport over time.
3 Public consultation. Three important forms of consultation were the following.
 (a) Formal meetings, with formally represented groups, such as local public authorities, community groups and school committees. This led to the creation of resource groups which took on responsibility to represent the views of people within their areas of concern.
 (b) Street meetings, held along the proposed route of the road and in the houses of residents. These were attended by local residents in groups of 10–40.
 (c) Special interest group meetings, with groups such as sports clubs, mothers groups and pensioners, to learn how they perceived potential impacts.
4 Interviews with retailers. The effects on the central retail area was assessed through retailer interviews. This information was then cross-checked with a shopper survey to investigate the possible effects on patterns of shopping behaviour.
5 Traffic survey data. This included known traffic problems, including accident trouble-spots, identified by local communities. Together with interview information on travel patterns this helped to identify the potential benefits of new road construction.
6 The number and membership levels of community groups, and media coverage of their activities. These sources helped to build the picture of the current organization of the affected community.
7 Historical research. The possible cultural significance of Maori sites was investigated by a Maori researcher specifically subcontracted for the project.
8 Written and oral submissions. These were collated through the regional council throughout the duration of the research.
9 Consultation on social impact reports. The draft impact assessments were sent to community groups for comment to ensure that their views had been accurately represented.
10 Feedback to engineers and planners. The social research team held meetings with professionals working on the project to relay community concerns and reactions. This enabled redesign of elements of the project proposals and other scheme improvements. These professionals also attended public and resident meetings themselves.

CONCLUSIONS ON SOCIAL IMPACTS

Such a pattern of research indicates the range of sources and methods than can have an input into the social impact assessment, and is by no means exhaustive. The outcomes of such research will be a combination of quantitative and qualitative insights into the development impact on communities. This is both an essential input into the decision-making process and, in addition, can be vital to the smooth running of the subsequent development phase (see Box 9.1). Although there is a tendency for decision-makers to focus on hard figures, the 'narrative' element of an SIA, describing potential changes in a community caused by development, can play an influential part in the overall assessment. Good reviews of SIA methodologies and outlooks can be found in Taylor *et al.* (1995) and Cernea (1991).

9.6 Summary

Environmental impact assessment is one of the few examples of a statutory requirement for consideration of environmental and social impacts of development alongside financial and economic criteria. Despite this process being widely used around the globe there are problems inherent in the EIA process, which assumes that all parties act in good faith.

The basic structure of an EIA follows a set of stages, leading from project definition, through scoping of the study and baseline assessment of environmental conditions, to estimation and assessment of impacts, and finally identification of mitigating activities and a monitoring programme. Within the estimation and assessment stage, a wide variety of techniques will be employed. Three possible tools were discussed, each dealing with data of different types. In general, most techniques require expert input and do not easily facilitate public participation. As an overall framework for undertaking an EIA, the AEAM has much to offer, but the classical reliance on a model of the system is perhaps overly restrictive. Recent research on methodologies in EIA has made considerable advancement in the development of computer-based assessment. It is important to remember that although mathematical

Box 9.1 Social impact assessment and the Niagara Community Agreement

An example of effective public participation is given by Ontario Hydro's development of a community impact agreement regarding the effects of a proposed hydroelectric scheme on the Niagara River (Smith 1995). Ontario Hydro is a public corporation with a long-standing involvement in the area, a major employer and operating a corporate policy which committed the company to the principle that 'no community shall suffer as a result of Hydro activities'. In 1991 it submitted plans for a new generating station in Niagara Falls, which involved a significant underground construction and associated transmission lines.

The company initiated consultation with affected communities from the outset of the planning process, and subsequently moved to formal negotiations with the affected municipalities in 1992. Impacts were discussed through a socioeconomic matrix table drawn up for the purpose. Interestingly, although the mandate of the consultation teams was limited to the new development, unrelated grievances with the behaviour of Hydro became an important issue. The Hydro teams became active within their own organization to sort these problems out alongside concerns with the project under assessment.

Following a year of consultation with both formal municipal authorities and other groups, a community impact agreement was signed at the end of 1993 between Hydro and the three affected municipalities of the Niagara region. The features of this agreement included:

- full access to information;
- a joint liaison committee;
- a citizen complaint process;
- a neighbourhood advisory committee;
- a process for monitoring and remediation;
- a transport impact management plan;
- a process for emergency services and response planning;
- a process to optimize local economic benefits;
- provisions for outside legal and consultant costs;
- an arbitration process; and
- an audit provision.

Financial figures were established for various compensation and mitigation programmes, payable upon approval of the project; these represented less than 1% of project costs. A liaison committee was also established to oversee the continuing management of the agreement. The local community, initially suspicious of Hydro activities, came to support actively the final proposal, which also incorporated a number of important design changes in response to public concerns.

techniques can help structure the analysis and aid in exploring the possible range of outcomes, ultimately the decision depends on human values at both an individual and political level.

The social impacts of a project are widely recognized as being important but historically SIA has not been well received within the EIA framework, often because its output has been more qualitative. The range of available techniques is again large, and assessment needs to reflect both objective changes in the affected environment and subjective responses to those changes. Ensuring the active involvement of affected groups will therefore always be central to the assessment process.

Further reading

Biswas, A.K. & Agarwala, S.B. (1992) *Environmental Impact Assessment for Developing Countries*. Butterworth Heinemann, Oxford.

Erickson, P.A. (1994) *A Practical Guide to Environmental Impact Assessment*. Academic Press, San Diego.

Glasson, J., Therivel, R. & Chadwick, A. (1994) *Introduction to Environmental Impact Assessment*. UCL Press, London.

10 Multicriteria Appraisal

'I have yet to see any problem, however complicated, which, when you looked at it the right way, did not become still more complicated.' [Paul Anderson]

10.1 Introduction

In this chapter we introduce two different styles of project planning and appraisal which represent widely different views on the process of analysis. In the first part we cover the formal planning methodology of multicriteria analysis (MCA), which is becoming a widely practised appraisal philosophy, particularly in the wake of advances in computer technology. It is a diverse field and the basic principles are covered with two simplified examples of popular programming methods. These examples represent what may be considered relatively 'hard-edged' appraisal techniques, integrating inputs from different disciplines through a formal assessment structure.

In the second part we introduce the principles of participatory rural appraisal (PRA), as a paradigm example of an alternative, 'soft-systems' approach to development planning. Participatory rural appraisal has developed primarily in the context of developing countries' rural development projects, but in recent years it has broadened its appeal, partly as a result of exemplifying a powerful reversal of the traditional 'reductionist' or 'positivist' approach to planning processes anchored in 'objective' scientific methodologies, and challenging it with an approach emphasizing uncertainty, adaptability, flexibility and context-dependence.

10.2 Philosophy of multicriteria analysis

When the US statesman Benjamin Franklin was faced with a difficult decision, he described his approach thus:

> '[I] divide a sheet of paper by a line in two columns; writing over the one pro, and the other con; then, during three or four days' consideration, I put down under the different heads short hints of the different motives that at times occur to me, for or against the measure. When I have thus got them all together in one view, I endeavour to estimate the respective weights [to] find at length where the balance lies.' (Parton 1864)

He added that this procedure was a great help in preventing him from taking rash steps.

The essence of Franklin's technique—determining the relevant criteria for decision-making and then weighting these criteria accordingly—remains very close to the core principles of modern MCA. Strictly speaking, MCA is not simply an alternative procedure to project appraisal techniques, such as cost–benefit analysis and environmental impact assessment. Rather, it provides a formal structure for integrating the results from all these other approaches in order to help decision-makers choose a plan or project that fits best with their own priorities and objectives.

In a nutshell, an MCA takes a set of plans or activities, a set of objectives to be achieved and a set of criteria by which to measure each of these objectives. It assesses the impact of each plan, or the activities that will compromise a plan, on each criterion. Then, through a formal programming procedure it compares these individual criterion impacts in order to evaluate and select the plan or combination of activities that most successfully meets the objectives.

10.3 The development of multiple criteria in decision-making

The general description above masks a very wide variety in the kinds of techniques that have been developed to deal with decisions involving multiple criteria. The oldest specialist field is multicriteria decision-making (MCDM), holding its first International Conference in 1974, but more recently this field has diversified into new branches, such as multicriteria decision aid (MCDA), multicriteria decision support (MCDS), multiple objective decision-making (MODM), and multi-attribute decision-making (MADM). For simplicity, in what follows we refer mainly to multicriteria analysis (MCA) as a generic title for techniques that involve multiple criteria. This should not disguise the fact that there is no universal procedure for multicriteria decision-making; what follows only picks out some of the most fundamental considerations in the use of these techniques.

Multicriteria methods were pioneered in the 1970s following the development of single-objective programming techniques, such as linear programming (LP). Linear programming is a method of mathematical analysis which calculates the optimal solution to a planning problem in which there is a single objective and several constraints on action. For example, a farmer may want to maximize his or her profits from farming, but he or she must work within constraints on the amount of land he or she can farm, the amount of water available for irrigation, the amount he or she can borrow, and so on. Given these constraints and the objective of the decision-maker, LP calculates the most profitable combination of crops to plant.

A multiobjective analysis extends the problem-solving approach by introducing additional objectives. A farmer may want not only to generate profits but also enjoy more leisure time, reduce his or her exposure to risks of crop failure, and perhaps enjoy a more attractive landscape. The problem situation modelled by this kind of multiobjective appraisal is generally closer to the real priorities of a decision-maker than a single objective approach. In fact, there are probably very few situations in which a decision can realistically be based on only a single objective.

While it might be possible to provide a financial valuation for some of these objectives (Chapter 6 introduces some of these techniques), it seems more sensible to measure each one by specific and appropriate criteria: leisure time in number of hours, landscape by acreage of tree cover and length of hedgerows, and so on. This introduces multiple criteria into the analysis. These criteria measure objective achievement on a range of scales from simple quantities to purpose-built indices. This freedom in data-gathering is a major advantage of MCA techniques, particularly in environmental planning situations where social and environmental impact assessments may score impacts on a number of different measurement scales.

Although MCA thereby avoids the problems associated with a single criterion approach, it presents analysts with three other problems:

1 how to compare criteria measured on different scales;
2 how to decide the relative importance of different objectives; and
3 how to compute optimal solutions once criteria have been made comparable and objectives weighted.

Whereas Franklin ultimately relied on intuition for his final judgement in the face of these problems, there are now well over 70 MCA programs available. The US Water Resources Council has utilized multiobjective planning techniques since the early 1970s and other international organizations, such as the United Nations Environment Programme (UNEP) and the World Bank have their own research programmes devoted to the field. All these techniques acknowledge that planning problems will inevitably involve the three dimensions of social, economic and environmental impacts. In the words of Keeney (1982), 'what has been lacking is not information but a framework to integrate and incorporate it with the values of the decision makers to examine overall implications'. The nature of this framework is shown in Fig. 10.1.

10.4 The elements of a multicriteria analysis

We noted above that the variety of techniques used within MCA is considerable. Some, such as goal programming, require complete information from decision-makers about their objectives and priorities before the start of the analysis. Others are designed to be interactive, working with the decision-maker to clarify his or her own priorities. Still others analyse problems without relying on preference information at all. Bearing this diversity in mind, the following sections introduce the basic elements that are involved in most MCA methods.

10.4.1 The plans or options

An MCA can be applied in either discrete or continuous planning settings. A discrete situation involves a set number of plans, and the problem is to select the best plan in the light of given objectives. Examples might be assessing several alternative sites for a new power station, or several alternative regimes for management of a water catchment.

In a continuous situation, planners can vary the levels of different activities they undertake to find a satisfactory mixture. Thus, managers can allocate their time, budget, personnel and physical resources in varying quantities between different activities. They want to find the most satisfactory regime which meets their management objectives.

The nature of the planning situation is obviously important in selecting an appropriate technique. Here,

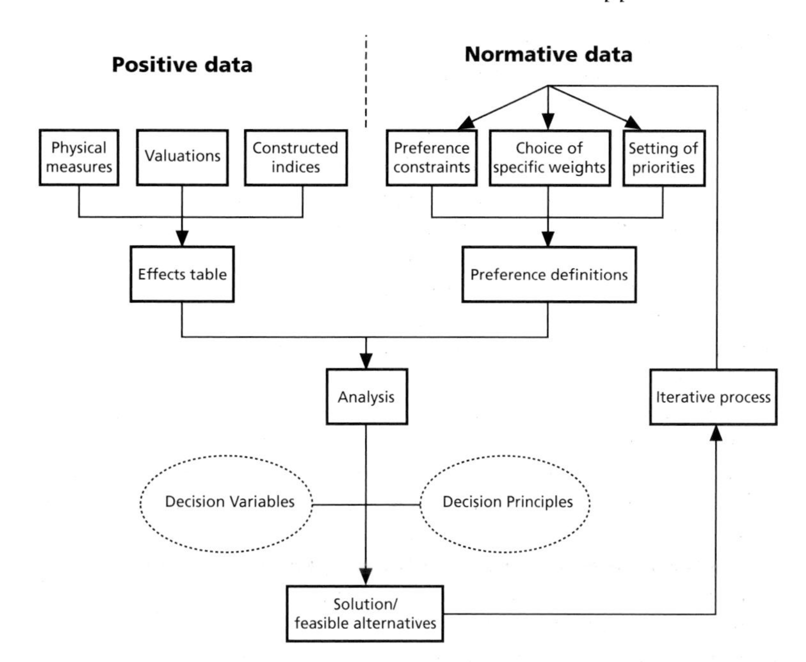

Fig. 10.1 A framework indicating the roles of both factual and value-based data in multicriteria analysis.

the very diversity of the field represents a problem. The range of techniques has expanded so far that Teckle (1992) even applied an MCA in order to assess the most useful MCA methods. In a review of past applications, he found that individual analysts were inclined to prefer the techniques with which they were most familiar, even when these techniques were not the most suitable for the problem they analysed. Unfortunately, there is little beyond common sense that a non-specialist can use to judge whether a technique has been well selected.

10.4.2 The objectives

The definition of objectives can take both a 'top-down' and a 'bottom-up' approach. In the case of a large infrastructure project, for example, a number of specific objectives are likely to have been set by a central planning body. On the other hand, the objectives for a local management plan might be generated through a grass roots process of public participation and discussion. Whatever the nature of the project, wide consultation is encouraged to ensure that important objectives are not overlooked.

A number of MCA methods have been developed specifically to help build consensus in planning solutions, often by ranking alternatives. The flexibility of MCA can therefore act as a conflict resolution tool in situations where there are strongly opposed interest groups. No views are systematically excluded and the analyst can work towards a compromise solution using information from all affected groups and individuals. However, it should be noted that these techniques are often time-consuming and the complexity of the programs can work against the need to gain acceptance from all those involved.

Where there is considerable argument over objectives, some analysts have stressed decision support and decision aid methods as more appropriate than decision-making. These techniques acknowledge that an optimal solution simply may not exist in cases where there are conflicts over objectives: that is, there is no single solution which will be the best from all points of view. The presence of competing objectives therefore means that there is no objectively right answer to the planning problem. Decision-making techniques introduce specific priority weightings to circumvent this problem (see Section 10.4.4).

10.4.3 The criteria

Criteria are measurable features of the planning environment which either singly or together indicate the levels of objective achievement. The relationship between criteria and objectives is shown in Fig. 10.2.

The choice of appropriate criteria for measurement in MCA is probably the most critical aspect of any analysis, and a number of guidelines should be followed

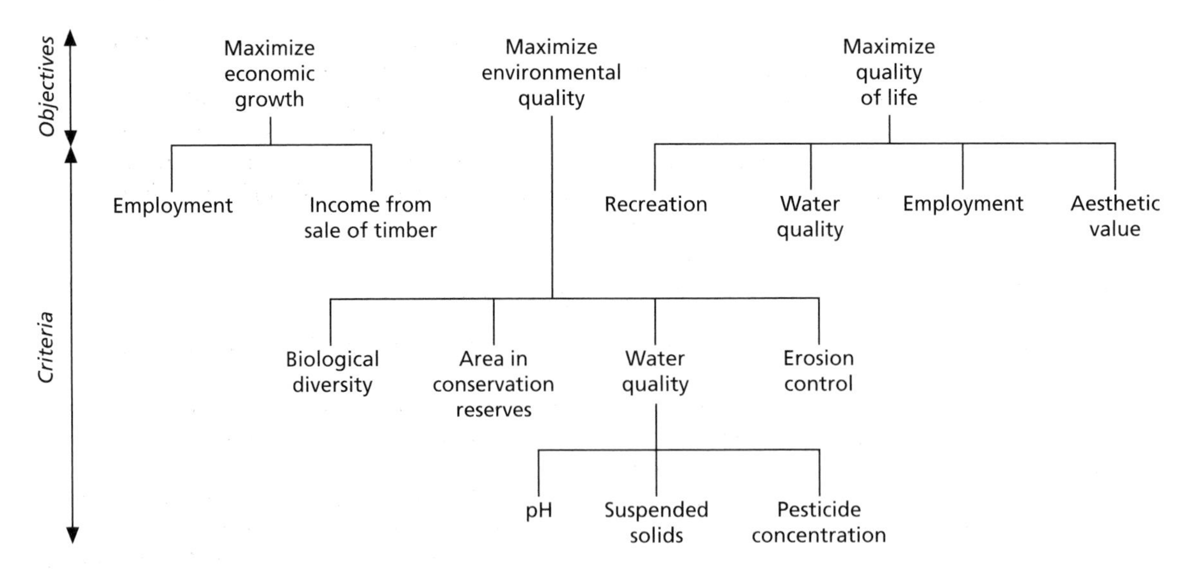

Fig. 10.2 The relationship between criteria and objectives.

(Keeney & Raiffa 1976). These indicate that the final set of criteria should have the following qualities:

Complete. If two alternative plans have the same overall score, then the plans must be considered equivalent. There should not be any further criteria which can be used to judge between them.

Operational. Each criterion must be capable of being measured in some significant way. This may require analysts to establish data limitations.

Decomposable. Two factors should not be in opposition in a single criterion. For example, if one objective is for habitat creation, and tree planting will be beneficial in some areas but damaging in others, tree planting cannot be used as a criterion.

Non-redundant. No aspect of the problem should be accounted for more than once. This is particularly important to avoid prejudicing the results in favour of a particular objective.

Minimal. No smaller set of criteria that satisfy the conditions above should be available. A minimal set improves efficiency in both data gathering and analysis, and also makes the structure of the final analysis clearer.

The choice of individual criteria themselves is a matter for the research team engaged in the analysis. In any public decision, wide consultation is encouraged to ensure that criteria reflect the views of the affected communities. There are no hard rules governing numbers or types of criteria, but the temptation to focus on easily measurable but relatively unsuitable criteria at the expense of more appropriate ones is an obvious one and should be resisted. Ideally, criteria should be linked as closely as possible to the specific objectives. The possibility of using purpose-built indices composed of several weighted measurements as criteria measures for such objectives as biodiversity or amenity can help mitigate the lack of good single indicators for these broad objectives (a review of such indicators in relation to nature conservation is given in Chapter 7).

In conclusion, it is worth noting Ackoff's (1977) warning that 'an optimal solution is not an optimal solution to a problem unless a model is a perfect representation of a problem'. Clearly, the divergence between model and problem can only be minimized by care taken in the definition of objectives and in particular, the criteria that measure them.

10.4.4 Prioritization of objectives and criteria

In many MCA techniques, decision-makers must decide on the relative importance of their objectives, even if this means only that all objectives are considered equally important. Romero and Rehman (1985) note that 'perhaps the greatest difficulty in the widespread use of the multicriteria decision-making paradigm is the availability of the substantial information required from the decision-maker on his objectives, goals, targets, weights and pre-emptive ordering of preferences'. Prioritization can be achieved either through relative *weighting*, or through *ordering*.

In an ordering method, objectives are ranked in order of importance and the MCA program searches first for the plans that meet the highest priority objective. Lower priority objectives are only considered once all higher ones have been fully achieved. This approach might well be appropriate for an objective such as safety, for example, where a minimum target level has to be reached for a project to be acceptable. In this case the highest priority safety criterion acts as a screening device, filtering out all plans that score below the threshold.

In a weighting procedure, each objective is assigned some fractional level of importance, with the total fractional weights assigned adding up to unity. Techniques for arriving at relative weightings can vary from relatively complex methods, such as Saaty's (1987) analytical hierarchy process, to simple scoring. Although weighting appears straightforward in theory, the process of weighting proves a major problem in many situations, particularly those involving multiple decision-makers, and ordering methods are often preferred as a result.

If the decision-maker is not sure of his or her priorities between objectives, some MCA techniques generate what is called the set of *non-dominated solutions*. These are the Pareto optimal plans or combinations of activities, where the level of achievement for one objective cannot be increased without reducing the level of another. The decision-maker is thus presented with a set of alternatives, all of which are in some way at the technical limits of the whole system. The number of possible plans for consideration is reduced thereby to manageable proportions. Whatever the decision-maker's relative preferences, it would clearly be irrational to choose a solution that was dominated by one which scored better on all criteria.

The problems associated with prioritization have also focused a good deal of attention on *interactive* programming methods; for example, the surrogate worth trade-off (SWT) method, first applied in water resource management by Haimes *et al.* (1979). These are methods aimed specifically to help clarify priorities by suggesting trade-offs between objectives. These pose questions such as, is a loss of 1 acre of tree cover more or less acceptable than a £5000 gain in profits? Or, is a 2% increase in air pollution acceptable for the creation of an additional 100 jobs? These techniques often offer the most helpful information on which to base a decision, when combined with a knowledge of total criteria scores. In a continuous setting, reactions to these trade-offs are used to identify a final optimal solution.

10.4.5 Evaluation

The specification of objectives, criteria and priorities set out the dimensions of a planning problem. Once these dimensions are in place, analysts can begin the work of evaluating possible solutions. This work involves two parts.

First, the analyst must determine the impacts of plans/activities on criteria. This will involve specifying a number of functional relationships. This information can then be used to build an effects table, or input–output matrix, detailing these relationships. The construction of the effects table for a large project is a highly detailed task, calling on the skills of specialists in a number of fields. The complexity should lead to a full understanding of the system, but areas where uncertainty persists should be noted for subsequent sensitivity analysis. All the problems associated with cost–benefit analysis (Chapter 8) and environmental impact assessment (EIA) (Chapter 9), themselves dependent on the tools of economic and ecological evaluation (Chapters 6 and 7), will confront the multicriteria analyst at the stage of building an effects table.

Once the functional relationships between criteria and plans have been specified, MCA programs must then use some search and selection procedure to identify possible solutions. The programming techniques available for this aspect of evaluation are diverse, but can be split into two distinct kinds.

1 Techniques that directly combine criteria scores together in some way, and find the best resultant combination score either algebraically or graphically.

2 Techniques that rely on building up out-ranking relationships between criteria, keeping criteria distinct.

From a decision-maker's point of view, these technical programming aspects of the analysis tend to represent a black box in the decision-making chain. This can lead to some distrust of the MCA methodology, particularly of complex methods that produce single solutions. Software packages that only require input–output data are also becoming increasingly available. Despite their easy-to-apply nature, it is useful to bear in

Table 10.1 Selected examples of multicriteria analysis applications.

Area of application	Alternatives considered	Criteria	Methods	Multiple weights?	Role and use in decision-making
Solid waste management in Ontario, Canada (Sobral *et al.* 1981)	10 disposal methods: landfilling, incinerating, composting, recycling, energy recovery and combinations thereof	7 criteria: water quality, air quality, land quality, social, political, legal, economics	Fuzzy set analysis; aggregation procedures; data consist entirely of qualitative evaluations of options provided by respondents to Delphi technique	Yes. Weights of individual respondents, and perceived weights of decision-makers	Technique applied to citizens advisory committee, which was addressing the waste management problem. No information on impact on decision
Solid waste management in the Netherlands (Mainome 1985)	8 disposal methods: composting, incineration, sanitary landfill, refuse-derived fuel, with minimum or maximum source separation for each	17 criteria, into four categories: cost, efficiency, recycling effectiveness, environmental consequences	Mixed data	Yes. Options assessed from three points of view: business; national goals; and environmentalist	Used in planning process. Two-stage evaluation procedure: treatment systems then siting. Regarded as successful, particularly in helping to organize the entire planning procedure
Freeway location in Italy (Guariso & Page 1994)	4 locations plus null hypothesis	12 impact criteria including land, aquifer, vegetation, fauna, recreation, landscaping, agriculture, historical site, housing, local economy	SILVIA software for three-stage process: preparatory (information gathering); analysis (effects tables) and decision (ranking)	Sensitivity testing of reliability of criteria weightings	Long-term effort by Italian Ministry of Environment. Used experimentally; in this case the preferred alternative was not accepted by the municipality in question

mind the value of specialists who are generally alert to the limitations of particular techniques.

Some selected examples of MCA applications are shown in Table 10.1.

10.4.6 Presentation of results

Finally, MCA models generally produce three kinds of outputs.

1 *A single optimal solution.* Single solution methods can be suited to discrete plan analyses where there are a large number of criteria but quite clear priorities. In these cases the program essentially uncovers the most desirable solution in a situation that is too complex for human analytical powers. However, it is particularly important that a sensitivity analysis is carried out to check the validity of the given solution.

2 *A feasible set of options.* The identification of non-dominated or Pareto optimal plans is generally preferred in cases where there are several interest groups and uncertain priorities. A number of techniques calculate the non-dominated set before narrowing down further to preferred alternatives.

3 *Specific trade-off relationships between options and objectives.* Trade-off analysis, often using graphs to show how two objectives trade off against each other under different plans, gives more information to decision-makers than a simple set of options. In particular, trade-offs can be established between financial costs and other criteria, presenting decision-makers with an idea of marginal gains and losses for comparison. Techniques may work with constraints on maximum allowable trades, offer specific trades to decision-makers, or ask decision-makers to indicate which criteria scores in a solution should be relaxed/improved in order to find a more acceptable non-dominated solution.

In all cases, a sensitivity analysis should always be carried out on the result(s) to investigate uncertainties in both input–output data, and in the prioritization of objectives.

10.5 Examples of multicriteria analysis methodologies

Finally, we introduce two popular MCA methods: goal programming and composite programming. There are numerous refinements and extensions to these techniques, but only the basic framework is introduced here.

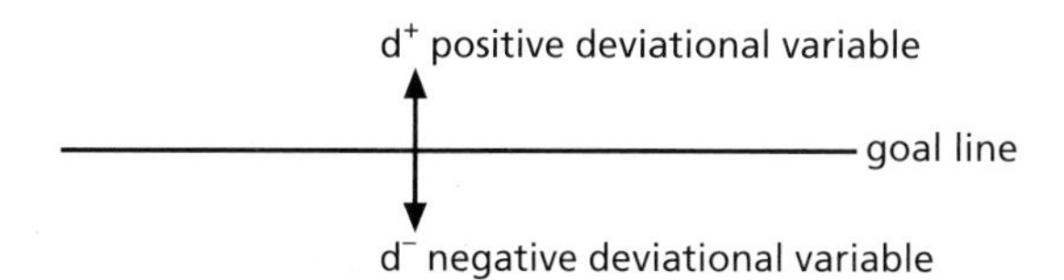

Fig. 10.3 Graphic representation of a positive and a negative deviational variable as used in goal programming. The variables measure over- or under-achievement of each criterion goal.

10.5.1 Goal programming

Goal programming (GP) was first developed in the 1950s (Charnes *et al.* 1955) and has subsequently been widely applied at both enterprise and regional level in natural resource management. In many ways its procedures match the basic approaches often used by managers—goal setting and prioritizing—which may partly account for its popularity.

Goal programming sets targets for the achievement of objectives, and then seeks for each objective to minimize the deviation between the target and the value that can be achieved in practice, given the need to satisfy targets for other objectives. Thus, each objective target has associated with it a positive and a negative deviational variable (see Fig. 10.3) and the decision-maker can weight the significance of under- or over-achievement of the target, according to his or her preferences, by attaching weights to these variables.

The GP problem to be solved through mathematical computation is therefore given by:

$$\text{Minimize}\ \Sigma d^- + \Sigma d^+ \qquad (10.1)$$

where d^- and d^+ represent vectors of deviations from proposed goal targets. The minimization process is usually achieved through either a weighting or a prioritizing procedure. Weighted goal programming (WGP) considers all objectives simultaneously and weights each goal according to its importance for the decision-maker. This gives a conventional linear programming problem which can be solved through computation. In lexicographic goal programming (LGP) the objectives are separated into sets, which are then ranked in importance. Each set is then dealt with according to its assigned priority level. There may be only one objective in a set, but if there are more, then each objective can be weighted relative to the other members of the set in the same way as in WGP.

The basic process of GP can be illustrated graphically with a simplified example. Consider a nature reserve

Table 10.2 Planning situation facing a reserve manager.

Criteria	Salary/ £10/wk	Training/ h/week	Reserve quality/ weekly score
Variable			
Qualified warden	5	1	5
Trainees	2	2	4
Manager's target	200	104	320

manager who can employ either qualified wardens or trainees for the summer season. These two types of worker have different salaries, different training needs, and different impacts on maintaining reserve quality (measured through a cumulative index—say, related to the number of trees prevented from damage). The reserve manager has targets for her budget, the amount of time her core staff spend on training, and the overall quality of the reserve. The data for this continuous planning situation is shown in Table 10.2.

The GP problem is formulated by drawing up a set of equations from this data which relate the variables to the criteria target through the inclusion of deviational variables:

$5W + 2T = 200 + d_s$ (salary objective)
$W + 2T = 104 + d_t$ (training objectives)
$5W + 4T = 320 + d_q$ (reserve quality objective)

where W stands for the number of qualified wardens, T the number of trainees, and d_s, d_t and d_q, the deviational variables for each criteria. To solve graphically, each of these three objective relationships can be plotted on a graph, which shows the relationship between the two decision variables, 'number of qualified wardens employed' and 'number of trainees employed', and each individual objective target (thus simplifying for the purposes of drawing by taking the deviations as zero initially). Each line on this graph therefore represents a combination of trainees and qualified wardens which exactly meets one of the three criteria, with no deviation. A graph of these three target lines is shown in Fig. 10.4.

The reserve manager is assumed to have some preferences about the importance of the targets for these criteria. In particular, she considers her targets for training hours and salary expenditure as maximum allowable values. Considering these two criteria, it is clear that the feasible set of employment combinations must lie within the area OABC. That is, a solution anywhere within this area satisfies the manager's budget and training objectives.

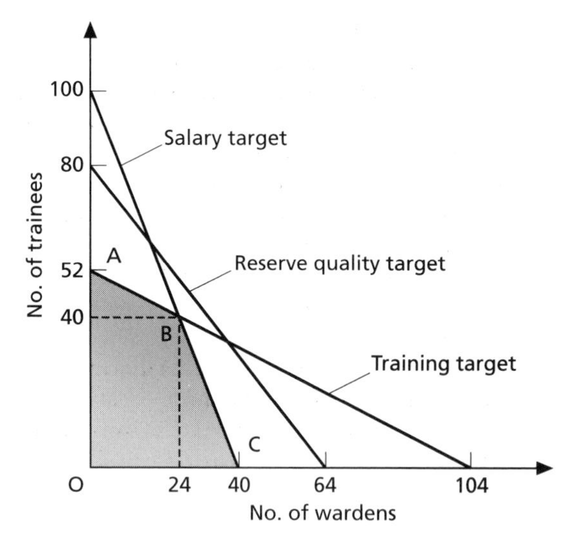

Fig. 10.4 A graph representing the planning problem in Table 10.2. Each line represents the combination of wardens and trainees needed to meet a target.

The target objective for reserve quality is an ideal, towards which she aims. Unfortunately for the manager, her ideal reserve quality is here unattainable without breaking one of the other two objective constraints. By searching the area of OABC, it is apparent that the point which minimizes the deviation from the quality goal is point B. With this combination, she meets the targets for salary and training exactly, while falling short of the reserve quality target by some deviation. Solving the two constraint equations simultaneously, the GP solution to the problem is to employ 40 trainees and 24 full wardens, with a shortfall in the reserve quality index of 40. At this simple level the procedure appears the same as that used in a conventional linear program, but the techniques diverge with problems of greater complexity.

If the manager had been flexible in the upper limit of the salary costs and training allowance, then locating a compromise solution would require mathematical programming, using precise weights assigned to the deviational variables for each of the criteria targets.

CONCLUSIONS ON GOAL PROGRAMMING

Goal programming is a computationally efficient procedure, because it works with closely defined objectives and priorities. It is also a logical and easily understood process of analysis, moving from goal definition to achievement in a logical progression. The process of constructing the model also forces the analysts to investigate the problem in great detail, which should lead to an accurate assessment of the interactions at work. A wide array of information can be made available to the planner through sensitivity analysis.

These strengths are in many ways also the technique's weaknesses. By narrowing down efficiently to a single solution, the final quantity of information supplied to the decision-maker is quite limited; in particular, it does not indicate the trade-offs being made between objectives. In addition, the requirements for information on the preferences of decision-makers are very exacting. In many situations decision-makers may want decision support to help clarify their own judgements, rather than locate solutions on the basis of predetermined values.

10.5.2 Composite programming

Composite programming is a technique developed by the United Nations Environment Programme (UNEP) for use in water resource management. It provides a relatively simple method for evaluating planning scenarios which gives a graphical representation of the outcome of the analysis and does not require complex mathematical programming. The technique builds on the principles of compromise programming (CP) developed by Zeleny (1976). Here, the basic principles of CP are outlined before describing the application of composite programming.

COMPROMISE PROGRAMMING

In CP the analyst defines a hypothetical ideal solution to the problem under investigation. This solution, called the ideal point (sometimes referred to as a benchmark plan), is defined by the values of all the objectives when they are each optimized individually. Thus, the ideal point is where all objectives take their best possible value given the technical features of the problem. It is highly unlikely that the ideal point gives a feasible solution to the problem, but it enables the programmer to define the best compromise solution as the plan that comes closest to the ideal point.

To find the ideal point, a pay-off matrix or effects table is constructed which includes the values of each objective when maximized. This matrix also gives the nadir points: the points at which objectives take their minimum or worst value. Table 10.3 gives a pay-off matrix for a two-objective problem.

Table 10.3 Pay-off matrix for a two-objective multicriteria analysis problem.

Objective function		
Profit/£	150 000 (max)	70 000 (min)
Leisure/h	40 (min)	80 (max)

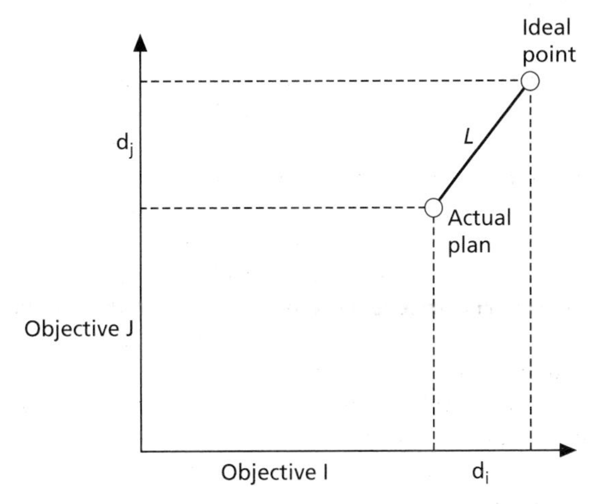

Fig. 10.5 Graphical illustration of a two-objective compromise programming problem. The performance of a plan is assessed against two objectives, I and J, and the distance *L* is a measure of how far this plan approaches an ideal solution.

The ideal and nadir points are used to standardize the values of the pay-off matrix. A multiobjective program is then used to generate all the non-dominated solutions to the problem; that is, all the solutions which cannot improve achievement of one objective without decreasing achievement of another objective. The ideal point now provides a criterion for choosing the best of these non-dominated solutions, by calculating the distance of each of the solutions from the ideal point.

This procedure is illustrated graphically for a two-objective problem in Fig. 10.5. The distance of the actual point from the ideal can be calculated by the generalized formula:

$$L = \left[\sum(\alpha \cdot s)^p\right]^{1/p} \qquad (10.2)$$

where L is the distance measure from actual to ideal point, α is a weighting expressing the relative importance of each objective, and s is the standardized value of each objective deviation from the ideal (d_i and d_j in Fig. 10.5). The index p expresses the importance of relative

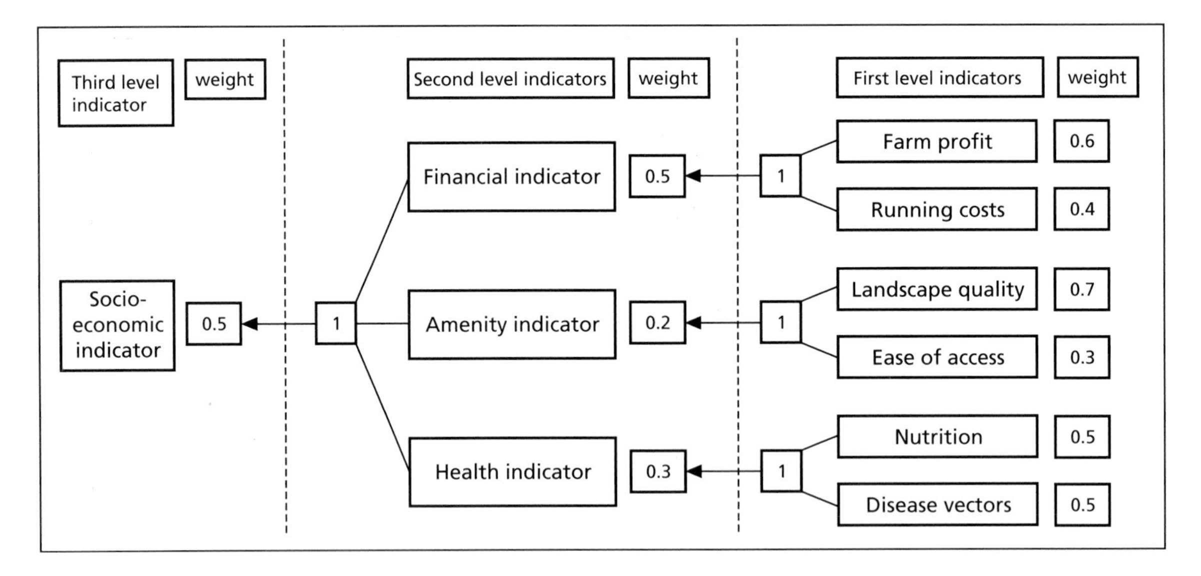

Fig. 10.6 Socioeconomic decision criteria arranged into a composite programming hierarchy. The weight given to the third level indicator is determined in combination with the equivalent ecological indicator (not shown).

deviations and its value, like α, must be determined according to the decision-maker's preferences.

COMPOSITE PROGRAMMING

Composite programming closely follows the basic approach of CP, but differs in the way it balances the different objectives of the decision-maker. In composite programming, the objectives or decision criteria are arranged into a hierarchy which compresses the total number of criteria describing the system down to two: a socioeconomic indicator and an environmental indicator. Each of these main indicators is thus an amalgam of lower level or basic indicators (the criteria). An example of this arrangement is given in Fig. 10.6.

Each set of indicators shown in Fig. 10.6 is combined according to the CP formula given as (10.2) above. Thus, the financial indicator for a given scenario is represented by a single standardized value, made up from the farm profit and running costs indicators. The weighting values shown are determined by the decision-maker, and may be set on technical or preferential grounds. Although the approach can theoretically handle any number of indicator levels, in practice three

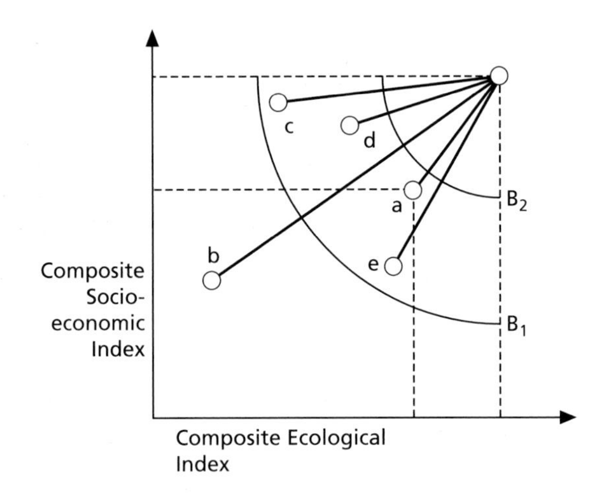

Fig. 10.7 Comparison of performance of five alternative plans (a–e) under a composite programming analysis. Plan a lies closest to the ideal point and is the preferred option.

levels seems to give an optimal balance between clarity and accuracy.

Once values for the final two composite indicators for each scenario have been calculated, the outcomes can be represented graphically (Fig. 10.7). In Fig. 10.7 we have added two boundary lines, B_1 and B_2, encircling the ideal point. These lines can be used to determine levels of acceptable and poor performance for each scenario: scenarios that fall outside B_1, for example, may be considered to be performing poorly and therefore

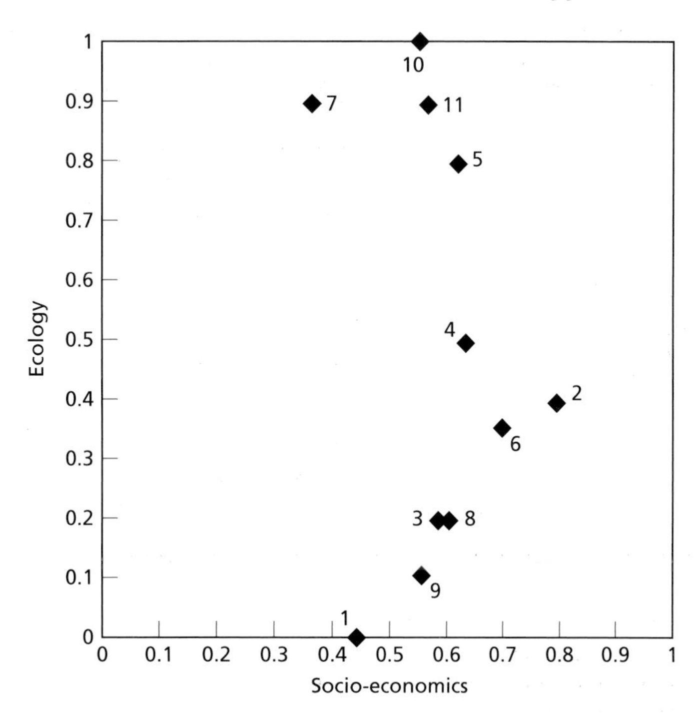

Fig. 10.8 Example of a composite programming analysis of a water management problem in Fife, Scotland. The potential effects of eleven alternative management systems to control eutrophication caused by soil erosion are shown in terms of their socioeconomic and ecological impacts on the study area. (Source: Bailey *et al.* 1995.)

do not constitute acceptable solutions. These boundary lines can be generated by considering specific achievement levels, or set according to political or other considerations.

The hierarchical arrangement of indicators enables planners to vary the priority they give to different objectives in the overall analysis. Thus, if economic objectives are particularly important this indicator can be weighted heavily at level 2 in relation to amenity and wildlife values. A new plot can be generated that shows the new levels of composite indicator achievement under these new priorities. The two composite indicators can themselves be combined to give a single overall score for each planning scenario, again using formula (10.2) given above, although this kind of 'grand index' number does not convey much information to decision-makers. The real advantages of the technique lie in making clear the relative trade-offs between major objectives. An example of a composite programming analysis is shown in Fig. 10.8.

10.6 Conclusions on multicriteria analysis

Multicriteria analysis techniques have some clear advantages over more restricted decision-making techniques, such as cost–benefit analysis. Their popularity has increased very substantially with improvements in both methodologies and computer power. Furthermore, their suitability to environmental and natural resource planning is increasingly being recognized. Table 10.4 gives a brief summary of the advantages and disadvantages of the approach.

It is not possible to give general rules for the use of a particular technique, but whichever method is chosen relies on the validity of two essential elements: a correct identification of relevant criteria; and a correct measurement of these criteria. Problems with individual programming techniques are likely to be minor in comparison with errors generated at these two stages. For these reasons, the success of an MCA relies ultimately on the skills of its research team. No amount

Table 10.4 Advantages and disadvantages of multicriteria analysis.

Advantages
Provides a systematic and comprehensive structure for problem analysis
Rationalizes problems with number of components beyond the ability of the human brain to process conventionally
Follows natural logic of problem solving, by enumerating impacts through attribute interactions within the effects table
Has flexible data requirements, incorporating quantitative and qualitative data
Makes specific provision for different points of view through the explicit use of weights to balance priorities
Through sensitivity analysis, enables exploration of alternatives and robustness of final solutions
Identifies where further data are required, where it will make little difference
Avoids need for single metric for decision trade-offs
Specifically identifies trade-offs that might otherwise be implicit
Can offer decision support through various scenarios, or single solutions
Disadvantages
Still compares what may be considered incommensurable properties
Offers many alternative analysis approaches without a clearly superior one
Complexity of analysis process can lead to mistrust or excessive faith in the results
Explicit weights can be falsely objective if other weights are still implicitly present (e.g. through the use of inappropriate indicators or absence of appropriate ones)
The effects table and appropriate weights require very considerable effort to develop accurately

of sophisticated programming will compensate for the omission of significant criteria. In the final analysis, it should also be remembered that like all programming applications, the results of an MCA still depend ultimately on the value judgements of those involved in the analysis. Good applications are simply procedures that enable those underlying preferences to be reflected most effectively in the final decision, in a way that is as inclusive and transparent as possible.

In conclusion, it is worth reiterating the comments in the opening of this section. To speak of a 'multicriteria approach' is to assume a level of similarity between techniques. While this assumption may be justified at a very general level, the field is more characterized by diversity than uniformity. It is therefore a field where the non-specialist is likely to continue to feel at best somewhat bewildered by what is on offer. That said, there is little doubt that MCAs will have an increasing role in natural resource decision-making in the future.

10.7 Participatory rural appraisal

The field of participatory rural appraisal (PRA) is a good example of an alternative planning and project appraisal paradigm to formal MCA methods. Rather than referring planning analysis to specialists and complex models, PRA centres its approach on the capacities and knowledge of ordinary people involved in the planning process. It is not really accurate to consider PRA as a directly substitutable process for a more formal MCA, because it is generally applied to very different kinds of problems; for example, local level or community-based agricultural developments. However, it is interesting to consider it here as it presents a very different outlook on the planning process.

In keeping with the fluid adaptive nature of participatory planning methodologies, in a review of the origins of PRA Chambers (1994a) notes that 'the approaches and methods described as Participatory Rural Appraisal (PRA) are evolving so fast that to propose one secure and final definition would be unhelpful'. It is within the philosophy of PRA as an appraisal methodology that it remains open to change, that rigid guidelines are avoided and that more emphasis is placed on experience, experimentation and action to discover what works, rather than on refining or justifying theory in an academic context.

10.7.1 Basic philosophy

Within the debate on planning sustainable development of agricultural systems, Pretty (1994) identifies five 'deviations' inherent in this context which separates them from conventional reductionistic planning approaches.

1 The ultimate end is unknown, therefore developing adaptability is more important than designing precise solutions.

2 All frameworks of understanding are relative, therefore there is no monopoly on designing correct solutions.

3 Uncertainty and problem-solving is endemic; that is, it is a continual process rather than an isolated event.

4 The key feature to successful development is therefore the capacity of individuals to manage change.
5 Systems of learning therefore need to be multiperspective, involving all sides in the construction of plans and processes.

The underlying philosophy of the PRA approach is perhaps best understood in contrast with the appraisal methodologies it has sought to replace. Conventional planning and appraisal methodology has been anchored in a reductionist scientific tradition, which is characterized according to several key features: objective, predominantly quantitative, and disengaged from the problems themselves. The orientation of PRA is, by contrast, towards empowerment of the individuals and communities involved in the appraisal, and their capacity-building in terms of their own abilities to analyse, design and implement their own solutions to local problems.

10.7.2 Participation and empowerment

The focus on participation within PRA has antecedents in the tradition of 'action research' exemplified in the Brazilian educationalist Paulo Freire's book, *Pedagogy of the Oppressed* (1968). 'Action research' is a style of social research that becomes actively involved in trying to address the problems it studies (Max-Neef's approach to the analysis of needs and wants is a good example; see Section 5.4.1). Part of Freire's work on adult education with the excluded sectors of Brazilian society emphasized the need for an educational approach that stimulated people's understanding of their own situation within society, through a process of 'conscientization'. Social development could only come about, in Freire's view, if disenfranchised people were allowed and encouraged to develop their own analytic abilities in relation to their place in the social system and their understanding of that system. Once this self-awareness and analytic capacity was developed, it became possible for those formerly oppressed by traditional power structures to work to change them, rather than passively accepting them as immutable.

Conventional education, with its emphasis on narrow disciplines, theoretical insights and tightly defined areas of subject matter was from the 'conscientization' perspective merely reinforcing the divisions between those with power and knowledge and those without, because it was defining knowledge and skills in relation to these academic categories. An educational elite thereby handed down its superior knowledge to inferior students, establishing a framework in which the personal skills and experience of students were by definition irrelevant or valueless. The Freirian 'pedagogy' (meaning 'teaching method') suggested instead replacing this 'containerization' view of education, in which students 'fill up their stocks of knowledge', with one of 'empowerment', in which students actively develop their own analytic skills in relation to issues and experiences that are of immediate social relevance and concern.

The role of the educator within this process thereby shifts from that of teacher to that of facilitator, whereby the relationship of superior–inferior is replaced by one of equality. Facilitators merely assist others in the process of constructing their particular vision of reality, rather than handing down a ready-made model of that reality developed by the teacher. All models of reality are seen as subjective interpretations, therefore no single model has greater validity than any other. The teacher thereby no longer has a particularly privileged position in the educational process resulting from possession of a more 'valid' or knowledgeable model; all models are equally valid. The facilitator is important only in so far as he or she has greater awareness of different possible models of understanding, and can help others to appreciate these alternatives from a position of greater experience. Once the 'conscientization' process is underway, however, students educate themselves by exploring social themes amongst themselves in relation to their own situations.

The planning process within participatory methodologies involves a similar principle to that described by Freire, although it is much less formal than his educational system and is orientated around a planning process. Where the two coincide is in the emphasis on facilitation rather than control; communities involved in PRA essentially take control of the assessment procedure themselves, rather than responding to a predetermined process designed by outsiders. In this regard Pretty (1995) isolates seven types of participation (Table 10.5), ranging from 'manipulative participation' to 'self-mobilization'. These types indicate the progression from the limitations, sometimes welcomed by planners, of more conventional notions of participation to the genuine empowerment of affected communities who come to manage the direction of development for themselves.

10.7.3 Tools and techniques of participatory rural appraisal

The methods and techniques of PRA focus on the capacities of local people to deal with these issues. In stressing

Table 10.5 Seven types of public participation. (Source: Pretty 1995, with permission.)

Typology	Characteristics of each type
1. Manipulative participation	Participation is simply a pretence, with 'people's' representatives on official boards but who are unelected and have no power
2. Passive participation	People participate by being told what has been decided or has already happened. It involves unilateral announcements by an administration or project management without listening to people's responses. The information being shared belongs only to external professionals
3. Participation by consultation	People participate by being consulted or by answering questions. External agents define problems and information gathering processes, and so control analysis. Such a consultative process does not concede any share in decision-making, and professionals are under no obligation to take on board people's views
4. Participation for material incentives	People participate by contributing resources, for example, labour, in return for food, cash or other material incentives. Farmers may provide the fields and labour, but are involved in neither experimentation nor the process of learning. It is very common to see this called participation, yet people have no stake in prolonging technologies or practices when the incentives end
5. Functional participation	Participation seen by external agencies as a means to achieve project goals, especially reduced costs. People may participate by forming groups to meet predetermined objectives related to the project. Such involvement may be interactive and involve shared decision-making, but tends to arise only after major decisions have already been made by external agents. At worst, local people may still only be coopted to serve external goals
6. Interactive participation	People participate in joint analysis, development of action plans and formation or strengthening of local institutions. Participation is seen as a right, not just the means to achieve project goals. The process involves interdisciplinary methodologies that seek multiple perspectives and make use of systemic and structured learning processes. As groups take control over local decisions and determine how available resources are used, so they have a stake in maintaining structures or practices
7. Self-mobilization	People participate by taking initiatives independently of external institutions to change systems. They develop contacts with external institutions for resources and technical advice they need, but retain control over how resources are used. Self-mobilization can spread if governments and non-governmental organizations provide an enabling framework of support. Such self-initiated mobilization may or may not challenge existing distributions of wealth and power

the underlying emphases of participatory planning approaches, Chambers (1994b) identifies several key principles. These can be summarized into four areas.

First, PRA attempts to reverse the learning process, so that outsiders/extension workers (those involved in 'extending' knowledge to local people) come to learn from local people. In the traditional planning model, outsiders analyse problems according to their own perceptions and merely 'extract' predetermined sets of information from local people. In contrast, PRA involves 'handing over the stick': allowing and encouraging the assessed to become the assessors, thereby planning and implementing their own analysis and subsequent actions.

Secondly, PRA emphasizes optimizing time and cost trade-offs in the research and planning process. This involves appreciating the concept of 'optimal ignorance'; that is, recognizing what is *not* worth knowing, and thereby balancing the costs of obtaining information with the usefulness of that information. The great economist Keynes endorsed a similar approach by being satisfied with 'appropriate imprecision' (as he put it: 'it is better to be approximately right than precisely wrong'). Participatory rural appraisal techniques are therefore rapid, flexible and encourage improvization to make the most of the opportunities inherent in any given situation.

Thirdly, the research orientation seeks to emphasize and recognize diversity, individual contexts and individuality, rather than aiming to develop general or

Box 10.1 Participatory rural appraisal techniques in development appraisal

A substantial number of techniques and methods have been developed for practising PRA. These tools and techniques have grown organically out of the field experience of researchers and rural extension workers, and there is an emphasis on the open sharing of new and developed techniques without concerns over attributing authorship. The following is only a small selection of some of the more widely used techniques.

Participatory mapping and modelling
People use the ground, the floor, paper or other readily available materials to make maps of social arrangements, demographic and health distributions, local natural resources, farm arrangements. The map is a public artefact, open to continual correction and verification by all members of the group.

Transect walks
A group walks with a researcher through an area, identifying different zones, soils, land uses, vegetation, new and traditional crop technologies and problem areas. Walks may follow slopes, comb through an area or loop around important sites.

Time lines and change and trend analysis
Individuals or groups build up chronologies of important local events, ecological histories, changes in land use and cropping patterns, trends in migration, education, health, population and customs.

Oral histories and ethnobiographies
People give oral accounts of local history and may trace the histories of single objects, such as a tree, a crop, an animal, a weed or a pest.

Venn or chapati diagramming
Groups identify the important individuals in local communities, the roles they fulfil and the important institutions within the community and their relationships to them.

Matrix scoring and ranking
Soils, crops and irrigation needs are analysed through scoring (using seeds, pebbles or other local media) across different characteristics on an open matrix constructed by the group. The group develops its own scoring system, important characteristics, etc.

Well-being and wealth grouping and ranking
Groups identify levels of well-being in the community, particularly noting the poorest members, but by ranking rather than in absolute terms. Key indicators of wealth or well-being may also be identified.

Team contracts
The research team agrees particular modes of behaviour with the local residents, the general manner of analysis, etc.

universal plans or solutions. It relies on the principle of triangulation—the constant cross-checking of impressions and assumptions through different techniques and viewpoints—to generate reliability and accuracy in any given planning or assessment exercise. By emphasizing diversity it also tries to offset analytical biases, such as inadvertently excluding groups such as the illiterate, the very poor, the old or sick, who in some situations may find it difficult to get their perspectives and needs recognized. The technique thus seeks new frames of reference, exploring unexpected and unusual avenues of inquiry.

Finally, there is a strong emphasis on self-awareness and personal responsibility on the part of PRA facilitators; being self-critical, and ensuring that facilitators constantly assess their own role and appreciate their mistakes as opportunities to learn ('failing forwards'), especially the tendency to dominate the process. Given the flexibility of the process, there is also a need for facilitators to take personal responsibility for the conduct of assessments and rely on their own judgements rather than following prespecified particular methods or guidelines. There is also a generalized culture of sharing of ideas and experiences (between and amongst both local people and outsiders), with the emphasis on improving and developing the effectiveness of planning rather than exclusively guarding or defending particular areas of expertise.

These broad characteristics of participatory planning approaches enable sensitivity to local contexts in the development process. Some of the techniques developed within PRA are listed in Box 10.1.

Chambers (1994b) identifies four important findings

Table 10.6 Importance of visual mode of expression in participatory rural appraisal. (Source: Chambers 1994b, with permission.)

	Verbal (interview, conversation)	Visual (diagram)
Outsider's roles	Investigator	Initiator and catalyst
Outsider's mode	Probing	Facilitating
Outsider's interventions	Continuous and maintained	Initial and then reduced
Insider's roles	Respondent	Presenter and analyst
Insider's mode	Reactive	Creative
Insider's awareness of outsider	High	Low
Eye contact	High	Low
The medium and material are those of	Outsider	Insider
The poorer, weaker, and women can be	Marginalized	Empowered
Detail influenced by	Etic categories	Emic perceptions
Information flow	Sequential	Cumulative
Accessibility of information to others	Low and transient	High and semi-permanent
Initiative for checking lies with	Outsider	Insider
Utility for spatial, temporal and causal analysis, planning and monitoring	Low	Higher
Ownership of information	Appropriated by outsider	Owned and shared by insider

from the operation of PRA in the field which reflect the particular strengths of the approach in relation to more formal assessment procedures.

1 Local people's capabilities are often far greater than outsiders are prepared to recognize. Chambers notes that 'local people have shown themselves capable of generating and analysing information far beyond normal professional expectations'.

2 The behaviour/attitude of facilitators and the rapport they develop with local people is critical to the success of participation. Rapport can be developed through simple actions, such as trying to learn basic tasks from local people and participating in their activities.

3 Sequences of techniques can be built up to provide richer information. For example, successive mapping exercises build up increasing levels of detail; preliminary mapping leads into transect walks which lead to identification of key problems, which in turn are analysed through matrices or ranking exercises.

4 The popularity of visually shared techniques, such as maps, diagrams and charts, in assessment is seemingly universal. Whereas questionnaires are seen to appropriate knowledge into the format and structure of the interviewer, diagrams developed by local people are owned and orientated by them.

The significance of visual communication is particularly emphasized. Chambers (1994b) identifies several aspects of this frame or mode of expression which makes it a particularly powerful tool in a participatory setting (Table 10.6).

SOFT AND HARD APPROACHES TO VALIDITY

In an impact assessment or planning context, PRA can offer advantages over more rigid or formal methods of analysis, most evidently in economy of time and effort. There remains a debate over the validity of data gathered through these techniques, however. Gill (1991) ironically entitled a review paper 'But how does it compare with the real data?' in commenting on rainfall patterns discovered in discussions with local farmers through PRA techniques when compared with meteorological station records. His conclusion was that local reports were more accurate of the situation on the ground, as the station reports from several kilometres away were averaged and not sensitive enough to take account of microclimatic differences in rainfall patterns.

Uphoff (1992) notes:

> 'Unfortunately, there appears to be an inverse relationship between rigour and relevance in most social science work. This may be because rigour always requires some reductionism, since certain aspects of phenomena are necessarily excluded by any classification and measurement. Moreover their changing nature tends to be ignored because taking this into account greatly complicates analysis.'

In reviewing past applications of PRA, Chambers (1994b) notes that PRA methods have frequently provided superior data to formally structured surveys, at lower costs and in much shorter time schedules.

He also identifies the openness and transparency of the appraisal process as a key strength of the PRA methodology. It can be impossible to check formal questionnaire-based surveys for enumerator bias after the event. However, the PRA process is constantly open to verification by both research teams and local people.

10.7.4 Conclusions on participatory planning

Although PRA is predominantly associated with local level development projects in the developing world, it has given rise to a multitude of related approaches that are constantly extending its boundaries. To date, PRA itself has been widely applied in four main areas: natural resource management; agricultural development; poverty and social programmes; and health and food security programmes. What these approaches share is an emerging consensus that new planning approaches must be flexible, adaptive, responsive to context, and that these features cannot be achieved without the full and active engagement of stakeholders within the planning and development process.

This emphasis on adaptive planning, building institutional resilience and flexibility, and a focus on context-specific and relativistic thinking rather than a search for universals has been identified as another example of a growing tendency towards *postmodernist* perspectives across disciplines. The postmodern approach stresses the importance of context, relative values, and the denial of absolutes not simply in value systems but also in analysis and in concrete knowledge. 'Reality' is, in fact, always perceived from some particular viewpoint or some particular frame of reference. The idea of an 'objective' reality which is separate from any individual's personal viewpoint, and uncovered through the scientific ideal of a 'dispassionate observer', is in this case simply an illusion.

Rosenau (1992) suggests that 'the absence of truth . . . yields intellectual humility and tolerance. [Postmodernists] see truth as personal and community-specific: although it may be relative, it is not arbitrary'. The PRA approach can be seen as in line with these general observations by its rejection of master plans, objective quantification and a reality that is separately determinable from people themselves. It suggests that it is only by turning the process of appraisal over to the people themselves that a model that is relevant to their needs can emerge.

10.8 Summary

Multicriteria methods model the decision-making situation by identifying a number of objectives and picking particular criteria through which to assess levels of objective achievement. Numerous methodologies have now been developed to help with this process.

An MCA analysis is structured, like an EIA, to work through the features of a project assessment in a methodical way: project definition; criteria selection; measurement; and analysis. Different techniques vary in the amount of information they require about the preferences of a decision-maker regarding levels of objective achievement. Outputs from the process may also vary, from identifying a single solution to providing several non-dominated alternatives. Goal programming and composite programming are two good examples of different styles of analysis, which vary in their input requirements and definition of 'best' solution.

In many local planning situations the hard-edged computer-based ethos of many MCA techniques is seen as inappropriate. The PRA approach stresses instead handing over the process of analysis to local people, and concentrating on capacity building for problem-solving at the local level. This style of problem analysis is increasingly identified with postmodern attitudes to science, which emphasize the importance of uncertainty, adaptation, relative values and context-dependent solutions. Highly mathematical and rigid planning methods may be ill-suited to these more uncertain environments.

Further reading

Bogetoft, P. & Pruzan, P. (1997) *Planning with Multiple Criteria*. Copenhagen Business School, Copenhagen.

Chambers, R. (1983) *Rural Development: Putting the Last First*. Longman, Harlow.

Pretty, J.N. (1995) *Regenerating Agriculture: Policies and Practice for Sustainability and Self-Reliance*. Earthscan, London.

Romero, C. & Rehman, T. (1988) *Multiple Criteria Analysis for Agricultural Decisions*. Elsevier, Amsterdam.

11 National Income Accounting

'Extrapolation reveals that human material demand now exceeds the long-term carrying capacity of Earth.' [Wackernagel & Rees]

11.1 Introduction

The national income accounts (NIAs) are conventionally how economists measure the state of the economy. Put simplistically, they represent a detailed breakdown of how much we consume; how much we spend; and how much we produce. The NIAs are a macroeconomic tool which can be used to analyse trends and make predictions about the performance of the economy. Further, they are often used as an indicator of the welfare of society.

Pure economics is concerned with the circular flow of scarce commodities between firms and consumers, as analysed in Chapter 3. The perspective of ecological economics is that this circular flow relies upon the ecosystem for the provision of resource inputs and the assimilation of waste outputs. Green NIAs attempt to accommodate this perspective.

In this chapter, in Section 11.2 we outline the mechanics of the conventional NIAs; the appropriateness of their use as a measure of social well-being is discussed in Section 11.3. We then turn in Section 11.4 to look at the application of supplements to the NIAs in the form of satellite accounts. Two fundamentally different alternatives to the neoclassical methodologies of the NIAs are presented in Sections 11.5 and 11.6: the index of sustainable economic welfare (ISEW); and an application of environmental space (ES) methodology by the environmental pressure group Friends of the Earth (FoE).

11.2 Conventional national income accounts

The reasons behind the current methodology used for the NIAs is more historical than systematic. Although the Commerce Department of the USA started reporting statistics on national output as early as 1934, it was the Second World War which stimulated the need for an accounting system, as politicians needed to know both how much military output was feasible and the sectoral effects of mobilizing resources for the war effort. With similar developments occurring in the UK and Canada during 1944, the League of Nations convened in 1945 to discuss NIAs methodologies. Although there were supplements and revisions to the system of NIAs in 1958 and 1965, the NIAs have been internationally standardized since 1947.

Economic production requires three factors of production, as discussed in Section 3.13: natural capital (NK); labour (L); and man-made capital (MK), or capital equipment. With reference to the NIAs, the circular flows of transactions involving these factors (referring back to Fig. 3.1) can be stated in three ways.

1 Households own the factors of production which they supply to firms, which in turn are used by firms to produce an *output* of goods and services.

2 Households receive *income* from firms in exchange for supplying these factors of production.

3 Households *spend* on the goods and services that the firms sell.

There are thus three methods of determining economic activity: the output approach; the income approach; and the expenditure approach. Theoretically, all three methods should arrive at the same 'answer' for national income (NI). In reality, there is some residual error and lots of caveats apply. For instance, what about firms' stocks and work in progress, firms selling to other firms as opposed to households, or households saving some proportion of incomes? These issues are resolved through a closer examination of the NIAs methodology.

11.2.1 Defining gross domestic product

We start by defining a benchmark statistic in the NIAs. *Gross domestic product* (GDP) measures the output produced by the factors of production located in the domestic economy, regardless of who owns these factors of production. Gross domestic product is widely quoted in the popular press, where it is considered synonymous to NI. In fact, NI is more precisely *net national product* (NNP), which is an adjusted GDP. We outline below the domestic to national adjustment, but we consider the GDP measure in detail first.

In terms of the NIAs methodology, it is perhaps useful to think in terms of one scenario as an example. Sony is a Japanese multinational which has set up a production plant in Wales producing electrical appliances, including televisions. The plant assembles *intermediate goods*; that is, partly finished goods which are inputs to another firm's (Sony's) production process, and are used up in that process. Thus Sony might source microprocessors, cathode ray tubes, electrical circuitry, loudspeakers, etc., either from subcontractors or other Sony divisions. What Sony sells is *final goods*, e.g. televisions. A final good is one purchased by the ultimate users—consumer goods purchased by households or manufacturing equipment (i.e. MK) purchased by firms.

It is important to differentiate between intermediate and final goods to avoid *double-counting*. This can be demonstrated using the scenario above. Assume that the firm that supplies the cathode ray tubes to be installed in Sony's televisions is paid £10 per unit. £10 is then the value of the transaction, and the *factor earnings* for the supplier is also £10, i.e. the sum payment to the factors of production in the vacuum tube manufacture. Assume that the television sells for £500; it would be incorrect for the NIAs to show factor earnings for Sony of £500 and for its supplier of £10, as the sum (£510) is greater than the expenditure on the final product (£500) because of double-counting. It is the *value added* by economic effort which needs to be recorded in the NIAs. The value added is the increase in the value of commodities as a result of the production process. If cathode ray tubes were the only intermediate good used by Sony, then the value added by Sony would be £490. The tube manufacturer's value added is £10, assuming that it does not source any intermediate goods from other firms. The sum of value added should be equal to both the sum of expenditure on final goods and the sum of factor earnings in the NIAs.

11.2.2 Savings and investment

Savings are simply that part of household income that is not spent on buying commodities, and investment is the purchase of new capital goods by firms. Savings are a *leakage* from the circular flow, in that the money does not directly come back to firms in the form of household consumption. Investments are an *injection* into the same circular flow in that they are spending decisions made by firms which are supplementary to the household spending on the firm's outputs. In accounting terms, the actual amount saved is exactly equal to the actual amount invested.

11.2.3 Inventories and work in progress

But what happens if Sony do not sell all the televisions produced in the accounting period? If the NIAs were not to allow for this, then the NI would be systematically different measured by the expenditure method as compared with the output method. Firms generally keep *inventories* of finished products and intermediate goods for future sale and production. Such inventories are treated as *working capital*, that is an inventory investment by the firm. In the NIAs, any increase in the stocks of televisions held by Sony in our scenario is classified as final expenditure by Sony. When the stocks are depleted, this is registered as negative investment.

11.2.4 The public sector

The public sector—both local and national government—raises revenues through taxation and spends it on behalf of society. For the purposes of the NIAs, the *transfer payments* element of public expenditure needs to be treated differently to other government spending. A transfer payment is a payment from the state that does not require the recipient to provide any good or service in return. Unemployment benefit, social security payments and state pensions are examples of transfer payments to households, and investment grants an example of transfer payments to firms. Neither transfer payments nor taxation should not be included in a measure of NI because no output is produced and no income generated. They merely *redistribute* income away from taxpayers to beneficiaries. Such payments are therefore taken out of the NIAs.

Other elements of public spending remain in the accounts. To justify this, consider state-funded education; expenditures include teachers' salaries, heating and electricity bills, books and materials. The provision of each of these goods and services implies the consumption of commodities. These therefore appear in the NIAs.

The other amendment that must be made to the NIAs as a result of the public sector is the *market price* to *factor price* adjustment. Gross domestic product at market prices is a measure of domestic output which includes any indirect taxes, such as value added tax or sales tax.

Gross domestic product at factor cost removes these taxes. If this adjustment were not made then an increase in indirect taxes would raise GDP (at market prices). This rise would not correspond with a rise in consumption levels and therefore GDP at factor cost measure is more appropriate. Gross domestic product at factor cost (GDP_{FC}) in a closed economy (without foreign trade, which we consider in the next section) is thus given by:

GDP_{FC} = consumption (C) + investment (I) + government spending (G) – indirect taxes net of subsidies (T_I).

If indirect taxes go up, then $(C + I + G)$ goes up but T_I falls by the same amount, keeping GDP_{FC} constant.

11.2.5 Foreign trade

We have thus far not considered foreign trade, and therefore have been considering NIAs methodology for a closed as opposed to an open economy. In order to include the foreign sector, we must differentiate between *imports* and *exports*. Exports are goods and services that are produced in the domestic economy but sold to foreign countries, whereas imports are produced in a foreign economy and sold in the domestic sector.

A useful rule of thumb for differentiating between imports and exports is to consider the currency exchange. Returning to our Sony scenario, Sony might import parts for producing televisions from the USA. Ultimately, the payment is in dollars, even if Sony pays a dollar equivalent in pounds sterling. This payment therefore represents an import. If Welsh employees of Sony were to go on holiday to the USA, they would pay their hotel bills and expenses in dollars and therefore this, too, would constitute an import, even though the money was spent abroad.

Gross domestic product relates only to the value added by domestic producers and so the NIAs must be adjusted to allow for the foreign sector. This is done by measuring total final expenditure on $(C + I + G + X)$, where X is exports, and subtracting imports (Z). Thus GDP at factor cost in an open economy is given by:

$$GDP_{FC} = C + I + G + X - Z - T_I.$$

11.2.6 Conversion from gross domestic product to gross national product

In the definition of GDP above, we stated that GDP does not depend on who owns the factors of production. This distinguishes GDP from *gross national product* (GNP). Gross national product measures total income earned by domestic citizens regardless of where the factor services were supplied.

To illustrate the distinction, consider again the Sony scenario. Assume that the multinational is owned by Japanese shareholders. Profits earned are then declared outside the UK, whereas wages are earned by the plant's UK workforce. The aggregate income earned in the UK is thus different from the value of the plant's aggregate output (which includes profits). However, as the output was produced in the UK, it constitutes part of UK GDP. In this case, profits are what is called a negative *property income from abroad*, as would be payments to foreign workers. Gross national product equals GDP plus net property income from abroad. This property income comes from interest, dividends, profits and rents from factor services supplied outside the domestic economy. Net property income is positive (and therefore GNP is bigger than GDP) if the inflows (from UK firms and workers abroad) exceed the outflows.

11.2.7 Conversion from gross national product to net national product

Net national product (NNP) is, by definition, the economy's NI, as opposed to GDP or any other measure. Net national product is simply GNP with the conversion from gross to net, which means subtracting *depreciation* from the GNP at factor cost. Depreciation is the rate of wear and tear, or obsolescence, on capital equipment; that is, the fall in value over time as equipment wears out. As it is a rate of capital consumption, it is a flow concept. It is an economic cost of producing output. Because all equipment has an expected working life that is finite, the depreciation adjustment corrects for this by factoring in a decline in the value of capital stocks each year.

11.2.8 Summary of adjustments

It is perhaps worth backtracking to check the various methodological stages that get us to NI, before we consider the use of the NIAs for welfare comparisons in the next section. A summary of the process is given in Fig. 11.1. We started with the simple circular flow between households and firms. In principle, the NI can be measured by measuring total output, total spending and total factor incomes. To avoid double-counting, only the value added appears in the output approach, and only the final goods in the expenditure approach.

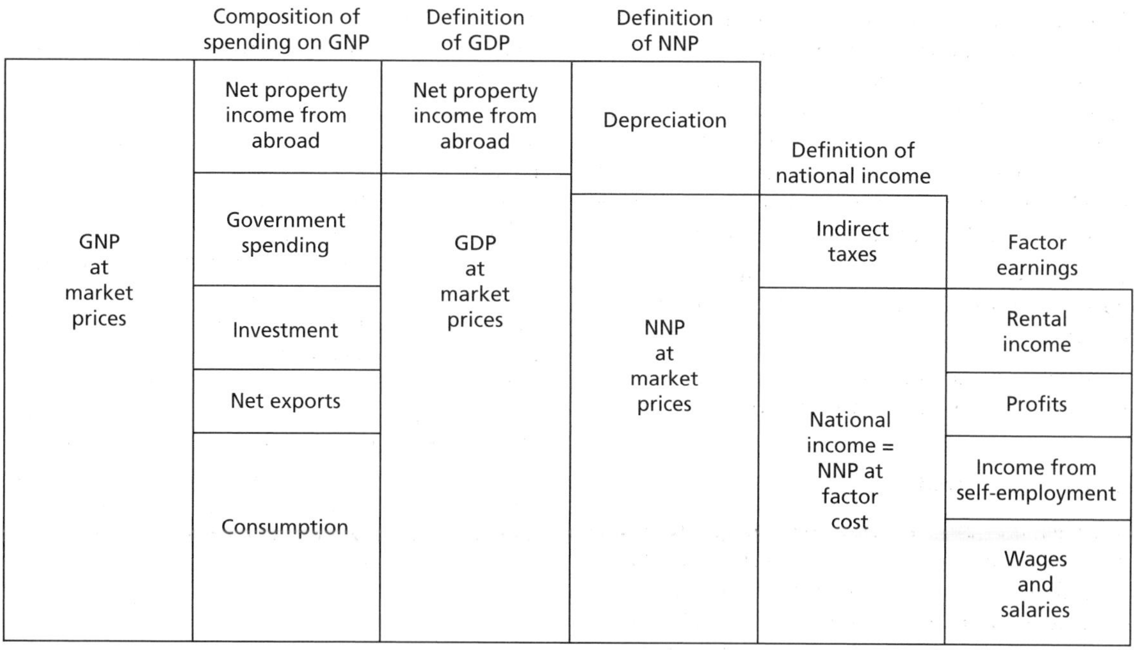

GNP at market prices

Composition of spending on GNP: Net property income from abroad; Government spending; Investment; Net exports; Consumption

Definition of GDP: Net property income from abroad; GDP at market prices

Definition of NNP: Depreciation; NNP at market prices

Definition of national income: Indirect taxes; National income = NNP at factor cost

Factor earnings: Rental income; Profits; Income from self-employment; Wages and salaries

Fig. 11.1 A summary of the national income accounting process. (Adapted from Begg *et al.* 1998, with permission of McGraw-Hill Publishing Company.)

We then dealt with various leakages and injections into this circular flow: investment and savings; inventories and work in progress. In considering the public sector, we found that transfer payments and direct taxes had to be factored out of the accounts as these only redistribute, as opposed to generate, income. The removal of indirect taxes is the conversion from GDP at market prices to GDP at factor cost. We then introduced international trade: exports are added to the NIAs and imports subtracted. This gives us GDP at factor cost for an open economy. The conversion to GNP requires the addition of net property income from abroad, and the final conversion in the NIAs to national income (NNP) requires the subtraction of depreciation from GNP.

The NIAs are a vital statistical tool for economists in general and macroeconomists in particular. They assist in general macroeconomic management issues, such as inflation, unemployment, interest rates and exchange rates, which lie beyond the scope of this book. However, apart from its use in macroeconomic policy appraisal, NI is often used as a measure of social welfare, by which valuations of the success of economic performance are made. This is the subject of the next section.

11.3 Social welfare and the national income accounts

The NIAs were never intended to measure either sustainable development or social welfare. Despite this, NIAs are used by development organizations, such as the World Bank, to measure a nation's wealth, and this can have repercussions in terms of the targeting of aid and investment. Such organizations are often aware of the pitfalls of using NIAs but are forced to rely on them owing to the lack of availability of other statistics and qualitative indicators.

The problems of using NIAs as a measure of welfare fall into two categories: those which can be addressed by amending the accounts, and more fundamental systematic ones. To give a flavour of the issues, consider how a road accident would be included in the NIAs. Most people would agree that a road accident is a bad thing, and that an increase in the number of accidents would decrease social welfare. However, in the conventional NIAs, a road accident would appear as an increase in NI. Economic activity is generated through the hospitalization of the injured parties, car repairs or the purchase of replacement cars and insurance administration. If the NIAs are treated as a welfare indicator without amendment or caution for cases such as these, then they can give the wrong signals. Some of the problems, and potential solutions, associated with using NIAs as a measure of welfare are considered below.

11.3.1 Conversion from nominal to real statistics

If an economy produces the same quantity one year as the next, but the price of these commodities increases by 10%, then the unadjusted (nominal) NIAs would show a 10% increase in GNP. However, aggregate consumption and production have not changed, so the nominal figure is misleading as an indicator of economic activity. Real GNP is deflated nominal GNP, a relatively straightforward adjustment to make.

11.3.2 Per capita adjustment

National income changes per capita are often more relevant than total changes. In theory, this is a simple calculation: divide the NI by the population size. However, population statistics are not readily available and/or inaccurate in many countries. To illustrate why it is important to make this per capita adjustment, consider two countries: Luxembourg and India. India's NI in aggregate might be higher than Luxembourg's, but the per capita NI significantly lower. Further, assume that Luxembourg has a real NI growth rate of 2% for one given year whereas the rate for India is 3%. India's economy is growing faster. However, if the population growth rate for that year is 1% in Luxembourg and 4% in India then the consumption level of the average citizen is rising in Luxembourg and falling in India.

11.3.3 Income differentials

Average changes in consumption do not imply that the change in every citizen's consumption is identical. A statistical indicator of the distribution of income in the economy—known as the Gini coefficient—can be used alongside the NIAs.

11.3.4 Quality of goods

The NIAs measure the quantity of goods consumed, which might bear no relation to the quality of the commodities consumed. Take computers as an example; if the real (inflation-adjusted) price of a 1980 computer were to be spent on a computer today then today's computer would be of a vastly higher quality. This would not register in the NIAs. Perhaps two more environmentally pertinent related issues are durability and reuse. If durability increases or commodities are reused, then less replacements are purchased, reducing NNP. If the NIAs are used as a welfare measure, then such environmentally friendly changes are registered as welfare reducing, simply because less material consumption is taking place.

11.3.5 'Black economy'

Certain outputs in the economy are not reported in the NIAs but do contribute to output. The 'black economy' is a term used for such non-reported economic activity, which generally is unreported in order to avoid taxation. Thus, 'cash-in-hand' payments to employees evade income tax and National Insurance contributions, and such payments to contractors evade value added tax. Such tax evasion is illegal. However, the fact that the contractors have carried out the work and the employees spent their incomes means that NI has actually risen—but this does not register in the NIAs.

11.3.6 Non-traded output

There are legitimate productive activities that also do not register in the NIAs. Subsistence agriculture contributes significantly to welfare in many developing countries. Indeed, it is often critical for survival. However, it does not appear in the NIAs as the output is not sold in the market-place. In a similar vein, assume that one adult in the household was previously responsible for cooking, cleaning and child-rearing. If a professional cook, cleaner and nanny are now hired, then these activities are marketed and they appear in the NIAs as output, thereby signalling an increase in NI. But there has been no change in aggregate output.

11.3.7 Exchange rate

For international comparisons of NIAs, an exchange rate must be chosen. If there is a *floating* exchange rate, then the demand and supply interactions determine the price of the currency. Many nations operate a *fixed* exchange rate system wherein the government sets a level which may or may not coincide with the market price. There are pros and cons associated with both systems. International comparisons of NI are more realistic with the floating exchange rate system. Assume that the exchange rate for the Indian rupee is fixed against the US dollar at twice the market clearing rate:

Indian NI converted to dollars would be twice the true level. If the 'true' market rate is known, then the NIAs comparison can be adjusted, but with a fixed rate system this can only be an approximation.

BASKET OF COMMODITIES

Diverse social and cultural norms and varying local conditions mean that different nations have different consumption patterns which are not figured into the NIAs. As an example, take fuel consumption. Does the fact that Canada consumes significantly more per capita on fuel for heating homes than Portugal mean that welfare is higher as a consequence? It could be argued that the expenditure on maintaining a pleasant indoor temperature in Canada simply need not be spent in Portugal. An international NIAs comparison between the two nations would not then represent respective welfare levels accurately.

DEFENSIVE EXPENDITURES

There are certain expenditures which, arguably, do not actually add to welfare. For instance, consider the purchase of antipollution masks worn by cyclists. If it were not for the increase in urban smog, resulting from industrialization, this expenditure would be unnecessary. Another example would be the installation of double glazing as a result of increasing noise pollution from the construction of a new airport runway; this defensive expenditure appears as an increase in NI, but only results in a return to that welfare level enjoyed pre-construction. The other side of this argument is that because individuals are not being forced into buying commodities such as antipollution masks, then they must be gaining some utility from consumption.

Despite the problems discussed above, the NIAs are widely used in economic analysis. There have been various attempts to modify the accounts, as well as the development of completely independent measures of social welfare. Some countries have generated satellite environmental accounts which are supplementary to the conventional NIAs; this is the topic of the next section, which considers the UK application.

11.4 Satellite environmental accounts

The function of satellite accounts is to provide a systematic and industrially disaggregated analysis of the impact of economic activities on the environment without interfering with the standard accounting procedure. Unlike the main NIAs, the procedure for satellite accounting is not standardized, although the United Nations (UN 1993) has developed a framework called the System for Integrated Environmental and Economic Planning. Several countries have prepared environmental accounts, including France, Norway, Canada, the Netherlands and India.

Here we focus on the UK satellite accounts. These are composed of:

1 a measure of the depletion of oil and gas reserves, converted to a monetary value;

2 atmospheric emissions by both UK households and industry; and

3 expenditure on environmental protection by certain designated industries.

The accounts are collected and administered by the UK Office for National Statistics (ONS) (Bryant & Cook 1992; Vaze 1996). They include both monetary and physical terms. As atmospheric emissions are, in general, non-marketed, the ONS contend that it would be methodologically unsound to attribute them a precise monetary valuation, and so the UK satellite accounts do not do so. Oil and gas, however, are marketed commodities and therefore a monetary figure is inputted to the satellite accounts. The three subsections are considered in turn below.

11.4.1 An adjusted measure of the 'real' value of revenues from oil and gas reserves

The depletion of oil and gas reserves is registered in the conventional NIAs simply as national 'income', but this is imprecise. This imprecision arises from the fact that there is a finite stock of oil and gas available. Oil and gas are non-renewable assets: we might locate more reserves, but the total amount of the fossil fuels on the planet is strictly limited. If we deplete these fossil fuels now, future generations cannot do so. In fact, the financial income from running down such stocks should not count as true income in the economic sense, as defined by Sir John Hicks: 'a man's income [is] the maximum value which he can consume during a week and still expect to be as well off at the end of the week as he was at the beginning'. Once some part of the stock of oil or gas is consumed, it will not regenerate in an economically relevant time-span, and therefore should not be treated as income in Hick's sense.

However, such consumption does, to the present day, register as income in the main NIAs. There are two principal reasons for the maintenance of this anomaly. First, historical precedence: any change would decrease the tractability and consistency of the NIAs. Secondly, although such consumption is not pure Hicksian income *per se*, its exclusion from the NIAs would lead to nonsensical statistics; for instance, Saudi Arabia might then be categorized as a 'poor' country.

Three methods can be used for valuing the changes in fossil fuel reserves and therefore correcting for the fact that asset consumption appears as income in the main NIAs (Vaze 1996): user cost; net price; and present value. Each is considered below.

USER COST METHOD

The user cost method (UCM) was developed by El Serafy. In order to calculate the proportion of the total income stream (R) from oil and gas extraction that is true income (X) using the UCM, the life expectancy of the reserves (n years) and the social discount rate (r)[1] must be estimated. If R and r are constants, then the following formula applies:

$$X/R = 1 - 1/(1 + r)^{n+1}.$$

If the social discount rate is 6% and the life expectancy of the oil reserves 50 years, then:

$$X/R = 1 - 1/(1.06)^{51} \approx 0.95.$$

If the social discount rate is 3% with the same life expectancy, then the ratio goes down from about 0.95 to about 0.78; with the original r of 0.06 and a life expectancy of 25 years, the ratio falls from about 0.95 to about 0.77. What the figure of 0.95 means is that 95% of the income from depletion can be treated as Hicksian income, and the other 5% is asset depletion and therefore not 'true' income. The 'true' or real income arises because of the investment that follows from the proceeds of oil and gas sales; this investment will contribute over time to providing other sources of income, such as, for example, the development of renewable energy resources. In general, the greater the life expectancy of reserves and the higher the social discount rate, then the higher is the proportion of true income in the total income receipts from oil and gas depletion.

[1] For a discussion on discounting see Sections 8.3.4 and 8.4.

NET PRICE METHOD

Robert Repetto at the World Resources Institute pioneered the second valuation methodology: the net price method (NPM). The calculation of the value of oil reserves according to NPM is *unit rent* multiplied by the change in the volume of proven reserves. The unit rent can be, simplistically, thought of as the profit per tonne of oil or per cubic metre of gas. More precisely, it is calculated as the revenue to the oil and gas industries from sales minus operating costs and minus the industry's previous investment in exploration and capital equipment expenditures.

Unlike the UCM, the social discount rate and the life expectancy of reserves do not affect the results. Any quantity of oil or gas reserves consumed are costed at full unit rental. The fact that the discount rate in no way factors into the NPM implies that a given quantity of the reserve extracted today is equivalent to that quantity being extracted some time in the future. But the revenue stream generated from the extraction today might be invested to accrue a positive rate of return. Indeed, this is one justification for a positive social discount rate. The NPM can thus be criticized for not taking into account when the stock is depleted.

PRESENT VALUE METHOD

If the volume of proven reserves rises in a given accounting period, that is new discoveries exceed extraction, Repetto's NPM proposes that a net positive contribution to NI appear in the satellite accounts. The NPM gives the same results as the present value method (PVM) if new discoveries are *not* treated as income.

The PVM does not simply impute an increase in reserves as income; instead it calculates the present value of the stream of income expected from reserves. Thus present value goes up with new discoveries. Thus expected real (inflation adjusted) price changes are taken into account in the balance sheet.

SOME RESULTS

The following estimates produced by the ONS are based on 1993 figures.

- Approximately 100 million tonnes of oil and 66 million cubic metres of gas were extracted.
- At these 1993 extraction rates, the reserves would last for about 40 years.

• 134 million tonnes of oil and 143 million cubic metres of gas were discovered, implying a gain in known reserves (net of consumption) for both oil and gas.
• The UK economy benefited by £2.2 billion from exploiting the fossil fuel resources. This figure is net of normal rates of wages and profits.

11.4.2 Atmospheric emissions by both UK households and industry

We now turn to consider environmental emissions in the UK satellite accounts. Atmospheric pollutants are split into three environmental themes:
1 Depletion of ozone layer.
2 Greenhouse effect (climate change).
3 Acid rain.
The list of atmospheric pollutants that are monitored is not comprehensive. Certain key pollutants have been selected. These are: ammonia; benzene; black smoke (from incomplete fossil fuel combustion); chlorofluorocarbons (CFCs); hydrochlorofluorocarbons (HCFCs); hydrofluorocarbons (HFCs); perfluorocarbons (PFCs); sulphur hexafluoride (SF_6); methane (CH_4); carbon dioxide (CO_2); carbon monoxide (CO); lead; non-methane volatile organic compounds (NMVOC); oxides of nitrogen (NO_x); nitrous oxide (N_2O); and sulphur dioxide (SO_2). Some of these pollutants impact on more than one theme; for example, CFCs contribute to both themes 1 and 2.

One benefit of this categorization is that there is reasonable awareness of the three theme issues. Further, it is simpler for those inspecting the satellite accounts to deal with three themes rather than 16 pollutants. However, methodological problems arise in deriving outcomes by themes. The impact of a particular emission of a pollutant depends on such variables as climatic conditions, susceptibility of local ecosystems and local demographics. The ONS has had to apply a weighting system to the pollutants and their combined impacts. As an example, consider theme 2 (greenhouse effect); the weighting of pollutants is referenced to one unit quantity (one megatonne) of CO_2. One megatonne of CH_4 has a relative impact of 21, and N_2O a relative impact of 310. These weightings are those applied by the Intergovernmental Panel on Climate Control (IPCC).

One issue that the ONS has had to resolve is that of pollution transfer to other countries. Consider the following scenario; a firm manufacturing microwave ovens switches its production plant from the UK to the Far East, as production costs are cheaper. In the satellite accounts this appears as an increase in environmental quality in the UK because all the pollution from production no longer occurs in the UK. But the problems—ozone depletion, greenhouse gas emissions, acid rain—have simply been exported to another country and have not been ameliorated. For CO_2 emissions, this effect is known as carbon leakage.

Moreover, if UK consumers continue to purchase the firm's ovens, then the UK global environmental impact is probably at least as large as before, but the UK satellite accounts show an improvement. On the other hand, commodities exported from the UK have the environmental impact of production attributed to the UK satellite accounts, even though they are not consumed in the UK. The ONS applies this methodology as it is consistent with standard accounting procedures. Many firms relocate not only because of reduced labour costs but also to take advantage of less stringent environmental legislation in the host nation. If this applies, then the impact from consumption would probably rise through relocation. Any additional transportation costs would also increase environmental impact.

The following results are taken from Vaze and Balchin (1998). The ONS disaggregate emissions by sectors. There are two distinct ways of considering environmental impact by sector: impact in absolute terms, and relative to the sector's contribution to aggregate NI. The ONS have produced summary outcomes based on 1993 estimates.
• 'Energy production and water' as a sector produced the highest pollution output in relation to contribution to NI. In absolute terms, the sector's share is 26% of greenhouse gas and 44% of acid rain emissions, whereas the sector constitutes only 3% of NI.
• 'Agriculture' ranks second in relative terms. However, in absolute terms the sector is not a major polluter: 4% of greenhouse gas and 10% of acid rain, with agriculture constituting 2% of NI.
• 'Education and health', 'distribution', 'construction' and 'finance' have a low relative impact, given that their aggregate contribution to NI is about 60%.

These results should be treated with caution. Just because 'energy production and water' has a high emissions impact does not imply that this sector has the greatest scope for improvements; that is, the marginal costs of pollution abatement might not be the lowest for this sector. Furthermore, some low-emission industries might purchase essential inputs from high-emission ones, the purchase of energy being a case in point.

11.4.3 Expenditure on environmental protection by certain designated industries

The third category of the ONS inquiry is expenditure on environmental protection by certain designated industries: extraction, manufacturing, energy production and water industries. Neither the environmental expenditure of other industries nor that of households is calculated by the ONS. The environmental expenditures being monitored are capital and operating costs that have been incurred by firms in pursuing an environmental objective. Capital costs include 'end-of-pipe' treatments of emission outputs and 'cleaner technologies' which deal with (or avoid) emissions earlier in the production process. Operating costs include management time spent to install and monitor environmental technologies, and payments to outside contractors for the purchase of environmental services, e.g. contracting waste management consultants.

Some summary results for 1994 were as follows.

- For the designated sectors, the aggregate industry environmental expenditure was £2340 million. This is approximately 0.5% of the aggregate gross output of these sectors (£416 437 million).
- Water protection constituted about 43% of this spending, air pollution control 28% and waste management 22%.
- 'Chemicals and man-made fibres' was the designated sector that spent the most on protection (£503 million), followed by 'pulp, paper and products, printing and publishing' (£367 million), and then 'food, beverages and tobacco' (£327 million). These three sectors alone constituted just over half of the £2340 million expenditure.

There are certain environmental expenditures by firms in the designated sectors which do not appear in the satellite accounts: if it is cheaper for a firm to cut back production rather than incur the environmental costs necessary for compliance then the firm's output in the NIAs decreases, but no environmental cost is registered in the satellite accounts.

11.4.4 Future developments in satellite accounting

Finally, it is worth noting some possible extensions to the UK accounts proposed by Vaze and Balchin (1998).

- The inclusion of supplementary new environmental themes (inland water and ground water; coastal water/marine; waste) with an associated expansion in the number of monitored emissions.
- Statistics on reserves and depletions of natural resources other than oil and gas (coal; timber; fish; water; soil).
- A new theme dealing with land (amenity values; habitats; soil).

Further, the ONS propose that the statistics from the satellite accounts could be used, with caution, to model changes in demand and emissions output arising from a change in government policy, such as the tax on carbon emissions proposed by the European Community.

Green accounting is very much in its infancy relative to the conventional NIAs. The work of the UK ONS and other comparable institutions is trying to move towards a systematic framework for analysing the environmental effects of economic behaviour, but significant reservations remain even regarding the satellite accounts. Herman Daly and John Cobb have developed a radical alternative to NIAs termed the index for sustainable economic welfare (ISEW). This is the subject of the next section.

11.5 Index of sustainable economic welfare

The index of sustainable economic welfare (ISEW) is an index calculated from a wide range of indicators which seeks to show changes in welfare according to specific monetarized criteria (Daly & Cobb 1991). The ISEW is an economic as opposed to a social indictor, because all its components are calculated in monetary terms. This means that changes in ISEW can be directly compared with changes in the NIAs, although it also confronts the analysts with very substantial valuation problems, as the authors acknowledge.

The full details of the elements of the ISEW are discussed extensively in Daly and Cobb (1991). Here we note some of the most interesting aspects of their approach.

1 *Income distribution* is included, under the premise that $1 extra to a poorer household is worth more than the same amount to a richer one.

2 Only changes in the stock of *fixed reproducible capital* are included. Changes in land prices are excluded because the quantity of land is fixed, and increased value is thus a result of increasing demand.

3 *Human capital*, the productive and entrepreneurial capacities of labour, is not included as there is no means by which it can reliably be measured.

4 *Net investment* in an economy is calculated. The source of this capital is considered—a reliance upon

foreign capital might not be sustainable—and whether the rate of net investment is sufficient to keep up with population changes.

5 The calculations account for *natural resource availability* because this is critical for sustainable production. For fossil fuels and minerals there is an estimate inputted for compensating future generations for current depletion in the form of a perpetual income stream, using the user cost method. Similarly, there is a depreciation allowance for wetlands and farmlands factored into ISEW to allow for erosion, compaction and shifts in land use.

6 The *costs of pollution and other environmental damage* are deducted, as well as an allowance for noise pollution. There is also a speculative estimate for the costs of climate change inputted into ISEW, which Daly and Cobb relate directly to energy consumption.

7 *Leisure time* is not valued in ISEW because of the lack of a clear definition of the term and the problems of measurement over time. For example, if individuals have been made redundant and seeking employment, should the time spent looking for work be termed 'leisure'? Does more time spent in this state increase their welfare?

8 *Unpaid household labour* is inputted into ISEW, although problems with definition similar to those for 'leisure' apply. Daly and Cobb (1991) cite the work of Berk and Berk (1979) who found through interviews that some activities—for example, child rearing and cooking—are *both* work and leisure. However, ISEW computes a value based on the average wage rate of household domestic workers.

9 Certain types of consumption (e.g. junk or convenience food, nicotine and pornography) which might be (arguably) welfare-reducing are not treated in ISEW.

The full set of calculations is detailed and complex, with much of this complexity arising from the desire to generate an index that is directly comparable with other monetary measures of welfare. Other alternative welfare indicators, such as the widely established United Nations Human Development Index, use statistics such as child mortality rates, literacy rates and calorific intake and combine these into a comparative index number but without monetary valuations. Although problems associated with valuation are avoided, these alternative indicators have their own problems. For instance, do obese individuals in the developed world who have a higher calorific intake than the norm have a higher welfare level? In fact, there may well be a negative correlation between food intake and welfare after a given level.

11.5.1 Index of sustainable economic welfare results from the USA

Daly and Cobb (1991) present estimates for US per capita ISEW from 1950 to 1985. The trends in ISEW are noticeably different to GDP. Measuring across this whole period, per capita GNP rose by an average of 102%, whereas ISEW only rose by 37%. The average US person consumed more than twice as much in 1986 compared to 1950, but welfare as measured by ISEW only increased by just over a third.[2] However, the overall changes mask the very distinct fluctuations in ISEW during the reporting period. Between 1951 and 1960, there was a modest but positive average annual rise of 0.84%. From 1960 to 1970, this figure had increased to 2.01%. However, the 1970s saw a slight average decline in ISEW (0.14%) whereas GNP grew by an average 2.04% per annum. From 1980 to 1986, the decline in ISEW sharpened to 1.26%, whereas average GNP still rose by 1.84% (Fig. 11.2).

Daly and Cobb suggest that ISEW is not an over-pessimistic assessment of welfare; the rate of increase in ISEW actually exceeded GNP growth for some of the 1960s. The reason for this was a change in the pattern of income distribution: whereas personal income increased by 37.5% between 1961 and 1968, the increase in the equality of income distribution implied a weighted increase of more than 57%. The opposite effect occurred in the 1980s because of enhanced income disparities; whereas measured per capita consumption grew by 24%, the weighting for the 13% increase in income inequality depresses this figure to about 10%. This weighting, along with resource use and environmental degradation and the decline in sustainable net investment, leads to a negative rate of change in ISEW during the early 1980s. Other factors in the ISEW include air pollution and car accidents; the former peaked in 1970 and the latter in 1978.

The improvements in the air pollution and car accident indicators show that state intervention can improve welfare as measured by ISEW. Other policies which would affect such an improvement are income

2 Real GNP per capita went up from $3512.2 in 1950 to $7226.0 in 1986; ISEW increased from $2488.0 to $3402.8 in the same period (Daly & Cobb 1991).

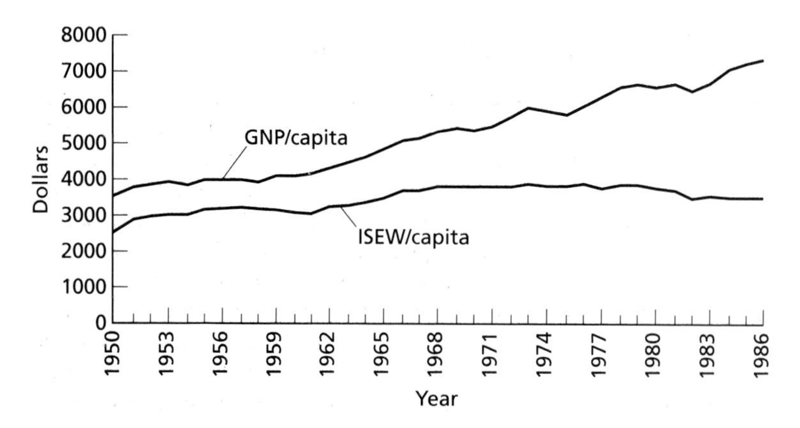

Fig. 11.2 The index of sustainable economic welfare (ISEW) graph for US ISEW vs. gross national product. (Source: Daly & Cobb 1991.)

taxation, resource use regulation and incentives for decreasing environmental degradation.

11.5.2 Some results from the UK

The ISEW has been calculated for the UK by the New Economics Foundation[3] (Jackson *et al.* 1997) using Daly and Cobb (1991) and the update, Cobb and Cobb (1995). The pattern for the UK was similar to that for the USA. Overall, per capita GDP in the UK was 2.5 times greater in 1996 than 1950, whereas ISEW only increased by 31% over the same period. Between 1950 and 1975, the annual increments in both GDP and ISEW are positive. After a period of stagnation in the late 1970s, ISEW actually fell. The fall of 22% from 1980 to 1996 compares with GDP growth of over one-third in the same period. The factors affecting this fall in welfare are the same as the US scenario: resource exploitation (in particular North Sea oil); environmental degradation; and increasing income differentials.

The UK study also compares the ISEW of the UK and USA with that for the Netherlands and finds a similar pattern. This leads to the authors' proposition that there is a threshold level of consumption beyond which further consumption is welfare-reducing: non-market institutions, such as the family and the local community, social values and the feeling of citizenship values are eroded by the market mechanism. In summary, the quality of life might be inversely correlated with the quantity of consumption. If this is indeed the case then a fundamental investigation into the worth of the NIAs is necessitated.

[3] The web page for the New Economics Foundation is: http://www.neweconomics.org/. Another related site is: http://www.sosig.ac.uk/.

11.6 Environmental space

Another perspective on the macroeconomic environmental performance of nations is provided by studies carried out by the environmental non-governmental organization, Friends of the Earth (FoE), using the concept of environmental space (ES). The FoE defines ES as: 'the total amount of pollution, non-renewable resources, agricultural land and forests that we can be allowed to use without draining the Earth's resources' (FoE 1995). The concept of ES is that there is a certain sustainable entitlement that every human being has to both natural resources and pollution sinks.

In general, on a per capita basis FoE suggest that the developed world is using more than its 'fair share' of this entitlement at present, and many developing countries less than their 'fair share'. The popular concept of an 'ecological footprint' (Wackernagel & Rees 1997) highlights this disparity. The ecological footprint is the environmental impact that an individual has as a result of his or her lifestyle, measured across a number of parameters related to resource consumption and waste disposal. The average North American citizen has an ecological footprint many times greater than that of the average Chinese or Indian. A second popular concept, the 'environmental rucksack', captures the waste aspect of this disparity. The environmental rucksack theoretically carries all the non-biodegradable or non-recycled waste that an individual produces throughout his or her lifetime. Not surprisingly, developed world citizens tend to have big rucksacks to complement their big footprints.

The ES calculation is not simply a case of dividing the total current use of resources and sinks of the planet by the number of people, as this would assume that current consumption patterns are sustainable. Sustainability is defined by FoE (1995) according to three environmental criteria:

1 the assimilative capacity of the environment;
2 the lifetime of reserves of non-renewables; and
3 the local and global carrying capacity.

The ES concept seeks to ensure that each nation's footprints are sustainable at national levels and, if not, to identify areas for potential trade (for example, Scotland has a surplus of agricultural land in its ES and this might be traded, say, with another country's ES entitlement for minerals). FoE (1995) propose three guiding principles for sustainability: the precautionary principle; the equity principle; and the proximity principle.

THE PRECAUTIONARY PRINCIPLE

The FoE definition of the precautionary principle is that caution should be observed when judging sustainability in the face of environmental uncertainty, with the assessment of sustainability of a resource being in terms of the three criteria mentioned above. For instance, the current consumption of coal implies emissions to air which must be assimilated, a decrease in the availability of this non-renewable resource for future generations, and an impingement on the global carrying capacity because of the global warming effects of the gases released in combustion. FoE's definition of 'precaution' in this case is to focus on this last issue—global warming potential—as FoE state that this is the most pressing (or least 'slack') of the three constraints on sustainability. The same type of assessment needs to be made for all resource use.

THE EQUITY PRINCIPLE

At the global level, the equity principle implies that a country's share of the sustainable level of resource use is determined by its population size. Thus, China and India are entitled to a greater amount of resources than they currently consume. The equity principle is certainly open to criticism. On a pragmatic level, it is politically unacceptable; the wholesale redistribution of resources from developed to developing countries, and indeed within individual countries, is certainly hard to conceive. The basis of the principle in population might also be questioned. As the average population growth rate in developing countries is much higher than in developed countries, each individual in, say, Europe, would have to reduce consumption levels constantly over time to compensate.

THE PROXIMITY PRINCIPLE

The proximity principle states that environmental problems should be solved as near to their source as possible. Thus, if pulp is required then it should be sourced locally; if hazardous nuclear waste is produced from energy generation then it should be disposed of locally. The application of this principle might well be inefficient in economic terms. What an economic analysis may consider is the marginal cost of waste disposal or resource exploitation. The marginal cost might not be minimized by choosing the local option. However, this economic argument can have some unpalatable implications, as became apparent in an incident described in George and Sabelli (1994): the 'Toxic Memo'. This infamous memo was intended as an internal one circulated in the World Bank by Lawrence Summers, the then Chief Economist, in 1991. The memo on 'Dirty industries' states:

> '[S]houldn't the World Bank be encouraging *more* migration of the dirty industries to the LDCs [Less Developed Countries]? . . . The measurement of the costs of health impairing pollution depends on the forgone earnings from increased morbidity and mortality. From this point of view a given amount of health impairing pollution should be done in the country with the lowest cost, which will be the country with the lowest wages. I think the economic logic behind dumping a load of toxic waste in the lowest wage country is impeccable and we should face up to that.' (Quoted in George & Sabelli 1994)

This is indeed impeccable logic, the same logic that values the loss of a life in the developed world more highly than one in the developing world. Not surprisingly, the leaked memo did little for the World Bank's reputation amongst those unconvinced that economic logic was the only decision-making criterion available, and encouraged the statement of alternative principles, such as those promoted by FoE.

11.6.1 Some environmental space results for Scotland

The ES results for Scotland provide both an inventory of Scottish resource use and a measure of by how much this resource use would have to change in order to achieve sustainability, as defined by FoE (1995) above. Because any move to sustainability is a lengthy process, FoE propose targets for 2010 and 2050. The research

Table 11.1 Friends of the Earth Scotland Environmental Space results for energy.

Year	CO_2 emissions (tonnes)	Tonnes per capita: Carbon	Tonnes per capita: CO_2	E-space 2010	Reduction needed (%)	E-space 2050	Reduction needed (%)
1993	49 178 823	13 412 406	9.63	5.4	44	1.7	82
1989	49 240 938	13 429 347	9.64	5.4	44	1.7	82

Abbreviation: E-space, environmental space.

Table 11.2 Friends of the Earth Scotland Environmental Space results for non-renewables.

Example resource	Present use (tonnes)	Present use (kg per cap.)	E-space 2050 (kg per cap.)	Percentage reduction needed by 2010
Cement	1 311 635	256.68	80	17
Pig iron	1 087 510	212.82	36	21
Aluminium	39 245	7.68	1.2	21
Copper	23 966	4.69	0.75	21
Lead	23 455	4.59	0.39	23
Chlorine	81 760	16	0	25

methodology acknowledges very considerable data limitations and availability problems. With these caveats, FoE (1995) presents preliminary results for various ES categories: energy; non-renewable resources; wood; water; and land. Summary findings for the first three categories are given below.[4]

ENVIRONMENTAL SPACE FOR ENERGY, WOOD AND NON-RENEWABLES

FoE Scotland rejects nuclear fuel as a non-sustainable option; nuclear power is assumed to be phased out by the year 2010. The global warming arising from the combustion of fossil fuels is therefore scrutinized with the assumption that the climate change recommendations of the Intergovernmental Panel on Climate Change for the protection of natural systems are to be adopted[5] (Table 11.1).

A 44% reduction in CO_2 output by 2010 is required for sustainability. FoE cite various means by which this 44% reduction in CO_2 usage by 2010 might be achieved: investment in renewables and energy efficiency initiatives. FoE claim that 7000 new jobs would be created in the construction and maintenance of these new projects.

As regards non-renewables, FoE focus on the waste arising from consumption rather than resource availability. A medium-term target of 50% reduction in global use is set, with a 25% reduction by 2010. This 50% global target implies an 80% reduction for the developed world, given the equity principle. Some results for Scotland are given in Table 11.2 for the non-renewables selected by FoE Scotland.

The reduction in non-renewables for Scotland by the year 2010 is in the range 17–25%. The suggested means to achieve this are recycling, reuse, and production of more durable and repairable commodities.

In considering wood, FoE (1995) finds that, if the UK as a whole were to meet its own current demand for wood, 172 197 km^2 of land (in 1993–94) would need to be dedicated to timber, i.e. over 70% of the UK's land area. Clearly this is an area where national self-sufficiency is simply not feasible. As starting assumptions, FoE propose that harvest should not exceed yield, that there should be no further logging of primary forest, and that 10% of existing European forests should be protected for nature. Given these assumptions, FoE estimate that Scotland must reduce its wood

[4] The results for water are not presented as for Scotland 'there is little likelihood of demand exceeding supply as a whole' (FoE 1995); those for land are subject to considerable unavailability of data.

[5] <0.1 °C change in climate per decade; less than a 2 °C change in total; a limit to global emissions of 2 gigatonnes of CO_2 per year by 2100.

consumption by 38% by 2010 to live within the available European timber resources.

ACHIEVING THE TARGETS

The FoE study documents targets for the medium and long term, with the caveats of imperfect data availability and quality already mentioned and, importantly, with the assumption that resources should be equitably distributed. The mechanisms proposed to achieve these various targets are closely aligned with those coming from a bioethical perspective: education; political and economic restructuring; and a change in social ethos away from the throwaway society. More direct recommendations include phasing out nuclear power for Scotland, and creating a coordinated and integrated public transport network.

11.7 Summary

The conventional macroeconomic tool for evaluating the performance of an economy is the system of NIAs. The NIAs provide a detailed breakdown of what is produced, what is spent and what is earned in an economy. The mechanics of the methodology involve making numerous accounting adjustments to provide a figure that represents national income. The accounts are also used as a measure of social welfare, but there are a number of limitations with their use for this purpose, only some of which can be overcome by adjustments to the accounts themselves.

Methodologies for inputting macro-level environmental changes have recently been developed. These data are developed as auxiliary satellite environmental accounts. One particular anomaly that the satellite accounts attempt to address is the fact that the proceeds of oil and natural gas sales should not be treated as income *per se*, although the conventional NIAs do so. The UK satellite accounts measure oil and gas consumption, pollution output, and defensive expenditures by certain key industries.

The methodologies used to generate the main and satellite NIAs are rooted in conventional economic theory. The ISEW is an alternative economic index which factors income distribution, natural resource availability, pollution costs and unpaid household labour into the assessment of welfare changes over time. The ISEW does not mirror NI, the former actually falling in the early 1980s as a result of increased income disparities in society. The second alternative macro-indicator is generated by the FoE environmental space methodology. The FoE estimate sustainability constraints for resource use and pollution levels, and then determine a national-level entitlement based on population. This vision of FoE is a fundamental restatement of how the economic system ought to operate in order to achieve a particular vision of sustainability.

Further reading

On calculating the standard accounts:
Begg, D., Fischer, S. & Dornbusch, R. (1998) *Economics* (5th edn). McGraw-Hill, Maidenhead, Berkshire.
On satellite environmental accounts:
Office for National Statistics (1998) *UK Environmental Accounts 1998*. Office for National Statistics, London.

Part 4
Applications: Theory and Practice

12 Resource Harvesting

'Invest in land. They have stopped making it.' [Mark Twain]

12.1 Introduction

The harvesting of biological resources is probably the oldest economic activity known to humanity. In this chapter we concentrate on some of the management issues that arise out of the directly productive assets of fields, forests and fisheries. These issues encompass the three central concerns of ecological economics: the economically productive use of assets, reliant on environmental quality and an understanding of the ecology of the resource being utilized, and often requiring the cooperation of diverse social interests with differing values and goals. We begin with a review of the concept of property rights, as these are often central to understanding the management problems inherent in resource harvesting. Property rights also feature importantly in the two following chapters on Nature Conservation, and Pollution and Waste.

Subsequently, we introduce some analytical insights into agriculture, forestry and fisheries. In agriculture, we introduce the idea of the production possibility frontier (PPF) as a means of exploring the development path taken by westernized agricultural systems, and consider what trade-offs the nature of this frontier presents. In forestry, we analyse the economic principles of harvesting in relation to the growth rate of the resource. We then explore two themes within forestry management based on criteria beyond the standard economic analysis: rural development forestry, aimed at assessing the viability of forest-based employment in rural communities; and the role of common property values in management systems in Nepal.

Finally, in the case of fisheries, we explain the economics of private and open access harvesting and the effects of fishing pressure on fish stocks under these two conditions. We then review a range of fishery management options, such as quotas, permits and gear restrictions, and assess their suitability as management tools in terms of their economic, ecological and social effects.

12.2 Property rights

Property rights have long been recognized as a critical element in the management of natural resources, and they have a particular heritage in environmental economics exemplified by the work of Coase (1960), Gordon on fisheries (1956) and Hardin (1968). Because much of economic analysis concerns the system of production and sale of goods, the legal and institutional conditions that determine the conditions for ownership, transfer and use of those goods is clearly important. However, neoclassical economists have tended to operate with very simplified models of ownership and exchange, and it has been institutional economists, such as Commons (1968/1926) and, more recently, Bromley (1989) who have tried to focus attention on the institutional complexities of the ownership system.

Economic analysis generally recognizes four basic categories of property rights.[1]

1 Private property, which is owned by individuals.
2 State property, owned on behalf of the citizens of a nation by the state.
3 Common property, which is managed collectively by a particular group.
4 Ownerless, or 'open access' assets, which are not owned by any group or individual.

Each of these categories of property provides people—both owners and non-owners—with different incentives for action. Some of the ramifications of these different incentives are explored in this and the following two chapters. Before examining these, we introduce two preliminary issues: philosophically speaking, what justifies ownership, and what are some of the factors

[1] An unusual fifth category—'property of all humanity'—has also been suggested. The 'property of all humanity' concept was an interesting attempt to justify the case in favour of preservation of certain unique world heritage sites in perpetuity, because no group or even nation could hold the right to destroy such assets.

that make up a right of ownership or property right? Understanding these basic positions provides a platform on which to assess the moral and political implications of different management strategies and the way in which they approach property issues.

12.2.1 The creation of property rights

The notion of private property is so familiar in Western society that its origins and justifications are rarely questioned. Indeed, it is asserted by no less an authority than the Fifth Ammendment of the US Constitution. In this section we go back to this very fundamental question—the justification of rights to property—and introduce the main arguments that have been advanced for recognizing some system of property rights in society.

There are broadly five established arguments advanced for recognizing rights to hold property (Becker 1977).

PRIOR APPROPRIATION OR FIRST OCCUPANCY

On the basis of prior appropriation, the first person to claim, or 'appropriate', unowned assets simply has the right to them. In the case of land, this is taken as being the first occupier of a previously uninhabited situation. Bromley (1989) notes that 'First in Time, First in Right' does often apply as a basic social ordering convention with regard to some scarce goods, such as bus seats and theatre tickets. However, prior appropriation suffers from several limitations. It is not clear exactly what can be appropriated ('all that one can see', 'all that one can use', 'all land as far as the next claimed portion', etc.) and it does not seem to have any particularly strong moral justifications why first arrival should hold the right of ownership in perpetuity, or the right simply to exclude others without using the land in any way.

LABOUR

The philosopher John Locke considered that people could take possession of what were free goods—say, the apples growing on a tree—by 'mixing their labour' with the goods: that is, in the case of apples, picking them. This is the essence of the labour theory of ownership—that as individuals 'own' themselves and their labour powers, then whatever individuals have worked on to change or create, they also have a right to own (Locke 1967/1690). The labour theory was predominant in the seventeenth century and provided the foundations to granting settlers' rights during periods of colonial expansion; title to land was achieved through working that land and unworked land along the frontiers was correspondingly open to appropriation. In some areas today land title is still created through the process of forest clearance and cultivation, which can create misincentives towards clearance. However, it is not clear why initial working of a piece of land entitles the worker to ownership in perpetuity; the labour theory would seem to justify ownership only as long as land is being usefully worked by the owner (a theme explored by the land reforming socioeconomist Henry George; see George 1979/1879).

SOCIAL WELFARE OR UTILITY

The argument for property ownership on welfare grounds is that social welfare is increased by recognizing an extensive system of individual property rights. There are really two aspects to this justification. First, without private property many aspects of life simply 'go less well' than with private property; we wish to have security in our own homes, our immediate possessions, tools for work and so on, and rights here seem to have a fairly strong utilitarian underpinning. Secondly, the utility argument suggests that private property provides the best incentives for good economic management of productive assets, leading to overall increases in social welfare. This is the philosophical justification generally adopted by environmental economists in favour of establishing stronger private property rights to goods, such as wild forests and fishing grounds. Privatizing these resources is seen as leading to their better management. This 'free market' approach can be criticized from a number of perspectives, outlined in Chapter 5 and later in this chapter.

NATURAL RIGHTS AND POLITICAL LIBERTY

The political liberty approach to property argues that the freedom to own property is a fundamental right of citizens. If a state fails to recognize and protect private property, it is therefore failing to secure the basic rights of its citizens. In effect, the ownership of possessions is taken as an expression of individual rights of comparable importance to freedom of thought or of belief. Two complications with this position are worth noting. First, how can clashes between this freedom to own property and what might be considered other fundamental rights of others (to food and shelter) be resolved? Secondly, because ownership of land is not generally absolute, how are the regulations governing ownership determined? These issues do not undermine the natural rights justification in themselves but they do qualify its absolute extent.

MORAL CHARACTER AND VIRTUE

The argument from moral character suggests that the ownership of property has a beneficial influence on the moral development of individuals. In effect, the ownership and management of property is regarded as a virtuous kind of activity that nurtures morally well-regarded characteristics. Such ideas have had some influence historically (e.g. Thomas Jefferson's views on farming land and moral development), but they have little current credibility. Counter arguments to a 'virtue' view can be identified, for example, in notions of the corrupting influence of excessive wealth. Overall, the issue of 'virtue' has in any case more or less subsided from public discussion of values.

The issue of justification for property ownership is particularly relevant in environmental management because of the lack of a clear concept of ownership for a number of environmental goods. Who 'owns' the ozone layer, the scenic view of Scotland's Ben Nevis, the air in the street, the deep ocean beds, the genetic material of the evening primrose, the last few hundred African black rhino? Some commentators argue that the lack of private property rights to these kinds of goods is leading to their economic mismanagement. Others argue that it is misguided to think of these goods in terms of 'property' at all; they suggest it is the very economic *commoditization* of these goods—treating them as property that can theoretically be bought and sold—that in itself creates the problem of mismanagement. They argue that we need to conceive of environmental management using a fundamentally different kind of language and concepts from ones of economic value; for example, issues of entitlements, duties of care, community stewardship and intergenerational equity. To follow this debate see Sagoff (1988).

Having noted these considerations, we now turn to consider more precisely the extent and operation of property rights.

12.2.2 Characteristics of property rights regimes

The property concept itself defines a triadic relationship (Bromley 1989): owner–owned property–non-owners. Ownership is popularly thought of as a 'bundle of rights', which gives the owner a certain freedom in how to use goods and property in relation to that property and to other people. The central aspects of the functioning of property rights include several features, of which the most important are the following:

Access/exclusivity. The key feature of property rights that concern economists is that of *exclusion*, or the conditions of access to a resource. Private property is assumed to be exclusive; that is, access is closed to others. Common property is exclusive to community members, and unowned property is referred to as open access, which anyone is free to enter and use or harvest.

Transferability. Access rights may be transferable, such as purchased fishing licences to a Scottish loch. They may, however, be non-transferable, such as the right to vote, or limited to location (i.e. only residents may use a forest for fuel, with current residents not allowed to lease this right to others).

Duration. Rights may exist for almost any time-scale, from hours (for example, in the case of parking tickets) to in perpetuity. Duration of rights is obviously as critical as access in influencing conservation or exploitation behaviour.

Formality. The formality of rights, which refers to their explicit legal status and codification or, perhaps more importantly, the lack of explicit codification, is a feature that simple economic models can easily ignore. Informal traditions and habits of behaviour can operate highly effectively as management systems without recognizing any particularly formal legal entitlements. Bromley (1991) identifies the neglect of these traditions by economic planners as a key factor in resource exploitation (see Section 12.4.4 below).

Prohibitions or use restrictions. Even where rights to private property are strongly identified, they are usually also strongly circumscribed by a substantial set of use conditions. Farm land, for example, may be ploughed and planted at will but only burnt at certain times, and not used for waste disposal or domestic development without consent. If it carries a 'right of way' it is not wholly exclusive, nor can the scenic view of it be exclusive even though the physical product of the land is.

Management system. Simple economic models generally equate property rights with management system; in effect either exclusive and private, or open and unowned. However, it is clear that legal title, access conditions and operating system can all be different. For example, a private house left deserted, housing a squatter commune, represents technically a communally managed, open access private property. In terms of resource management, analysis needs to focus on the operating system and prevailing access conditions, and not simply the legal framework.

The systems of functioning property rights arising from these factors can be extremely complex. As an example, the Chesapeake Bay in North America contains a fishery, which is a fugitive renewable resource; it is also migratory. The migrations cover several different State jurisdictions. The coastal waters are in areas controlled by local community systems, with some exclusive and non-transferable restrictions. Different gear and landing restrictions operate in other areas subject to State control, which are operated under qualified open access conditions. State, regional and local organizations all have elements of control over certain aspects of the bay's use. The behaviour of fishermen is itself also influenced by local traditions, length of experience and faith and trust in the efficacy of the various arrangements and their perceived validity.

Having considered some aspects of the ownership of natural resources, the following sections introduce related issues within agriculture, forestry and fisheries in more detail.

12.3 Agriculture

Since the 1950s, the agricultural systems of the world's advanced economies have been focused more or less exclusively on the single target of increased food production. Agricultural policies have had multiple objectives, including national food security, maintenance of farmers' incomes and rural employment, but policymakers simultaneously assumed that farm land itself would, and should, be managed to maximize food output. This has led to more and more intensive agricultural systems, using increased levels of fertilizers, pesticides and other inputs, and increases in farm size and mechanization.

12.3.1 A multivalue approach

Environmentalists have long argued for an approach to agriculture that reflected a much broader appreciation of the value of agricultural lands than food production alone. A well-known representation of this wider consideration is given by Traill (1988) in Fig. 12.1. Traill's figure indicates the relationship between agricultural intensification, plotted horizontally, and the 'value' of a number of aspects of the agricultural systems, including food yield, environmental quality and rural employment. Although it is not possible to specify the vertical scale directly because many of these aspects are cur-

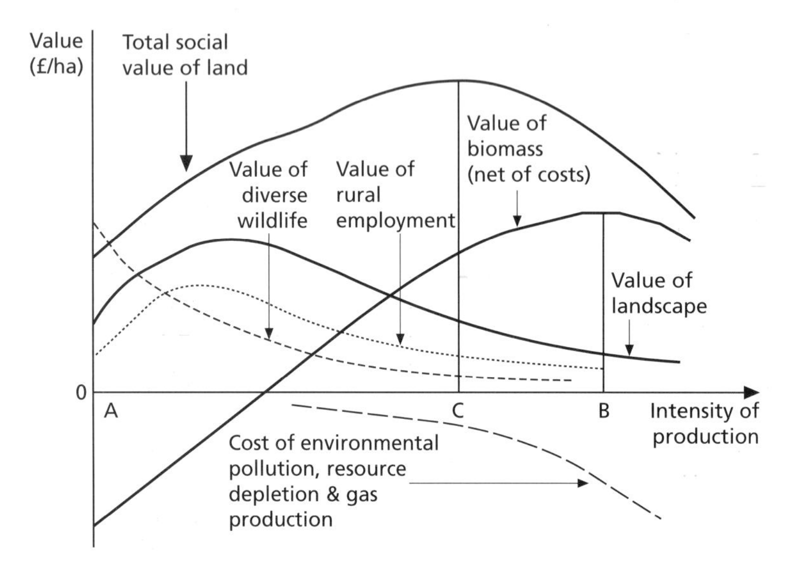

Fig. 12.1 Traill's (1988) schematic analysis of multiple values and agricultural production (reproduced with permission).

rently unpriced, it indicates the kind of analysis that might be undertaken in order to determine a social welfare maximizing level of intensity of agricultural production. From the viewpoint of the current generation, farming output would be ideally set at C, where the vertical sum of the farm output and all the features of the wider ecological system is at its greatest.

It is important to bear in mind that the social optimum identified at C in Fig. 12.1 will almost certainly not be the optimal output level for the owners of the resource, the farmers themselves. This is because they bear the full financial cost of lost crop sales if output is cut back, but they gain only a fraction of the total social welfare gain from environmental improvements. In this case, a free market in food will not alone maximize social welfare because food output decisions will not reflect the numerous external effects that arise as a result of the choice of agricultural output system. In fact, food producers would be expected to intensify up to point B, because this maximizes their returns overall.

12.3.2 A production possibility frontier for agriculture

The issue identified by Traill is essentially one of balancing competing interests. The production possibility frontier (PPF) for food and other agricultural goods is a useful way to conceive of this problem. The PPF shows the relationship between all the non-market (environmental) outputs of agriculture, and the marketed food or fibre output (Fig. 12.2). The food output is measured on the horizontal axis, and the possible output of environmental and other goods on the vertical. For each possible level of food output, the PPF curve identifies the corresponding level of environmental goods that can be produced, assuming that managers maximize the output of these goods subject to the constraint of maintaining a particular level of food output.

To understand the PPF more clearly, consider point A. Here the output of food (F_f) is very high, meaning that the agricultural system is highly intensified and almost all available land is being used for such intensive production; correspondingly, the overall output of environmental goods is low (E_f). As we move along the frontier towards B, food output, and thus the resources devoted to that output, are being reduced, which correspondingly generates more opportunities for providing environmental goods. Each point on the PPF thus identifies a combination of these two goods—food and 'environment'—which is at the limit of the total production capacity of the overall agricultural system. Note that no sensible society would chose to produce any combination of goods that lies inside the PPF, such as C, because it is technically possible to produce at a point such as B,

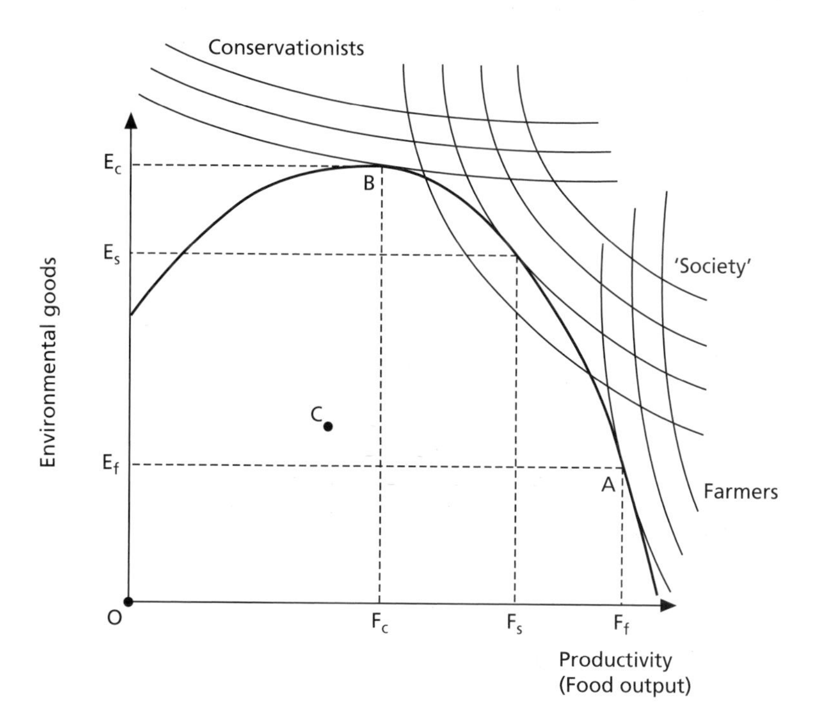

Fig. 12.2 Graph showing a production possibility frontier for agricultural output and environmental goods, and indifference curves for conservationists, farmers and society. The preferences of these three groups will lead them to favour producing different combinations of environmental and food outputs around the frontier (F_c E_c, F_s E_s, and F_f E_f respectively).

and this provides society with more of both kinds of goods. The PPF defines the limit of possible outputs for any two goods for any given economic system with a particular state of technology, hence the name 'production possibility frontier'.

Establishing the PPF opens up a number of interesting questions about the relationships between outputs of the ecological and the economic system, many of which analysts and managers are currently trying to answer. Four issues are particularly interesting.

Where are we now on the frontier? Currently, we are only aware quite roughly of some of the trade-offs implicated in the PPF model. Monoculture cropping techniques are acknowledged to have poor biodiversity, for example. However, it is not possible to say with any precision at what position on the frontier we currently lie, partly because the frontier itself is only an approximation, and partly because of the level of variation at the local level. For example, some areas, such as uplands, may already have high environmental outputs and low yields, whereas lowland arable farms produce the opposite kind of mix.

Which direction do we wish to move? Despite uncertainty regarding the current position on the frontier, strong lobby groups exist in favour of moving towards more of one type of good or the other. These issues are political in that they reflect different fundamental beliefs or preferences about the 'best' state of agriculture and the environment amongst different groups.

How do we achieve a movement around the frontier? While deciding on which direction to shift systems is mostly political, determining how best to achieve that shift is both political and economic. The use of economic instruments to achieve environmental objectives is covered in more detail in Chapter 14. We note here that a system of strong private property in combination with free markets for food will tend, *ceteris paribus*, to move towards greater intensification. Thus, the form of the institutional structures within which choices are made—private property, food markets—leads to a particular point on the frontier. Alternative institutional structures (for example, the SSSI system of conservation, see Section 13.3.2) are needed to shift to a new point on the frontier.

What changes the shape of the frontier over time? It is important to remember that the PPF is a representation for a particular economic system with a particular state of technology. The shape of the frontier thus changes over time as that system and its technology changes. Historically, the PPF has expanded outwards considerably, but almost entirely horizontally. This reflects the market incentives for technical innovation, which have generally always been for increasing food output. This path looks likely to continue unless policies adjust incentives accordingly. If serious long-term ecological damage arises from over-intense production methods, however, it is possible that there would be an eventual contraction of the PPF, leading to a corresponding reduction in social welfare. This phenomenon is apparent in localized cases (for example, eutrophication of fishing lakes); the effects of globally significant environmental problems, such as ozone depletion, would seem to make this contraction in the PPF a possibility.

12.3.3 Concluding comments on agriculture

The comments above are intended only to put the environment–agriculture debate within a broad context. It is worth noting here that issues facing developing world agriculture are, in general, very different. Food security to assure adequate access to food, fair access to land, protection of indigenous rights, pressures for intensification for growing populations and species loss are giant and complex themes, beyond the scope of this text. A good starting point for these issues is Pretty (1995b).

12.4 Forest resources

The set of issues that are raised by the theme of forest resource management is potentially vast. The roles of forests and their associated products are as diverse as for any ecosystem, and we will focus here on just a few illustrative examples. We begin by discussing the principles of forest harvesting for a single product, namely timber. We take this example under a private property arrangement to illustrate the contribution of economic analysis to a simple harvesting problem. Subsequently, we investigate two other issues. We introduce a preliminary analysis of rural development forestry (RDF) in terms of its employment and community development potential and we review an example of a failure to accommodate informal property rights regimes in the case of Nepal, and its impact on forest management.

12.4.1 Biological growth and optimal harvesting strategies

If we ignore for the moment the issue of externalities, which are likely to be extremely significant in many

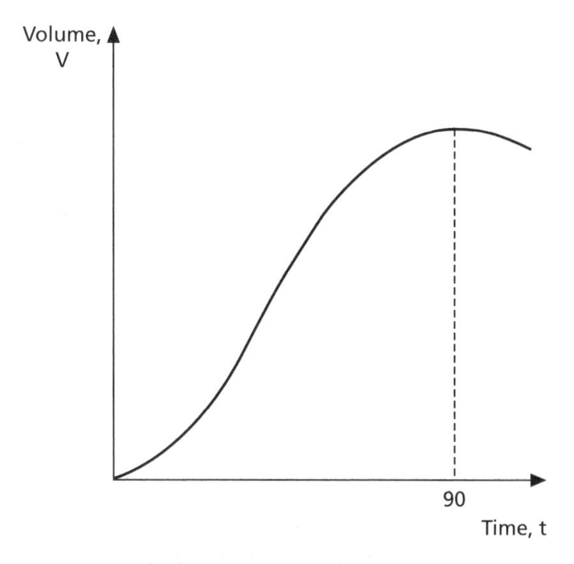

Fig. 12.3 Standard growth function for a tree.

forest ecosystems, the basic problem facing a single owner of a forest resource is how to maximize personal returns from the timber that it contains. To begin with a simple model, we assess the biological growth function of the resource.

Each tree grows according to a standard growth function, and thus the timber volume of an even-age stand is dependent on the time for which it has been growing. A conventional growth function for a single tree is represented in Fig. 12.3. Clearly, for an even-aged stand this represents on average the growth of the whole resource. The volume of timber rises relatively slowly at first, then grows very quickly as the trees approach maturity, at which point the rate of growth slows again and eventually actually declines as trees die and the wood starts to decay. A wide range of factors influence this relationship, such as rainfall, wind, spacing and silviculture techniques, but the basic principle remains unchanged.

If the only management consideration is volume of timber from this one batch of trees, without regard to any other criteria, clearly the maximum value of the resource is when t = 90, when the total volume has grown to a maximum. However, woodland managers are generally more interested in maximizing the output of a series of plantings, in which case waiting for each stand to attain full maturity is not optimal. They are concerned to get the greatest timber returns over an indefinitely long time period, and to do this they find a rotation period based on the highest average growth rate of the forest, or the highest mean annual increment (MAI). This is the point at which the ratio of timber volume to age of the stand is greatest, known as the culmination of the MAI (CMAI). It is clear that if we intend to maximize timber volume over a number of years, we should harvest each cohort of trees when the average growth rate of each cohort is at its highest value, because total timber yield over time is simply the number of years of growth multiplied by that average growth rate.

Note that this is not when the trees are growing at their greatest rate. The rate of growth is called the current annual increment (CAI), which is the marginal growth rate of the forest. This is shown in the lower half of Fig. 12.4, together with the MAI curve. The CAI curve crosses the MAI curve when the latter is at its maximum, because the average growth will be increasing if the marginal growth is higher than average and decreasing if marginal growth is lower than average. In practice, the CIA is very difficult to estimate accurately, and may often be taken as the average rate over a set period, say 5 or 10 years. An example for the Douglas fir is given in Fig. 12.5.

Does the MAI provide a good economic reason for harvesting? From the financial viewpoint its major failing is that it fails to take account of any *discount* rate. That is, it does not consider the opportunity cost either of alternative uses for the land currently forested, or for the money that is realized whenever the timber is harvested. Even if there are no alternative uses for the land, and it is not going to be replanted, it still might be the case that the trees should be harvested earlier rather than later because the money raised could be invested in something else that would yield a higher rate of return than the additional growth of the volume of timber. Clearly, the slower the biological growth rate, the more likely that other investments in the wider economy will yield better returns. Consequently, a financial appraisal will try to maximize the net present value (NPV) of the timber resource, which is dependent on opportunities for profit in the rest of the economy as well as the growth of the resource itself.

Again, keeping the analysis reasonably simple by focusing on the single crop case, the financial criterion for harvesting must be when the NPV of the timber is greatest or, alternatively, when the rate of financial return from the wood is equal to the rate of return of other possible investments, indicated by the prevailing interest rate. An example of some values for these calculations is given in Table 12.1. The rate of return on investment (ROI) is given by taking the CAI as a

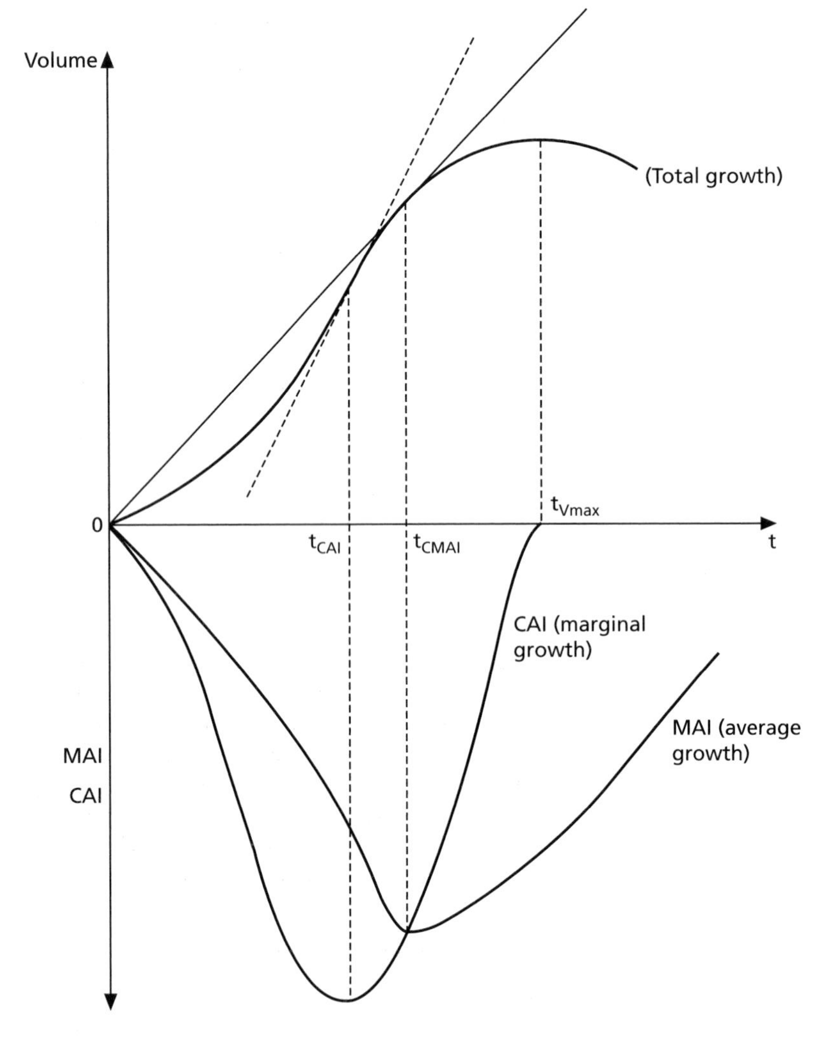

Fig. 12.4 Curves specifying tree growth, mean annual increment and current annual increment relationships. The highest growth rate is achieved at t_{CAI}, but the highest average growth at t_{CMAI}.

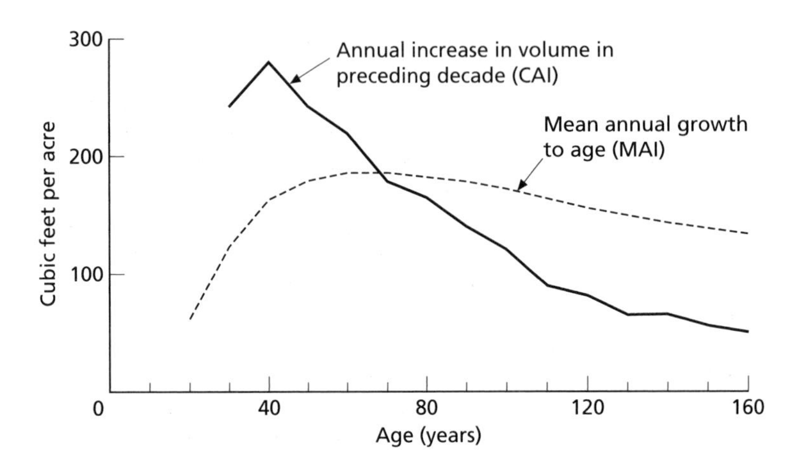

Fig. 12.5 Mean annual increment and current annual increment for a specific stand of Douglas fir. (Source: Clawson 1977, with permission.)

Table 12.1 Calculations for net present value and related factors for determining optimal forestry rotation period. (Source: Kuuluvainen & Tahvonen 1995.)

Age	Vol. (m^3/ha)	MAI (m^3/ha)	CAI (m^3/ha)	Timber value (ECU)	ROI (%)	Discounted value ($r = 3\%$)
20	4.0	0.2	6.9	600.0	172.3	−4 667.8
30	96.5	3.2	11.3	14 467.5	11.7	960.4
40	222.7	5.6	13.7	33 408.0	6.2	5 241.5
50	363.0	7.3	14.1	54 442.5	3.8	7 418.7
55	432.1	7.9	13.5	64 814.1	3.1	7 753.3
56	445.5	8.0	13.3	66 829.1	3.0	7 766.8
57	458.8	8.1	13.2	68 817.0	2.9	7 763.7
60	497.3	8.3	12.5	74 592.0	2.0	7 660.7
69	596.7	8.6	9.4	89 507.7	1.6	6 643.7
70	605.9	8.7	8.9	90 877.5	1.5	6 477.6
71	614.5	8.7	8.4	92 178.8	1.4	6 120.1
80	668.8	8.4	3.4	100 320.0	0.5	4 427.8

Rotation for Norway spruce (assuming a regeneration cost of 5000 ECU/ha at time zero, harvesting without costs and timber price 150 ECU/m^3) using maximum MAI and maximum present value of harvest income over one rotation.
Abbreviations: CAI, current annual increment; MAI, mean annual increment; ROI, return on investment.

percentage of the total value of the stand; this indicates by how much the forest is appreciating in value as a result of growth. The discounted value is calculated using the standard NPV formula (given in Section 8.3.4) applied to the timber value in the relevant end year. In this example, it can be seen that the highest NPV under a 3% discount rate occurs at year 56, which is also the time when the ROI is at 3%.

In a stand of mixed ages, the same basic harvesting logic applies. If we look to manage the forest from a pure volume perspective, we can design a maximum sustainable yield approach that harvests selectively those trees that have attained their maximum average rate of growth. If the CMAI is 40 years, we can design a forest system of 40 hectares such that one hectare attains this CMAI each year and is consequently harvested. By taking out these specimens we allow room for replanting. However, it should be noted that there are very substantial management issues involved in bringing an unmanaged stand into an optimal harvesting strategy. This will depend on the current age mix and may never come to resemble an even-age strategy (see Hartwick & Olewiler 1986).

One further economic problem that we note here is that of the appropriate harvesting strategy for a series of even-age rotations (the problem posed by German forester Faustman in the last century, and solved later according to the still-current Faustman formula). This requires the maximization of a series of NPVs, which occur at discrete time intervals (T). The calculation of T is beyond the scope of this section, but we can note that it is likely to be shorter than for the single rotation example because delayed planting postpones all subsequent plantings, encouraging more faster-growing rotation periods. However, costs of replanting and harvesting, which we have ignored so far, will tend to increase the rotation period *ceteris paribus*.

12.4.2 Extensions of forestry analysis

So far we have focused on a very simplified financial appraisal of forestry economics. This is useful to indicate how economic analysts determine optimal management regimes for a renewable resource, and to note in passing under what conditions it may be optimal to eliminate that resource altogether, even though it offers a sustainable income stream into the future. If alternative investments or alternative land uses offer higher income streams, the current use will not continue.

In economic terms, the fact that alternative uses or investments offer higher yields is only to recognize that the allocative mechanism of the market is drawing resources away from one particular use to another one which has a higher value to society. This higher value is represented by the higher prices paid for alternative land uses, or the higher interest payments made by investors who need capital to invest in other projects. There are several problems with this limited perspective.

1 Positive externalities generated for society from forests will not be considered by owners of forest resources if they cannot benefit from them. Similarly, external costs, such as flooding caused by loss of water retention capacity following felling, will be ignored.
2 Rates of tree growth are generally far slower than rates of interest in the economy, the latter driven partly by technological innovation. It is interesting to note that most UK forestry planting has been State run and an unusually low discount rate applied to forestry projects.
3 In many developing countries there is often a large self-sufficient subsistence economy which has no access to capital but whose welfare is dependent on forest products. The needs of these forest-dependent people may not enter government financial appraisals of potential revenues from timber yields, yet their welfare is wholly dependent on forest maintenance.
4 Logging, in many timber-producing countries, operates on an unsustainable basis, in which natural capital is depleted but still counted as income (see Section 11.4.1). The going market price for timber therefore fails to reflect the real price of maintaining a sustained yield in the future.

From a resource management perspective, it is likely that in many cases these complications are a very significant challenge to the basic financial rationale for harvesting. The rest of this section therefore explores two elements of these other issues: community and rural development forestry, and property rights mismanagement in Nepal.

12.4.3 Rural and community development forestry

In the section above, we derived first the biological and then the financial (basic economic) optimum forest timber management strategy. In Chapter 6 we introduced other techniques that might theoretically calculate a full economically optimal management strategy in terms of the total economic value (TEV). In this section we introduce yet another consideration: what is the community, or 'socially' optimal strategy?

What do we mean by 'community optimal'? Essentially, we are asking how the woodland resource can be managed for the maximum benefit of the people who are most directly affected by it. This is likely to be those who work in or near it, and those who live nearby. In developing countries, these are people who may well be part of the subsistence economy, dependent on the forest for many products and for whom the financial optimum if the timber is sold on international markets is irrelevant: (a) because they will not receive the revenues, or indeed possibly any compensation; and (b) have no alternative livelihoods. For such people the TEV is similarly of no practical interest unless they receive some form of transfer payments for maintaining many of the non-market values that the TEV records, and forgoing harvesting options.

Given the caveats above, forest managers and planners in developing and, more recently, in developed countries have been increasing attention to woodland management based around the principles of rural development forestry (RDF) and community development forestry (CDF). These approaches have recognized that forest management should, and in many cases must, involve local people if it is to be successful. We might summarize these different approaches below.

Objective	*Decision Criteria*
Biological optimum	MIA (MSY)
Financial optimum	NPV of timber
Economic optimum	NPV of forest TEV
RDF	NPV of forest TEV with weighting for local employment/local retention of benefits
CDF	Local management plan (implicit community TEV).

The basic rationale of RDF and CDF management approaches is fourfold.
1 They recognize the multiple uses of forested areas: for timber, other minor non-timber products, recreation and conservation, among others.
2 They concentrate on small-scale business development, preferably at the local level.
3 They require local involvement in the planning of management strategies.
4 They look to secure the benefits of local strategies for the benefit of local people.

Some preliminary economic investigation of RDF have been developed by Slee & Snowdon (1995) in Scotland, UK. They investigated the potential for alternative management approaches in Scottish forest areas. Extensive consultation was carried out with local communities in these areas, which led to the development of three management strategies: one employment-centred, one recreation- and amenity-centred, and one a combination of the two, recognizing that there was some potential conflict between these two centres of interest. Employment creation was encouraged via policies that included employing local people rather than contractors for felling, using local support services, tree nurseries

Table 12.2 (a) and (b) Results for a prelimary analysis of rural development forestry options in Moray and Wester Ross, Scotland. The options considered were employment-centred (RDF a), recreation- and amenity-centred (RDF b), and combined employment and recreation/amenity (RDF c). (Source: Slee & Snowdon 1996.)

(a) Financial and economic returns per hectare in terms of net present value for control and rural development forestry (RDF) forest systems in Wester Ross (at 8% discount rate).

	Net present value (£/ha)			
Forest system	Financial value	Recreation (including wildlife)	Carbon sequestration	Economic value
Control	–1 489	49	225	–1 215
RDF a	–1 626	49	225	–1 352
RDF b	–2 093	148	180	–1 765
RDF c	–2 187	148	180	–1 859

(b) Financial and economic returns in terms of net present value for control and rural development forestry (RDF) forest systems in Moray (at 8% discount rate).

	Net present value (£/ha)			
Forest system	Financial value	Recreation (including wildlife)	Carbon sequestration	Economic value
Control	–1 172	223	196	–753
RDF a	–1 355	223	196	–936
RDF b	–1 669	668	186	–815
RDF c	–1 789	668	186	–935

All carbon sequestration values used in Table 12.2 (a) and (b) are adapted from Pearce (1991) (in *Forestry Expansion: A study of technical, economic and ecological factors, Paper* No. *14*) who used a 6% discount rate. All other values used (financial, recreation and wildlife values) are calculated at 8%.

and processing mills, and utilizing smaller scale working methods, such as motor manual felling. Conservation policies included more diversity of tree species and active conservation management.

Using Forestry Commission data, the changes in harvesting costs under these three strategies were compared with a control. The results of their preliminary calculations indicated that none of the alternatives could generate returns equal to the conventional silviculture system, though they point out that in these cases the current system is always likely to have cost advantages because it is already established. The preliminary figures are given in Table 12.2(a) and (b). It should be noted that these figures report only a limited financial analysis.

In addition, they considered the multiplier effects of local jobs on the local community: these are the additional jobs that are created for the original forest jobs in ancillary and related industries, such as processing, mechanical servicing, etc. These were around 1.2–1.3 (meaning roughly one extra job outside forestry for every three jobs within it; this is an average figure for land-based rural industries). Also important for a fuller analysis is the cost of creating other jobs in forest areas, which is a concern of government. Rural development forestry was estimated to provide employment at a cost of £50 000–80 000 per job, falling to £35 000 if multiplier effects were considered within this.

Community development forestry retains the local focus of RDF, with a wholly bottom-up approach to the management planning process. Less emphasis is placed on the direct employment benefits as opposed to facilitating the development of a locally based plan that is agreed communally. Approaches such as participatory rural appraisal (PRA) (see Section 10.7) have been utilized in both developing and developed forest areas as a part of the CDF process. Where ecological regeneration is important, such as reforestation projects, PRA is now a statutory part of many development organizations' planning approach.

The approaches exemplified by RDF and CDF are interesting primarily because they link forest resources with locally orientated management. In developing countries, involvement of local people in all aspects of planning has been recognized as a critical factor in the success of projects. Community management is also seen in all contexts as an alternative to decision-making systems reliant on cost–benefit analysis and other more formal planning appraisal techniques. In effect, forest-based communities can reflect the value of many of the forest's functions in their plan, without necessarily quantifying all of these individually.

12.4.4 Common property rights and local forest use in Nepal

We noted in the Introduction to this chapter the significance of property rights in relation to problems of resource use management. The simplified typologies of rights which are used in economic analysis emphasize the incentive structures that different systems of property rights present to those involved in conserving or exploiting the resource. The system of incentives hinges on the opportunities for making profits and sustaining losses, given the behaviour of others and the nature of the resource. Many incentive problems relate to the management of open access resources, most obviously forests and fisheries—those areas that are not subject to either privately or collectively enforced forms of management. These often provide the strongest incentives to free-ride at the expense of the population as a whole.

Despite the economic logic of these analyses, there are empirical studies of local resource use behaviour in developing countries suggesting that the economic assumptions of free-riding behaviour are questionable. Bromley and Chapagain (1984) present an interesting example of forest management in Nepal subsequent to its independence in 1951.

The pattern of local resource use in the Nepalese highlands centred on the village. Village members, in addition to any land they may own privately, also have access to other designated areas for purposes of gathering fuelwood, fodder and other products, under a system of rights called *nistar*. Areas covered by *nistar* include waste land, such as road verges, ditches and inaccessible areas, in addition to forested lands under community control.

On achieving independence, the Nepalese government initiated a series of development plans, drawn up by the central state authority. One key initiative, from the perspective of forest resource use, was the nationalization of all forested lands, transferring ownership from local communities to the state. Three main reasons seemed to support this move:

1 population was increasing rapidly, putting pressure on local resources;

2 programmes to control malaria had opened up previously inaccessible hill lands to cultivation, accelerating the rate of land clearance; and

3 the perception at state level was that national management of Nepal's natural resources would be superior to that of many isolated individual villages (Bromley & Chapagain 1984).

This reallocation of property rights was unfortunately confronted with two fundamental problems. On the one hand, there was no reduction in the local demand for forest products and villagers were therefore deprived of an important local source of these products. On the other, government had no effective way of enforcing its jurisdiction over the nationalized lands. In addition, combined with this inability to apply national management was a growing distrust of the central authority on the part of many rural people. This probably had its roots both in unrealistically high expectations about the speed with which economic development would move following independence, and in the recognition that the national forestry policy was targeted more towards the potential for national revenues from timber sales than meeting local and domestic demand in the traditional subsistence section of the economy.

As a consequence of nationalization, Bromley and Chapagain suggest there was an initial surge to convert forest to agricultural land, thus retaining it within the local community. Secondly, a sense of individual responsibility for observing the use of *nistar* lands was weakened and the nationalized lands were opened up to greater exploitation as a consequence.

In effect, concentrating on a single policy initiative—nationalization—had the very undesirable consequences of both undermining faith in government and, most crucially, of removing a sense of local responsibility by the act of reallocation. Coupled with this, there was no effective policy to tackle either the basic supply shortages that were driving exploitation, or the demand for resources, by providing alternative supplies or technologies. The lack of state enforcement, which had been less critical when an ethic of personal responsibility was maintained within the village, was a critical failing when those responsibilities were denied by state appropriation.

As a part of the field research on this project, Bromley and Chapagain also tested the community orientation of village members through a questionnaire asking individuals about their willingness to make donations to a village development fund. Their conclusions were that many villagers did not show a tendency to free-ride, even if they knew others in the village were doing so. The investigation was, of course, hypothetical, and it is possible that those questioned would behave differently in real life than in response to a questionnaire. Nevertheless, a clear understanding and acceptance of a fairness ethic was apparent amongst those questioned, and insisted upon as the appropriate model for behaviour.

In conclusion, Bromley (1991) suggests that
1 informal rather than legally defined community values may operate to achieve effective resource management;
2 changing institutional arrangements can change the operation of these values; and
3 formal institutional changes without supporting norms are potentially problematic, particularly when such changes are unenforceable.

12.5 Fisheries

The issues surrounding the management of fisheries have been developed to great mathematical sophistication. In this section we introduce some of the basic principles related to the harvesting of fish. In addition, we also consider some technical and socioeconomic aspects relating to fisheries management policy.

Although fish are often considered to be an open access kind of resource in economic modelling, labelling of property rights is not a simple task. As Buck (1989) points out:

> 'A fish caught on the high seas, which is no one's property, or *res nullius* (Grotius 1972/1608), becomes its captor's property by virtue of his labour; no nation or individual could lay claim to it prior to the capture. This lack of ownership, however, resides not in the fish but rather in its location. The same fish caught in a Scottish lord's salmon stream is private property before the first line is cast, which is why fishing rights in salmon and trout streams may be leased. Had the fish travelled instead into American waters to spawn, it would have belonged to the state (*res publica*) in which it was found and its capture bound by numerous state imposed restrictions such as season, licences and gear. However, our peripatetic and biologically unlikely fish may have swum into waters whose fishing is assigned to a tribe of Native Americans, in which case it would become the property of the entire tribe (*res communes*) rather than of one individual. Thus we cannot casually label a fish as a common property resource: it may be *res nullius*, *res communes*, *res publica*, or simply private property, depending upon where it is found, how it is caught, and by whom.' (Buck 1989)

12.5.1 Population growth and sustainable yields

By a fishery, we mean a geographically discrete fish population: that is, a group of fish that interbreed but whose population size is limited by the carrying capacity of a particular area. A typical growth pattern of such a population is shown in Fig. 12.6 (known as the Schaefer curve). Note that the vertical axis plots the *marginal* or *instantaneous* population growth; that is, the net additions to the stock dependent on the overall size of the biomass. At k, the carrying capacity of the habitat, new births equal deaths and a relatively stable population is reached.

It is the net additions to the biomass that provide the opportunity for a sustainable harvest. If only the net additions to the stock are harvested over any given time period (the instantaneous growth), then the size of the total biomass will be maintained at a constant level. For any given size of biomass, therefore, a sustainable catch can be harvested that exactly equals the instantaneous or marginal growth of the biomass. Given this observation, we can identify the maximum sustainable yield (MSY), the largest catch which can be harvested indefinitely. This is catch h_m, with biomass b_m (Fig. 12.6). This maintains the population at its maximum *biological* productivity. An example of an estimated MSY for the Pacific halibut is given in Fig. 12.7.

What happens if the catch rate differs from the growth rate? We can examine this situation using Fig. 12.8. For any catch below the MSY, such as h_s, there are two sizes of biomass (b_l and b_h) where marginal growth exactly equals catch size. These harvest/biomass combinations are in equilibrium. Consider now what happens if the biomass is to the right of b_h. In this case the instantaneous growth curve lies below the harvest rate h_s. The size of the biomass will therefore be reduced, increasing its productivity until it reaches b_h,

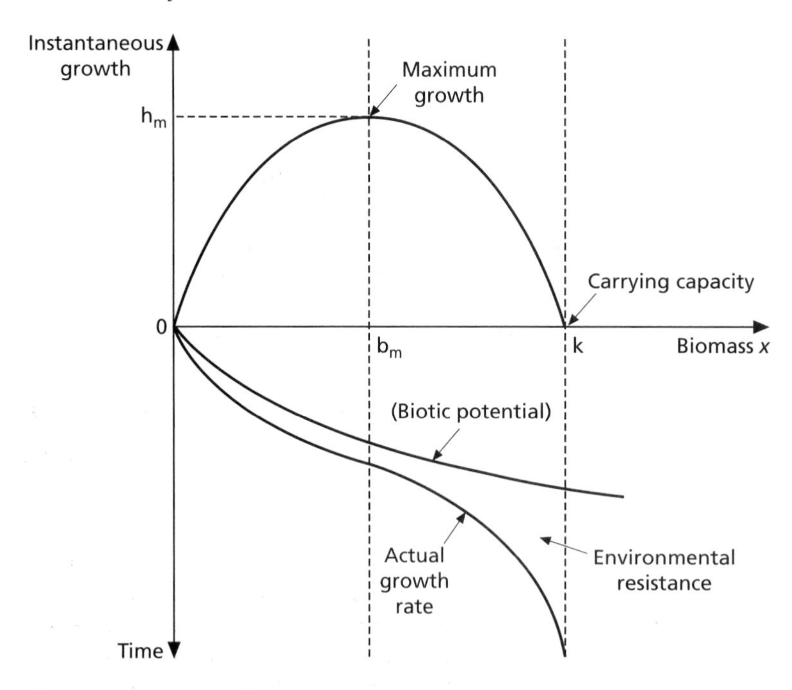

Fig. 12.6 The Schaefer curve, showing how instantaneous stock growth varies with total stock biomass. The potential for continued exponential growth in biomass is restricted by environmental conditions or 'resistance'.

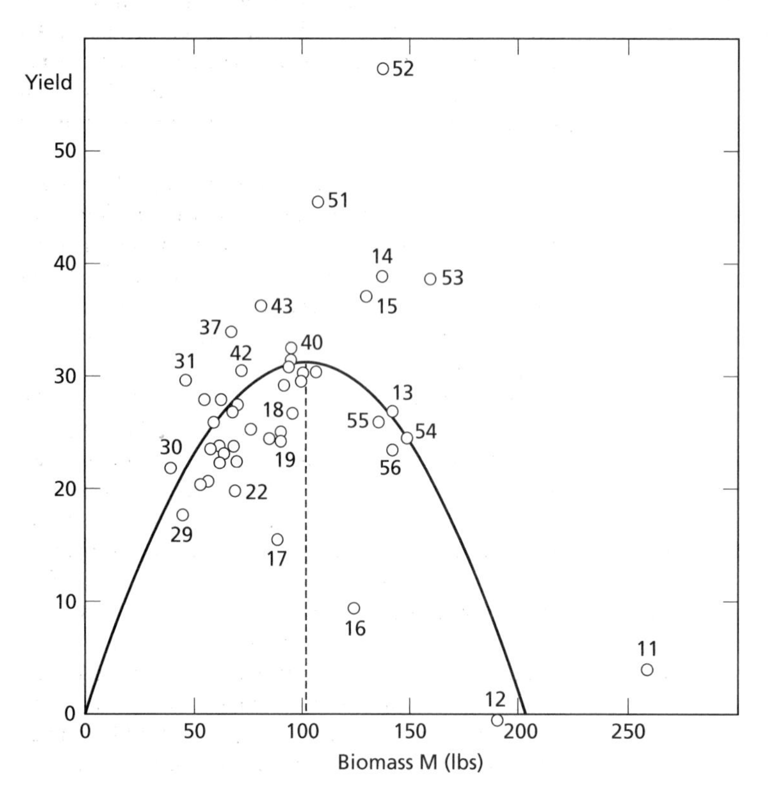

Fig. 12.7 Yield data for the Pacific halibut catch over the years 1910–1957, with an estimated marginal growth curve fitted. (Source: Ricker 1975.)

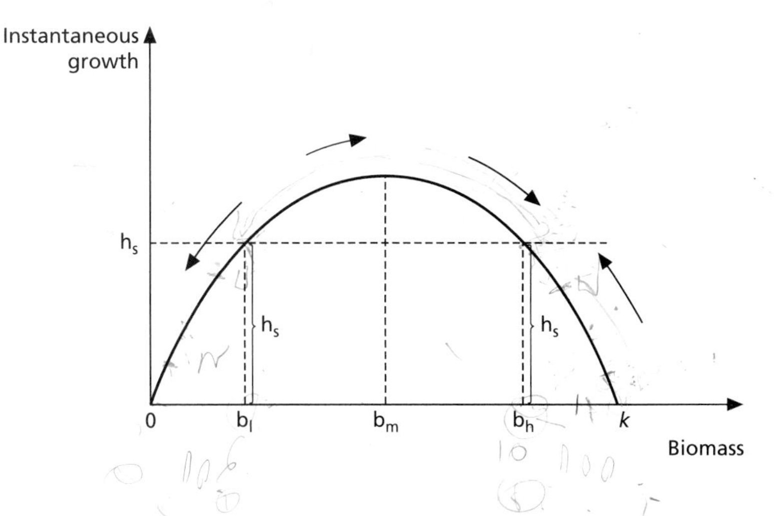

Fig. 12.8 Harvesting dynamics for different population sizes. A sustainable harvest, h_s, can be drawn from two levels of biomass, b_l and b_h. This harvest is in stable equilibrium with the biomass at b_h but in unstable equlibrium at b_l.

at which point the situation attains equilibrium. What happens to the left of b_h? Here the marginal growth exceeds the catch, so there are net additions to the stock. The biomass therefore expands, reducing productivity until it again reaches a stable equilibrium at b_h.

The situation at b_l is quite different. If the biomass is to the right of b_l, the stock will expand since marginal growth is greater than catch: it will expand until it reaches the stable equilibrium at b_h. To the left of b_l, however, harvest exceeds marginal growth: the stock size is therefore reduced, eventually causing extinction. These dynamics are indicated by the arrows in Fig. 12.8. From a precautionary principle, it is therefore always preferable to operate with a biomass above median carrying capacity, and with catch rates set somewhat below the MSY.

12.5.2 Private owner harvesting problem

Having established the basic biological form of the population, we turn now to the human effort required to harvest the stock. Here we find that the maximum biological yield may not equate with the optimum *economic* yield. This is because of the *fugitive* nature of fish stocks; as fish get scarcer, more effort is needed to catch them (we are assuming here that we are dealing with so-called 'search' fisheries, where the fish are more or less evenly distributed across the habitat, and not a 'schooling' fishery which has somewhat different properties). In this section we explain the thinking behind the economist's calculation of the economically optimal harvesting rate for a privately owned fugitive resource.

The economic objective of a fishing boat is to maximize *total profit* from fishing. Total profit is simply the total revenues from fishing minus the total costs of fishing. In this case, fishing effort incurs costs (in terms of boat hire, crew wages, fuel, etc.), and the fish catch landed provides the revenue when sold at market. The fisherman wants to fish up to the point where the revenue from fish sales, less fishing costs, is maximized. Here we introduce an important point. The total profit for any firm is maximized when marginal revenue exactly equals marginal cost. What does this mean exactly? We explain it below with reference to the fish problem.

First, the *marginal revenue* from fishing is the revenue from each successive fish that is landed, in addition to those already caught. This is a constant value—namely, the price of a fish in the fish market. The total value of the catch goes up with every fish caught, but each one of those fish contributes the same marginal revenue: the price of a fish.

Now consider the *marginal costs* of fishing. The marginal costs of fishing are the costs of landing each fish, in addition to the fish already landed. In other words, marginal fishing costs are costs per fish caught. Importantly, as more fish are caught, fewer remain in the sea, and therefore those remaining take more effort to catch: the costs of catching additional fish therefore increase as more fish are caught. The marginal cost of fishing therefore increases steadily as fish get scarcer.

Now consider the relationship between costs and revenues. Marginal revenues from fish caught are constant, whereas costs per fish caught are steadily increasing. At some point, it is not worthwhile to incur any

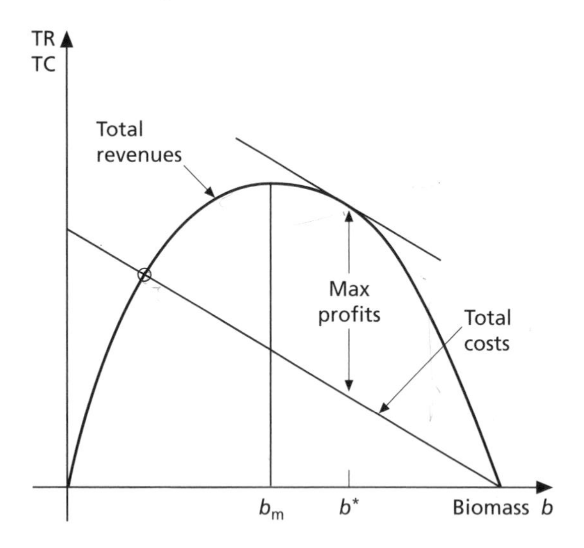

Fig. 12.9 Total costs and total revenues for fishery harvesting. Profits are maximized when the biomass is at b^*, and the difference between total revenues and total costs is at its greatest.

extra (marginal) cost in fishing effort, because the extra fish caught, which provides the marginal revenue when sold, is worth less than the effort expended (in extra fuel, time, etc.) to catch it. We have gone beyond the point at which marginal cost equals marginal revenue. Why, then, does this point maximize profits? Up until this point, the revenue from each fish is higher than the cost of catching it. Therefore each fish caught is contributing to total profit: it is contributing the difference between its market value and its cost of capture. Beyond this point, the cost of catching the fish is actually greater than the value it yields when sold; trying to land more fish is now reducing total profits, because fish are simply too expensive to catch. In order to maximize profits, the boat owner should therefore catch fish up until marginal fishing revenue equals marginal fishing costs; there are simply no more opportunities to make money on any fishing trip beyond this point.

We can represent this graphically in Fig. 12.9, where we show both total costs and revenues. Note first that the harvest curve is itself the total revenue curve: total revenue is simply size of harvest (fish caught) multiplied by fish price. Interpreting the total cost curve, and ignoring for simplicity fixed costs such as the fishing boat itself, it lies theoretically at zero when the fish stock is at full carrying capacity. This is because in a full sea, catching a fish is, theoretically at least, very easy. As more and more fish are taken out, the costs incurred in catching what remains rise until a maximum, the total cost required to catch the last fish in the ocean.

Having established these relationships, we now need to identify the point at which the difference between total costs (TC) and total revenue (TR) is maximized: this yields the greatest possible profit. Geometrically, we can see it lies at the biomass stock where the tangent to TR is parallel to the TC line, given by b^*. This therefore defines the economically optimal level of fishing effort, or what we can call the optimum sustainable yield.

12.5.3 An open access analysis

The theory above was a general case based on the optimal harvest under private, or state-controlled, management. Private management can maximize profits by maintaining the fish stock at an appropriate level. But what happens in an open access situation? This classic situation has been analysed exhaustively by fisheries economists, and a particularly interesting case study is presented by Wilen (1976) in relation to the North Pacific fur seal.

At the turn of the century the North Pacific fur seal fishery was found along the north-west coast of North America, where the seals occurred in large numbers and migrated long distances along the coast. The harvesting of seals on land during the mating season was governed by international treaty, but during the coastal migrations the situation was one of open access. Following a number of years of high yields and increasing fishing effort, two further international treaties were signed in the early 1900s effectively ending the fishery altogether, in the interests of preserving the seal stock from collapse. Wilen examined the harvest pattern during the last years of the nineteenth century, when harvesting efforts increased significantly, in order to assess the likelihood that extinction would have occurred had an open access situation continued.

The full details of the study are quite complex but it rests on two essential relationships. First, firms (fishing boats) enter an industry because they see potential profits to be made, and they exit when they cannot cover their costs. Secondly, the opportunity to make profits depends on the available seal stocks and also on technical factors (cost of crews, available boats, and so on).

Given these conditions, potential fishermen sum up their chance of making money. If potential profits are high, because there are a lot of seals, many new boats will enter fishing. However, as more boats enter, the

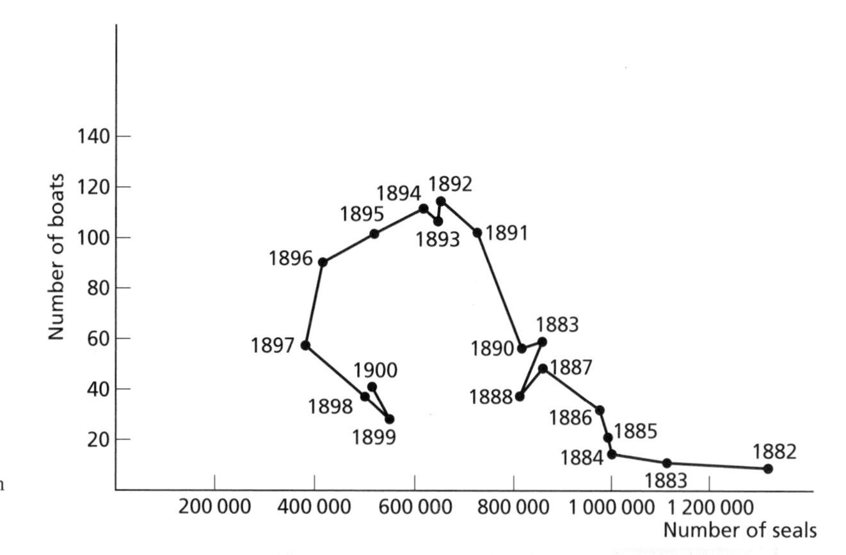

Fig. 12.10 The variation in North Pacific seal stocks and fishing effort, 1882–1900. (Source: Wilen 1976, in Neher 1990, with permission.)

seal stock goes down, average catches go down and marginal costs go up, and profits per boat thereby go down. Eventually the stock depletes so far that profits for the worst fishermen, who have slightly higher costs than others, become negative, and they are forced to leave the industry. The important result of this relationship is that the fishery may be self-regulating: as seals become scarcer, it simply becomes unprofitable to land them beyond a certain aggregate catch, given the prevailing state of fishing technology.

In reality, because there is imperfect information about total numbers of seals, about fishing costs and about boats, including those already fishing, those intending to leave and those considering entering, the self-regulation aspect is likely to be relatively imprecise. What is important to Wilen's analysis is that given a particular average state of technology and a particular fugitive biomass, there may be a point of equilibrium where boat profitability and available biomass balance. Fishermen respond to profit incentives and fishing stocks respond to the fishing effort, and the two interact dynamically to determine a catch–biomass equilibrium point for the industry.

By painstakingly examining archival sources, such as ships' logs and traders' inventories, Wilen pieced together the catch and boat numbers for the seal fishery between 1882 and 1899. These are shown in Fig. 12.10. The path shown does indeed bear a striking resemblance to the predictions of the theoretical model. A more recent example is given by Conrad (1995) (Fig. 12.11) for the North Atlantic herring stock, which shows a similar

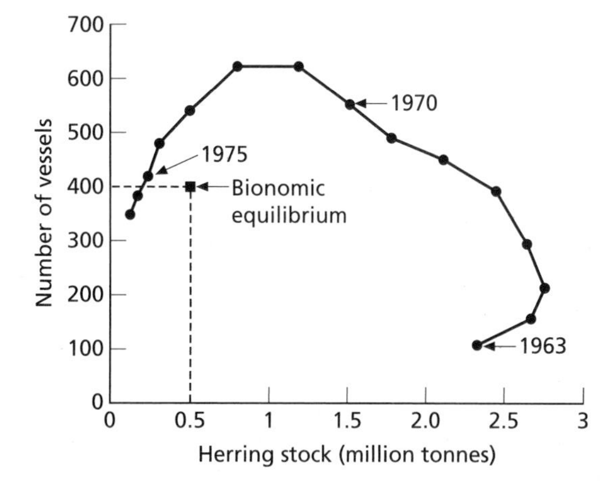

Fig. 12.11 North Atlantic herring stocks variations over time. (Source: Conrad 1995.)

situation, though here again it is interesting that government action was taken to limit fishing in 1976.

Although these analyses seem to support a *laissez-faire* approach to fisheries management, there are several important qualifications that should be made.

1 Most significantly, improvements in technology shift the catch–biomass equilibrium point to the left. Less effort has to be used to find fish, or alternatively, larger harvests can be achieved without an increase in fishing effort. The risk of stock extinction is therefore increased as technology improves.

2 Estimates of current growth may not be good for the

future. Biological uncertainty makes it very difficult to estimate the MSY correctly.

3 Each individual boat has an incentive to maximize its own profitability, even if this leads eventually to population extinction. This is an example of the 'fallacy of composition'—what seems good for the individual is bad for the whole group. Without some form of agreement to control access in some way, this state will persist.

4 Small-scale artisan fisheries are part of the subsistence economy in many developing countries which sustain numerous small communities, largely outside the main monetary system. Commercial fishing fleets, which are highly capital-intensive, can eliminate these subsistence livelihoods.

12.5.4 Sustainable ocean fishery management options

Ocean fishery management has been likened to a game between fishery managers and the fishermen themselves, with the fishermen always left the last move (Anderson 1995). Whatever the intention of the managers with regard to resource conservation, individual fishermen will always have incentives to work around the restrictions in other ways that thwart the initial intent. In this section we review the economic, ecological and social implications of the range of main management approaches.

The central pillar of most fisheries management is the total allowable catch (TAC). This must not exceed the MSY, and ideally would be set to ensure a safe minimum standard of biomass growth. The TAC for a number of pelagic (ocean) fisheries are set by international boards each year, with nations bargaining with each other for their share of the allowable catch. Despite the need for keeping within biological limits, this may be exceeded for two obvious reasons. First, scientific uncertainty may simply lead to a misestimate of the MSY, and also weakens the hand of policy advisers because political and economic pressures are always in favour of less restricted access. Secondly, negotiation failure can prevent an acceptable allocation of TAC being reached, in which case individual nations may allow their own fleet a greater catch than is required to maintain sustainability.

Four main management instruments are discussed below: input orientated measures (tackle; boats; time); area-based measures; financial measures (taxes; subsidies); and licence arrangements (quotas; landing limits).

INPUT ORIENTATED MEASURES

Restrictions on inputs into fishing are, in economic terms, ways of reducing the effective effort of fishermen. Examples are restrictions on minimum mesh size of fishing nets, maximum engine power or haulage power for raising nets, and the maximum storage capacity of individual boats. All these measures reduce the fishing impact of individual boats; they make the capital employed in fishing relatively inefficient. Put another way, they make it more expensive to catch fish. The result of this is twofold: there is a deadweight loss to society because more people are tied up in fishing than theoretically need to be; and fish is thereby more expensive than it would be if the system was centrally planned. As it is, input restrictions effectively lock in a level of inefficiency into the fishing fleet.

From a socioeconomic perspective the issue is not so clear-cut. Where fishing communities are small and reliant on fishing income, deliberate limits on capital efficiency can form part of an organized, long-term adjustment strategy that retains people in meaningful work while the local social and economic infrastructure is developed to provide alternatives; an applied example of Schumacher's principle of appropriate 'small-scale' activity. This is likely to be a long-term process. It is also a distributive justice issue, concerning to what extent consumers should have to pay more for fish, and how government money is allocated to support 'inefficient' economic sectors.

On an ecological note, mesh sizes are clearly a positive restriction, particularly by allowing immature fish to escape (see Fig. 12.12). Trap size limits are a central element of lobster harvest management in Maine and Novia Scotia, for example, enabling more females to reach breeding age. However, problems remain if stocks are multispecies. Limits on engine power and haulage carry additional risk burdens for what is already a risky industry; fishermen have an incentive to underpower larger boats, or overload them, making them riskier in high seas.

In fishing terms time is also an input, and a number of fisheries operate closed seasons to allow stocks to recover. These are not without problems; supply is consequently variable, whereas demand is relatively constant, leading to unnecessary price fluctuations. Capital is tied up in dock for parts of the year, which is an example of an inefficient allocation of resources. However, fishermen are free to take other jobs during

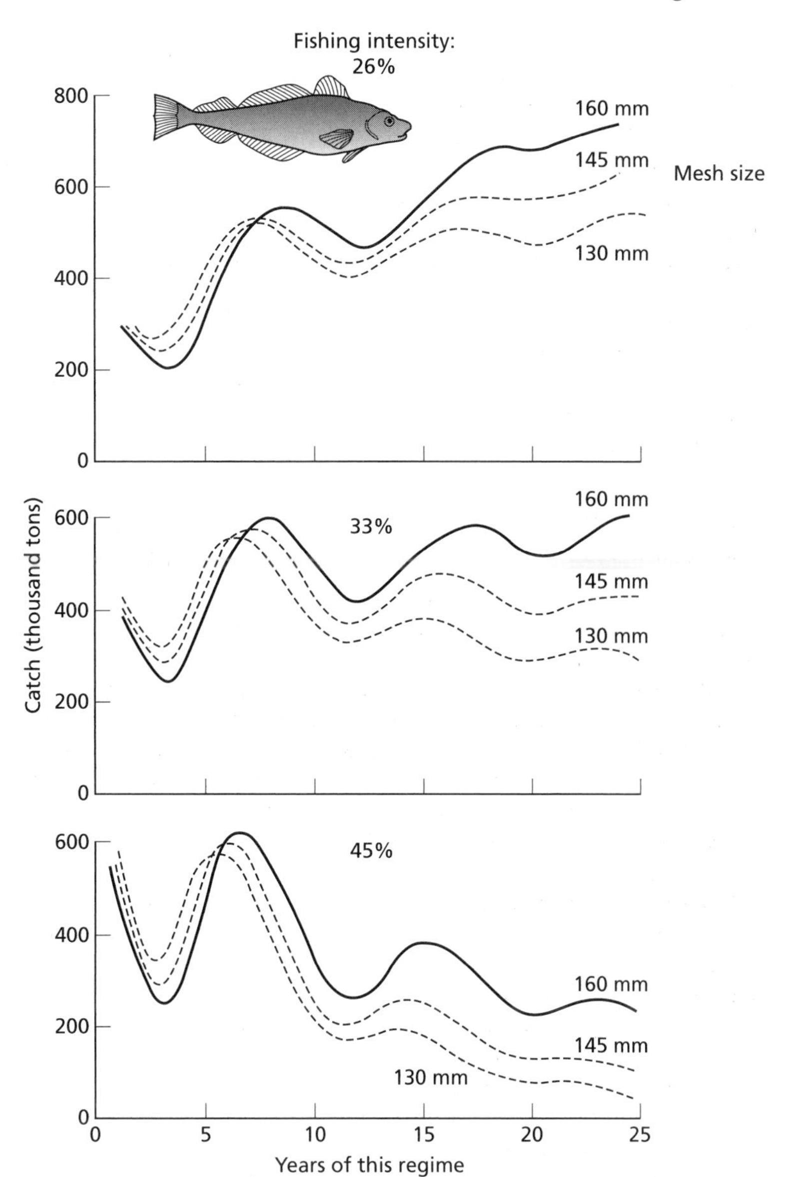

Fig. 12.12 Predictions for the Artic cod stock under three fishing intensities and using three alternative mesh sizes. (Source: Begon *et al.* 1990.)

closed seasons, so this is less inefficient than reducing the catch per trip. If fishing times are heavily restricted and stocks deplete quickly once opened, boats have an incentive to 'get in first' as this maximizes their profit potential. This adds to risk—for example, fishing despite bad weather—and to the bumpiness of supply. From an ecological point of view, sudden depletions and recovery are generally less satisfactory than a stable state, and less prone to sudden shocks.

AREA RESTRICTIONS

Area restrictions take three basic forms: protection of key habitats (for example, spawning grounds); national territories (economic exclusion zones (EEZs)); and local restrictions (established for the protection of indigenous or small-scale artisan fisheries).

Area-based measures can have obvious benefits in ecological terms, where there is little dispute over their usefulness. As ever, problems are chiefly political in establishing to universal satisfaction that restrictions should be in place. An advantage of exclusion zones is their ease of monitoring; it is far easier to detect the presence of a violating boat, than to stop and investigate mesh size or other input regulations at sea.

Local restrictions are particularly important where the interaction of local systems with commercial fleets is a major equity problem. In instances where foreign fleets pay for licences to fish in another nation's waters (e.g. Japan) there are serious conflicts between the point of revenue collection, at national government level, and the local level impacts on welfare of indigenous subsistence economies, which may well be both severe and remain uncompensated. As local systems are generally coastal, they can be effectively protected by exclusive territories, though these coastal waters may also be particularly abundant in fish, creating the pressure for incursion.

FINANCIAL MEASURES

Financial incentives for general environmental management are covered in detail in Chapter 14, and only a few of their effects are noted here. Financial measures are either positive (subsidizing preferred activity) or negative (punishing unwanted actions).

Taxes can be applied to any material input into fishing, making it more expensive to fish, and to output by taxing catch landed. In all cases they act to raise the price of fish to consumers and to reduce the income to fishermen, redistributing it to central government. In an industry with low incomes, this can be a politically difficult move to make.

Subsidies in relation to fishing can act to encourage or reduce effort. The reduction can be achieved through government retirement incentives, ship buy-out schemes and other financial inducements for effort to leave the industry, or simply not to fish. These subsidies are transfer payments by the state to provide fishermen with alternatives to fishing. The difficulties arise in setting these incentives at effective levels when government spending is limited. Fishermen, like some other occupations, such as farmers, that represent a well-defined way of life, are often very unwilling to move into alternative jobs that lack the same social identity, even when alternative jobs are available in their locality. This may partly reflect a personal preference but also personal aptitudes and local traditions.

The alternative subsidies are those that encourage industrial activity, giving grants for boat-building, investment in new tackle and so on. These tend to be made in times of economic boom and when biological stocks undergo unexpected increases. Some subsidies are one-off, others ongoing. There is evidence that direct and indirect subsidies over time have led to vastly over-capitalized fishing fleets, which are difficult to manage and which create additional political pressure for more open access (Huppert 1990).

LICENSING ARRANGEMENTS

Fishing licences of various forms can operate to restrict the fishing fleet to within TACs. In their simplest form, a licence may involve agreement to observe certain restrictions on activity. More usually, licences are restricted to those who qualify for them under some set of conditions, such as belonging to a fishing family or having already served on a fishing boat. These help to reduce the numbers who may obtain licences, though their effect on achieving the TAC obviously depends on many other factors.

Other than acting as a filter to deter large numbers of potential fishermen, licences may operate as quota systems. The quota may apply to the total catch landed, to the number of boats (see proviso above), to the number of days allowed at sea, or to the number of trips made per year. Each of these may be appropriate in certain circumstances, and each sets different incentives in front of fishermen. Landing and trip limits encourage high-grading, when boats throw away low-grade fish at sea so that they land as many high-value fish as possible. Limited days fishing also encourages riskier practices, staying out at sea for longer periods to maximize time spent in good fishing waters. Boat limits can be undermined by up-grading gear, as noted above.

Individual transferable quotas

A relatively recent innovation is that of individual transferable quotas (ITQs), which in theory provide for the most efficient allocation of effort in landing any given TAC. The theory behind ITQs is similar to that for industry pollution permits, explained in Section 14.3.2. Each boat is allocated an individual allowance or quota which allows it to land a certain catch, with the sum of all such quotas adding up to the TAC. As the quota is transferable, the fisherman has the option either to catch exactly up to his own quota, or sell some (or all) of it and catch less fish, or indeed try to buy more in order to land more fish than he was originally allocated.

The principle of a transferable quota is that each fisherman will consider his marginal costs for fishing, the revenue from the fish he catches, and the revenue from selling some portion of his quota. The key factor is the difference in marginal costs between boats; those that are highly efficient (better equipped, faster, using sonar, etc.) can fish more cheaply than others, and therefore make more profit per fish. If they are sufficiently efficient, they can compensate the more costly boats for not fishing by buying their quota, and still make a profit themselves. Everyone should gain as a result.

Quotas have been introduced in various forms into New Zealand, America, Australia and Canada. Note that they rely on exclusion of other boats and a reasonably fluid market (if it is too inflexible the transaction costs of locating willing buyers or sellers are too high and no trades will occur, in other words no market for permits is established). Resistance is generally high amongst fishermen as quotas limit their potential earnings; setting an acceptable initial allocation is therefore important. Note too that announcement of a quota or licence system can lead to a surge in landings as uncertainty exists about future catches and operators look to maximize their short-term returns.

In New Zealand, where the system does appear to have had some success, three problems were particularly noteworthy. Initially, there was some suspicion about how long the ITQ regime would last, which encouraged fishermen to over-catch/cheat where possible, because the long-term benefits were questionable. Secondly, the authorities auctioned permits to raise capital. This led to two problems. It effectively taxed the fishery, as everyone who fished had to pay to do so. This caused resentment amongst fishermen, particularly those who already had low incomes. Furthermore, the permits quickly became concentrated in relatively few hands: an effect described as the giants getting bigger and the dwarfs getting smaller. This shifted the traditionally mixed fishing fleet to one with a few large operators. Whether or not this was considered socially desirable, it created a new strongly focused lobby group that increased the political effectiveness of the fishing lobby, pressuring the government for ever-larger quotas.

12.5.5 Fisheries: concluding comments

The management of ocean fisheries involves a complex array of economic incentives, biological relationships and political concerns. As in other primary industries, such as farming and forestry, many of those involved in fishing are often dependent on the resource, from small communities with few job alternatives. Management must take into account the monitoring/enforcement costs of any policy, in addition to its stated objectives. In many instances, this requires at least the cooperation and preferably the active participation of those affected by the policy, in design as well as enforcement.

Further Reading

Berkes, F. (ed.) (1989) *Common Property Resources: Ecology and Community-Based Sustainable Development*. Belhaven Press, New York.

Bromley, D.W. (1991) *Environment and Economy: Property Rights and Public Policy*. Blackwell Scientific Publications, Oxford.

Hill, B. & Ray, D. (1987) *Economics for Agriculture: Food, Farming and the Rural Economy*. Macmillan, London.

Price, C. (1989) *The Theory and Practice of Forest Economics*. Blackwell Scientific Publications, Oxford.

13 Nature Conservation

'It took Britain half the resources of the planet to achieve its prosperity; how many planets will a country like India take?' [Mahatma Gandhi]

13.1 Introduction

In this chapter we focus on issues concerning the preservation of species and habitats. This debate has essentially two aspects: first, how to determine which species or habitats to protect and, secondly, how to organize that protection once the preservation decision has been taken. To put the urgency of this debate into some context the well-known ecologist E.O. Wilson (1980) suggests that of all the aspects of modern environmental problems, the extinction of species is the one for which future generations are 'least likely to forgive us'. The increased rate of extinction in all taxa during the last century are well documented, and the causes are largely anthropogenic, often being related to habitat destruction and the introduction of alien species into areas. Some evidence for this comes from data presented in Fig. 13.1 which show that the a disproportionate number of the recorded extinctions of mammals which have occurred since 1600 have occurred in the lands most recently colonized by Europeans.

It is worth noting that the preservationist orientation of the conservation movement has come under strong fire from a number of prominent economists for utilitarian ethical arguments. These arguments are that species and habitat destruction frequently occur because alternative human wants and needs have simply

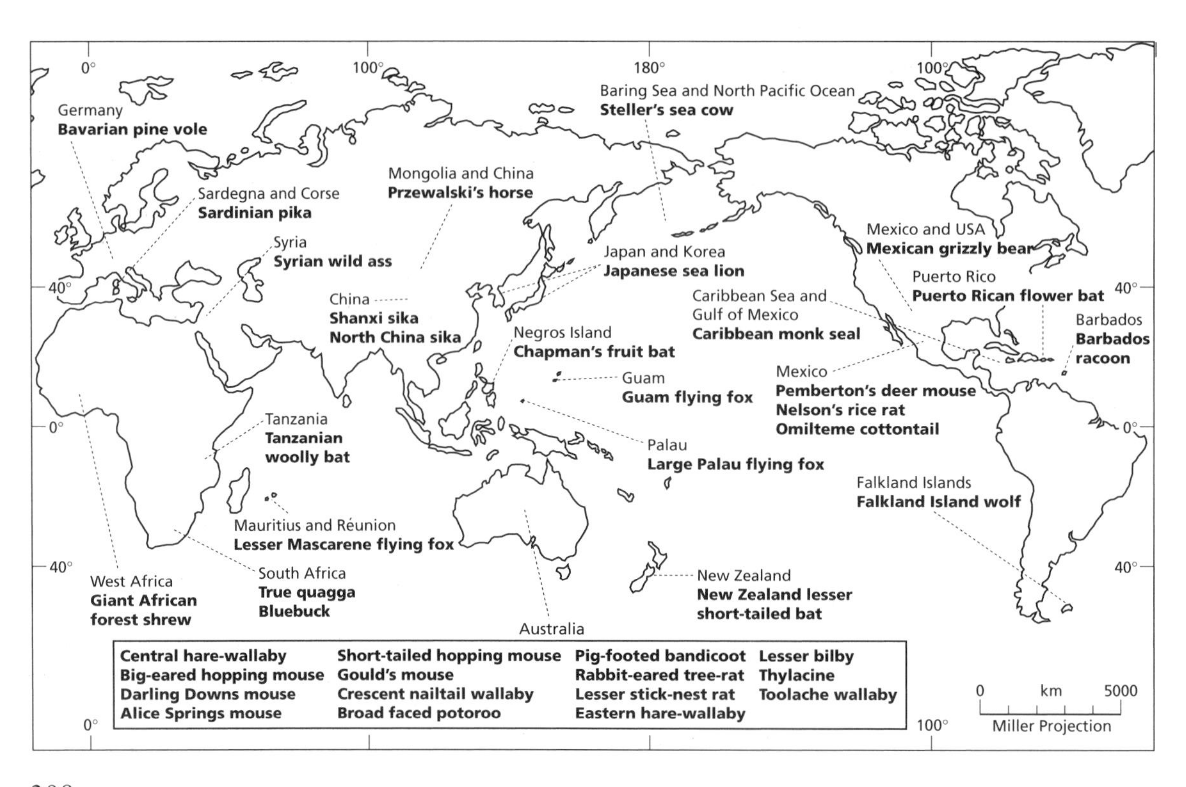

Fig. 13.1 Distribution of mammal extinctions since 1600. (Source: Caughley & Gunn 1996.)

outweighed, in significance, the preservation values of a species or habitat. Beckerman (1992) is characteristic of these critics and attacks what he perceives as overly zealous efforts at species or habitat preservation, particularly in the face of pressing human needs in developing countries. In his book, *Small is Stupid* (an ironic title in reference to E.F. Schumacher's (1973) seminal 'green economics' title *Small is Beautiful*), Beckerman calls for rational appraisal of the costs of preservation as against the benefits of development. He suggests that preservation is often a 'middle-class' Western obsession carried out at the expense of the poor, both between the classes within nations, and between rich and poor nation states. As Beckerman acknowledges, the forces threatening and driving biodiversity loss are vast and complex. In the face of this complexity, simple decision rules such as 'All organisms are of equal value' are quite simply worthless when there are a multitude of competing reasons for pursuing environmentally damaging development projects.

In this chapter we expand on previous discussion of ecological and ethical aspects of the conservation debate presented in Chapters 4, 5 and 7 and consider the economic, social and institutional aspects of the debate. In order to achieve this we first consider issues surrounding species, through four frames:

1 attempts to include them in economic valuation;

2 modifications to decision criteria based on valuation that arise as a result of uncertainty: the concept of the safe minimum standard (SMS);

3 two decision rules for conservation under budget constraints; and

4 the application of marginal analysis to preservation/development trade-offs in the case of the Northern Spotted Owl.

In the second part of the chapter, we consider protected areas and habitat conservation measures, beginning with two examples of economic valuation drawn from Africa. We then consider some institutional issues arising out of two UK conservation policy instruments, sites of special scientific interest (SSSIs) and environmentally sensitive areas (ESAs), specifically implementation problems surrounding SSSIs, and a cost-effectiveness comparison of the two instruments.

13.2 Species conservation

13.2.1 Economic considerations

One seemingly simple criterion for deciding whether or not a species is worthy of conservation is to consider the monetary benefits that the species provides to humankind and to compare these with the costs of maintaining its existence. A conventional economic analysis would conclude that if the expected benefits of conserving the species were greater than the expected costs of conservation, then it should be conserved.

The costs of species preservation are of two kinds: the *opportunity costs* of not proceeding with developments which threaten extinction, such as the forgone power output from a hydroelectric dam, which has a direct market value; and the *expenditure* on continuing protection of the species, such as policing reserves and prosecuting those endangering the species. This is also a directly measurable cost. As both these costs are directly valued in markets, placing values on costs (expenditure and opportunity costs) of preservation is relatively straightforward.

Estimating the benefits from preservation is a contentious task, and some of the problems with available techniques were introduced in Chapter 6. Nevertheless, it is worthwhile examining where these techniques have generated monetary estimates, and we review three examples below. These correspond to two of the components of total economic value (TEV): use values and existence values.

DIRECT USE VALUES OF MARKETABLE SPECIES

While food and shelter are the most obvious direct use values of plant and animal species, the pharmaceutical value of flora is less widely appreciated. In fact, their use for such purposes is widespread. Farnsworth and Soejarto (1985) estimate that around 25% of all US prescription drugs are based on plant material and, of those, three-quarters were already being used by indigenous peoples before being adopted for commercial development. Given the direct benefits that these drugs provide, one way to assess the value provided by a plant species is in terms of the drugs it provides. Three methods have been used to do this:

1 the market value of the plants when traded;

2 the market value of the drugs for which they are the source material; and

3 the value of the drugs in terms of their life-saving properties.

If we consider the market value of the drugs for which the plants provide source material, then evidence suggests that during the 1980s about 40 plant species accounted for the plant-based drug sales in USA. On the basis of prescription value only, each of these species was worth $290 million p.a.

However, if we value the drug according to the value of avoided deaths arising from the use of drugs, then assuming a statistical life to be worth $4 million this method suggests a value per plant of $6 billion p.a. Given that many of the plants from which these drugs are derived originate in developing countries, it is not surprising that the traded value of the plant material suggests a much lower value for the plants than the above two methods.

Not only can these sorts of calculation highlight the value of certain plant species to humankind, but they can also provide a basis of valuing habitats rich in plant species which may prove to be useful in the future (the option value of a habitat). For example, 65–75% of all higher plant species are indigenous to moist tropical forests, so by estimating the probability that a certain forest will provide one or more drugs in the future, and assuming that current values for drugs are valid in the future, it is possible to place a monetary value on an area of forest or any other habitat or group of species, in accordance with its probability of supplying a drug. Pearce and Moran (1994) undertook such an analysis and suggested that the value of 1 hectare of tropical rainforest ranges from $0.01 to $21 per hectare, yielding a net present value (NPV) for tropical rainforest of around $420 per hectare. This value is an option value for pharmaceuticals alone and does not include the other elements of TEV.

The issue of plant-derived pharmaceuticals raises a number of issues beyond the immediate values associated with the drugs themselves. One such issue is concerned with the ownership of base genetic material and the ability to patent drugs based on plant and even human-derived material. A second is concerned with the distribution of benefits from drug discoveries between sources of plant material and centres of drug development.

The patenting of pharmaceutical knowledge derived from genetic material gives rise to 'intellectual property'. Naughton-Treves and Sanderson (1995) note that US objections to early drafts of the United Nations Conference on Environment and Development (UNCED) Biodiversity Convention centred on the issue of intellectual property rights. The President of the USA (1993) asked for 'fair and equitable sharing of benefits arising out of the utilization of genetic resources' and supported 'the concept that benefits stemming from the use of genetic resources should flow back to those nations that act to conserve biological diversity and provide access to their genetic resources'.

In reality, the division of labour falls such that the advanced economies of the Northern Hemisphere possess the 'prospecting skills' to uncover and commercialize potentially valuable plant-based drugs, while the poorer developing economies possess the base material and often valuable knowledge about medicinal properties at the local level. How the benefits from commercial exploitation should be shared is a divisive issue. Private companies in the North may favour fixed prospecting fees on a per hectare or time basis, whereas developing world governments are inclined to regard royalties on subsequent discoveries as more equitable.

INDIRECT USE VALUES OF SPECIES

Species provide tangible benefits to humans in other ways besides being the source of food and pharmaceuticals. These benefits include recreational hunting and fishing trips and many other touristic activities. Estimates have also been made of the value individuals place on engaging in such activities, and these values represent estimates of the direct benefits derived from maintaining these species (Tables 13.1 and 13.2).

EXISTENCE VALUES

In addition to their direct and indirect values to humans, many species are considered to possess an existence value (reviewed in Chapter 6). This is the value that humans derive from simply knowing that a certain environmental good exists, regardless of whether or not that individual will ever use the good directly or indirectly. Existence values are currently only accessible through the use of contingent valuation methodology (CVM) and are the source of much debate. Despite this debate, the list of species for which monetarized existence values have been obtained continues to grow. Table 13.3 provides some examples of these.

Probably the most significant criticism of estimated existence values is that valuation of this kind provides the wrong kind of 'reference conditions' for making decisions about public goods values (Sagoff 1988; Jacobs 1996). Whereas individuals may make decisions about their private budgets in isolation from everyone else and with regard only to their personal tastes, they may adopt a different, more publicly responsible attitude towards species preservation or other public good issues. Sagoff (1988) describes this attitude as reflecting

Table 13.1 Annual total resource values for each of six case study locations in 1988 $US. (Source: Shafer *et al.* 1993.)

Good	Technique	Net economic value per visitor day ($US)*	Annual total resource value ($US)†
Catch and release trout fishing	TCM	44.50	70 515 per stream mile
Trout fishing with fly	TCM	16.10	14 076 per stream mile
Viewing waterfowl		3.57	270 963
Watching elk		20.43	147 096
Watching migration of raptors	CVM	12.53	577 633
Wildlife viewing in an educational setting		3.70	166 500

Abbreviations: CVM, Contingent Valuation Methodology; TCM, Travel Cost Methodology.
* The amount a typical visitor would have been willing to pay per day over and above actual expenditure.
† Total number of visitor days multiplied by net economic value per day for that activity.

Table 13.2 Willingness to pay (WTP) for wildlife-related activities. All values converted to 1988 constant dollar using gross national product implicit deflator indices.

Good	WTP values ($US)	Technique	Reference
Elk hunting	72.00	TCM	Duffield (1988)
Antelope hunting	67.64	TCM	Loomis & Cooper (1988)
Deer hunting	60.00	TCM	Brooks (1988)
Elk viewing	20.43	CVM	Brooks (1988)
Sport fishing in all Montana's streams	111.89	CVM	Duffield *et al.* (1987)
Sport fishing in all Montana's lakes	95.11	CVM	Duffield *et al.* (1987)
Sport fishing in all Montana's waters	75.94	CVM	Duffield *et al.* (1987)

Abbreviations: CVM, Contingent Valuation Methodology; TCM, Travel Cost Methodology.

Table 13.3 Examples of willingness to pay bids for threatened and endangered species (all figures in 1993 $US). (Source: Loomis & White 1996.)

Species	Willingness to pay		
	Low value ($)	High value ($)	Average all studies ($)
Studies reporting annual WTP			
Northern Spotted Owl	44	95	70
Pacific Salmon/Steelhead	31	88	63
Grizzly Bears			46
Whooping Cranes			35
Red-cockaded Woodpecker	10	15	13
Sea Otter			29
Grey Whales	17	33	26
Bald Eagles	15	33	24
Bighorn Sheep	12	30	21
Sea Turtle			13
Atlantic Salmon	7	8	8
Squawfish			8
Striped Shiner			6
Studies reporting lump sum WTP			
Bald Eagles	178	254	216
Humpback Whale			173
Monk Seal			120
Grey Wolf	16	118	67
Arctic Grayling/Cutthroat Trout	13	17	15

'citizenship' rather than 'consumer' values. Contingent valuation methods are not designed to ellicit such 'citizen' responses. Sagoff (1988) suggests that debates about environmental preservation should therefore be addressed through some form of public or open forum, rather than determined by privately assessed valuations. Economic values will clearly be important within this debate but the final outcome will reflect a wider range of criteria than economic efficiency alone.

13.2.2 Species conservation decisions under uncertainty

Estimating the costs and benefits associated with any good into the future is an imprecise process, and such estimates are frequently bounded by some measure of variability. In a situation of risk, a probability distribution of all likely future states of the variable exists. Usually within a cost–benefit analysis (CBA) it is possible to construct such probability distributions for all costs and benefits over the lifetime of a project. In this way CBA can be applied to projects where there is an estimate of the variability surrounding the future state of any of the input variables.

In a situation of uncertainty, however, it is impossible even to estimate the likely variation surrounding the future state of a variable. Decision-making in this case occurs when there is simply no knowledge of the probability of any of the potential outcomes. In these cases the concept of a best choice cannot be made objectively, but is dependent on the attitudes of the decision-makers to the various risks involved in the choices available. Several decision-rules have traditionally been utilized to deal with these kinds of uncertain situations. These rules of thumb have some relevance to species conservation. The basic theory is discussed in the next section and then extended to the issue of species preservation.

DECISION STRATEGIES UNDER UNCERTAINTY

As an example of making decisions under uncertainty, consider the situation faced by a company executive who must decide on adopting one of three advertising strategies, S1, S2 and S3, below:

S1 involves an expensive combination of television, magazine and newspaper advertisements;

S2 is limited to a single television special and a few selected magazines; and

S3 involves only limited newspaper and local radio advertisements.

Table 13.4 The pay-off table. (Adapted from Cooke 1985.)

	State of nature		
Strategy	E_1 (£)	E_2 (£)	E_3 (£)
S1	500 000	200 000	−300 000
S2	300 000	100 000	−100 000
S3	100 000	50 000	25 000

Table 13.5 The pay-offs received from each of three decision criteria. The shaded figure indicates the recommended strategy for each decision criterion. (Adapted from Cooke 1985.)

	Pay-off		
Strategy	Maximin (£)	Maximax (£)	Minimax regret (£)
S1	−300 000	500 000	325 000
S2	−100 000	300 000	200 000
S3	25 000	100 000	400 000

The executive knows that there are three possible future economic scenarios: E1, E2 and E3. Under E1 the economy is rapidly expanding, under E2 it is growing steadily, and under E3 it is contracting. Only one of these future scenarios will actually occur, but the executive has no idea which.

However, the executive is able to estimate the annual profit that would be realized under each advertising strategy in each of three possible future scenarios, and these estimates are shown in Table 13.4, the so-called 'pay-off' table. Under conditions of certainty the executive can easily decide which strategy to choose from this table; however, under uncertainty the choice is more difficult. There are several criteria available to influence the choice.

Maximin

One of these is the maximin criterion. This is based around the assumption that the decision-maker is risk-averse and pessimistic, and suggests that the executive should choose the strategy which will maximize the minimum pay-offs. The minimum pay-offs for all strategies are shown in column 1 of Table 13.5 and it is clear that the maximum minimum pay-off is associated with strategy 3. But, as seen in Table 13.4, this strategy will limit the maximum pay-off to £100 000.

Table 13.6 The regret table.

	State of nature		
Strategy	E_1 (£)	E_2 (£)	E_3 (£)
S1	0	0	325 000
S2	200 000	100 000	125 000
S3	400 000	150 000	0

Maximax

An alternative criterion is the maximax criterion. If the executive is interested in maximizing the company's maximum profit and is generally a risk-taker, optimistic and can stand any losses, then the maximax may be the criterion to adopt. This is based on the same idea as the maximin but the strategy selected is the one which provides maximum pay-off, here strategy S1.

Minimax regret

What if the executive makes a mistake, chooses S1 and state of nature E3 occurs? With hindsight the executive may regret having chosen S1 and would wish she had chosen S3. If the executive had done that she could have a made a profit of $25 000 rather than a loss of $300 000. In fact, a measure of her regret is $325 000; that is, the difference in the profit associated with the best and worst strategy in that scenario. This concept leads to another criterion, that which minimizes the maximum regret or, in shorthand, the minimax regret criterion.

The minimax criterion is calculated by subtracting every number in each column from the maximum number in the column and putting these figures in a regret table (Table 13.6). A value of 0 means there is no regret, the executive chose the best strategy for that state of nature. So, for example, choosing S2 in E1 makes $300 000 profit rather than $500 000. This leads to a regret of $200 000. If the executive wants to minimize regret, she should choose the strategy with the lowest possible regret overall, here S2. This means that, regardless of the state of nature, the worst regret the executive will suffer is $200 000.

Note that in this example, depending on the criterion used—maximin, maximax or minimax regret—the executive could justifiably have chosen any one of the three strategies available. Selecting the correct criterion to use is a function of the executive's personality (is she risk-averse or a risk-taker?) and the state of the company (can it stand a loss?). The correct decision-rule will therefore vary according to changes in these factors.

13.2.3 Endangered species and the safe minimum standard

The concept of the safe minimum standard (SMS) for nature conservation was first proposed by Ciriacy-Wantrup (1968) and developed further by Bishop (1978). In philosophy the concept is simple: adopt a level of environmental protection/preservation that safeguards against the extinction of valuable, or potentially valuable species, unless such protection results in 'unacceptable' losses of social welfare through preventing potentially beneficial developments. In Bishop's (1978) words: 'adopt the safe minimum standard unless the social costs are unacceptably large'.

Two themes within the SMS are worth separating. The first concerns the uncertainty surrounding potential future benefits to be derived from species; the second associates the SMS with guidelines for harvest management of already valuable species. In the harvesting case, the SMS is a guideline that suggests setting yields taken from renewable stocks, like fish and game, within 'safe' limits, which ensure the maintenance of viable populations given scientific uncertainty about their recovery and growth rates under harvesting pressure. In effect, the SMS operates here as a rule of thumb that favours lower but safer yield levels over higher ones that risk population extinction (Section 12.5.1 introduced this idea more fully).

In the face of complete certainty about values of all the future benefits associated with preservation and development, and public acceptance of these values as a basis for making decisions, there would be no need for an SMS approach. Values could be estimated for the two available options: develop (causing extinction) or conserve (preventing development); and a cost–benefit analysis would provide a utilitarian justification for choosing one option, within the usual caveats on CBA noted in Chapter 8. The SMS is important because uncertainty exists over the preservation value of species regarding the future. It is worth noting in passing that the current value of preservation is often overlooked in arguments for SMS (Crowards 1998). Invoking the SMS decision-rule does not mean that no estimate of current benefits is necessary. Rather, including current preservation benefits is important

Table 13.7 Pay-offs for different states of nature.

	Sy	Sn
D	Db	Db
P	Pb	0

Table 13.8 Regrets for each outcome and greatest values.

	Sy	Sn	Greatest regret
D	Pb – Db	0	Pb – Db
P	0	Db	Db

in order to get a true value for the potential benefits of development.

In order to address the uncertainty issue, we consider two options: development (D) which causes extinction, and preservation at an SMS (P) which will prevent development; and two future states of nature, Sy, in which the species under consideration turns out to have very valuable economic properties, say as a medicinal source, and Sn, when it turns out to have no particular value.

We can now assess the outcomes under these four possible scenarios in a simplified pay-off matrix. We call the future benefits of development Db and assume that this figure is a net figure representing the gross development benefits (Dgd) minus the costs of development (Cd) and minus the estimated current benefits of preservation (Pkb), which is the baseline against which development benefits should be compared. In other words, Db is the net overall gain to society of development at the current moment. Similarly, we call the benefits of preservation (Pb) a net figure, and this represents additional benefits over and above the known baseline benefits (Pkb). We can now construct a pay-off table in the spirit of the business example above, based on the possible states of nature and the costs, including opportunity costs, of pursuing different strategies (Table 13.7).

The important factor in considering this table from the SMS point of view is that potentially Pb may be very large under future state Sy, though of course we do not know if state Sy will occur or not. In this case the greatest possible loss across all scenarios would be Pb, occurring if development goes ahead. According to the minimax rule, we should look to minimize the risk of maximum losses, i.e. we should therefore favour preservation, under which option the most we can lose would be the forgone benefits of development, Db.

We can note in passing that the maximax strategy would yield the same result, though on an alternative logic. As Pb is potentially very large, it can be viewed not as the greatest loss that must be avoided but as potentially the largest gain that can be achieved. High-risk-takers might therefore also favour preservation, on the off-chance that they would reap a big pay-off under Sy.

A final way to regard this analysis is through the minimax regret criterion, and we can view the results of this in Table 13.8. We can see again that in the face of potentially large preservation benefits, even if uncertain, the minimax regret strategy favours preservation, under which the greatest loss is simply that of forgone development.

Although the arguments above suggest that when faced with uncertainty, preservation may be a preferable option under a number of well-accepted decision-rules, there are two important caveats. The first is the precise specification of what constitute 'unacceptable' social losses, or benefits forgone, from forestalled development, which we cover below. The second related issue is the level of uncertainty attached to the preservation option. This is a twofold uncertainty: will the preserved species have any value at all and, if so, how great will that value be? If neither of these questions can be answered, then the pay-off tables cannot even be constructed because few of the potential outcomes are known. It cannot make sense to make huge expenditures to preserve every single species without any knowledge of the possible outcomes, when the effects of forgone development may be substantial.

The issue of 'unacceptable' social losses through forgone development has been regarded as one 'with which economics is ill-equipped to deal' (Bishop 1978). Essentially, it is a politically and/or socially determined issue which will reflect society's judgement about the relative merits of the two alternatives. Interestingly, such a debate would provide an opportunity for pluralist ethical standards to come together. The development/preservation decision in this case is not characterized as a simple utilitarian trade-off between certain current benefits and uncertain future ones, but may expand to address interspecies interests and philosophies of rights to existence. Those supporting species survival ethics must thereby confront the direct social costs entailed in

Table 13.9 Genetic distinctiveness of the species of stork for species existing in Southeast Asia and Southern Africa. (Source: Eisworth & Haney 1992.)

Region	Latin name	Common name	di
Southeast Asia	*Anthropoides virgo*	Demoiselle crane	1.4
	Grus vipio	White-naped crane	1.3
	G. monachus	Hooded crane	1.3
	G. grus	Eurasian crane	1.0
	G. japonicus	Japanese crane	1.4
Southern Africa	*Balearica regulorum*	South African crowned crane	3.7
	Anthropoides paradisea	Blue crane	1.5
	Bugeransus carunculatus	Wattled crane	1.5

Abbreviation: di, the genetic distinctiveness of species as measured by DNA, the method of DNA–DNA hybridization of Krajewski (1989).

Table 13.10 Species richness of stork and relative genetic endowment (G) for two regions, Southeast Asia and Southern Africa. (Source: Eisworth & Haney 1992.)

Region	Species richness	Relative genetic endowment (G)
Southeast Asia	5	6.4
Southern Africa	3	6.7

G, sum of di in a region.

these positions, arguing at the limit that no development benefits can ever provide justification for species loss. The deliberately ill-defined measure of 'unacceptable' social losses can therefore be construed as a strength of the SMS position, requiring the decision-making process to acknowledge arguments beyond direct utilitarian considerations, even if utilitarian arguments in the end carry the day.

MAXIMIZATION OF GENETIC DISTINCTIVENESS

An alternative to basing conservation decisions on species is to consider the genetic distinctiveness of each species and to allocate expenditure in order to maximize the diversity of conserved genes. This analytical framework assumes that rather than concentrate on conserving the diversity of species in existence, we should conserve the existing diversity of genes. Given the advances in molecular genetics that have occurred in recent years it is now relatively straightforward to estimate the degree of similarity between the genetic make-up of two or more species. If we could measure the degree of similarity and, inversely, the degree of distinctiveness between species, we could allocate the conservation money in order to conserve the maximum amount of genes.

An interesting study that demonstrates the consequences of allocating conservation money to genes rather than species is given by Eisworth & Haney (1992). This example is concerned with the conservation of cranes in two regions: Southeast Asia and Southern Africa. There are five species of crane resident in Southeast Asia and three in Southern Africa, and the authors assume that the overall threat to these species is the same in both regions. They also assume that there is a finite budget for conservation of cranes in the two regions, and this budget is not sufficient to guarantee the conservation of all eight species in the two regions. The crane species occurring in each region, and their genetic distinctiveness, is shown in Table 13.9.

If we sum the degree of genetic distinctiveness for each region we get a relative genetic endowment (G) which is shown in Table 13.10. If we based conservation expenditure according to species richness we would allocate the money Southeast Asia: Southern Africa as 62.5 : 37.5. Alternatively, if we allocated the money according to genetic distinctiveness then it would be allocated 49 : 51, with the majority of the money going to Southern Africa.

This framework does not take specific account of the benefits and costs of preservation of a species, but the net benefit is assumed to be greater the more unique the species (genes) and the more imminent the loss. The specific issue of costs in relation to benefits is addressed from a different perspective in the case of the Northern Spotted Owl below.

13.2.4 Marginal analysis of expenditure on conserving a particular species: the Northern Spotted Owl

In some cases the development/preservation decision is in essence a binary one: either the development goes ahead, or the species is saved. In such a situation, there is no compromise position that can appease both sides. In many other cases, however, the issue is not so clear cut. Particularly in the case of less dramatic land use changes, such as new agricultural developments, there can be compromises between competing uses. These compromises entail trade-offs in costs and benefits as above. This section summarizes some elements of one such analysis applied to the case of the Northern Spotted Owl (Montgomery *et al.* 1994). It shows some interesting aspects of what has become a highly polarized debate.

The Northern Spotted Owl is endemic to the old-growth forest areas of the northwestern USA and, in particular, western Washington and Oregon. In 1990 it was designated an endangered species under the US Endangered Species Act, which precipitated various government plans for its protection. The owl is dependent on the characteristics of old-growth forest for its survival, and in 1992 the US Fish and Wildlife Service designated 6.1 million hectares of critical natural habitat in prime timber country to be protected.

Numerous estimates of the cost in terms of timber revenues forgone have been made, as well as the likely impacts on the local communities that are mostly dependent on timber revenues for their economic survival. In addition, contingent valuation studies attempted to estimate the existence value of the owl. It should be noted that US government policy requires positive action for the protection of endangered species, although relevant authorities must also take account of the costs involved and the likelihood of success.

The key issue that Montgomery *et al.* (1994) sought to address was twofold. First, it should be recognized that no protection programme could ensure a probability of survival of 1. Plans could only have a greater or lesser chance of success. Secondly, given that survival was only more or less probable under different recovery plans, Montgomery *et al.* were interested to find out not only the overall probability associated with any particular plan but, crucially, the *marginal* costs associated with achieving that particular probability. In other words, how did the probability of survival change with the change in expenditure on protection? Their analysis is followed through here before noting some objections to it.

First of all, we need to know what affects the probability of survival. For the owl, this is dependent on the number of breeding pairs, which in turn is dependent on the total area of suitable habitat available. In this case, therefore, the costs of aiding species survival are directly related to the total area protected: the costs incurred are the revenues forgone from the lost opportunity to harvest the timber which is now under protection. In other cases, costs might be incurred in active management for recovery, but the same style of analysis could be applied.

One point can be noted immediately; whereas the marginal cost of not logging each hectare of timber is constant because each hectare, roughly speaking, yields the same amount of timber, the marginal effect on species survival is almost certainly not constant; it depends on the amount of habitat already protected. If half a billion hectares of forest are protected, a single further hectare is unlikely to affect the probabilities much either way. But the timber from it will fetch the going market price regardless of how many hectares remain to be logged (though we note in passing that harvesting rate and market values are most certainly affected by the total available supply of timber, both now and in the future).

Using an econometric model for timber production, the relationship between timber supply, market demand and supply restrictions as a result of conservation policy can be plotted, as seen in Fig. 13.2. This indicates the welfare loss, in terms of lost revenue, for any given reduction in the timber harvest. The points A, B and C refer to three different species management strategies.

Having established the financial costs associated with protected areas, we now turn to their ecological significance. As stated above, owl survival probability is dependent on the breeding population, which is a function of the available habitat. It is assumed that the population will expand to the carrying capacity of whatever area is available. In Fig. 13.3 we show an estimated population survival probability model plotting probability against habitat capacity in terms of the number of pairs that any given area can support.

This probability and population data can now be plotted against the marginal physical cost of provision in terms of timber volume: that is, what volume of timber harvest is forgone in order to support each particular population, and its associated survival probability. This relationship is shown in Fig. 13.4.

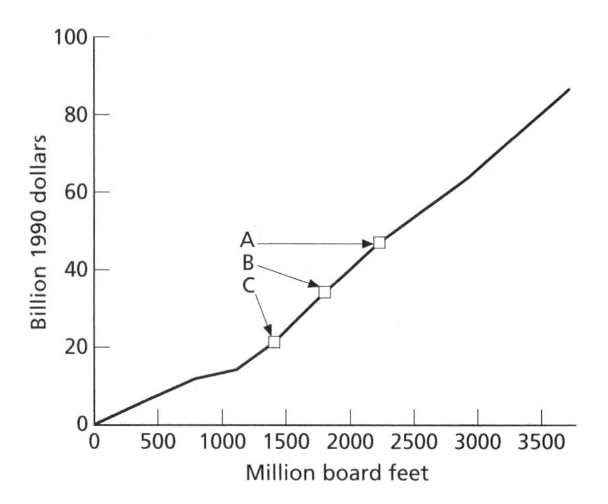

Fig. 13.2 The welfare loss, in terms of lost revenue, estimated for any given reduction in timber harvest. (Source: Montgomery *et al.* 1994, with permission.)

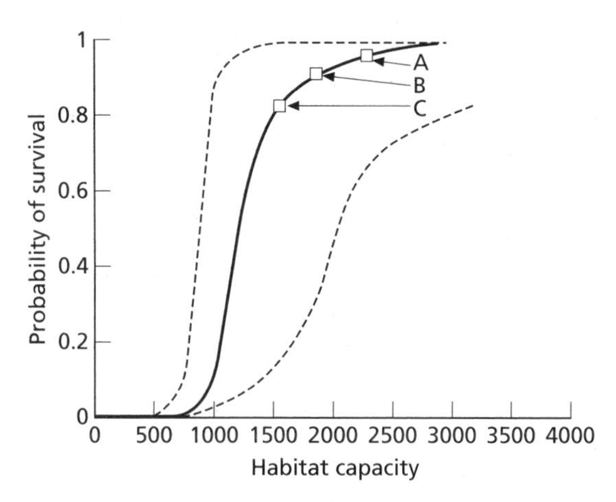

Fig. 13.3 The estimated probability of species survival associated with any given habitat capacity (measured in terms of breeding pairs supported). (Source: Montgomery *et al.* 1994, with permission.)

Having established this relationship, we can now combine it with the financial data on revenues forgone to provide a graph of the marginal cost, in terms of forgone revenue, per unit increase in the probability of species survival. This is shown in Fig. 13.5. The three possible preservation plans are again indicated by A, B and C. As might be expected, the marginal cost of an additional one percentage point increase in survival probability begins to rise very steeply once survival probability is slightly above 90%. Put another way, it becomes increasingly expensive to ensure higher levels

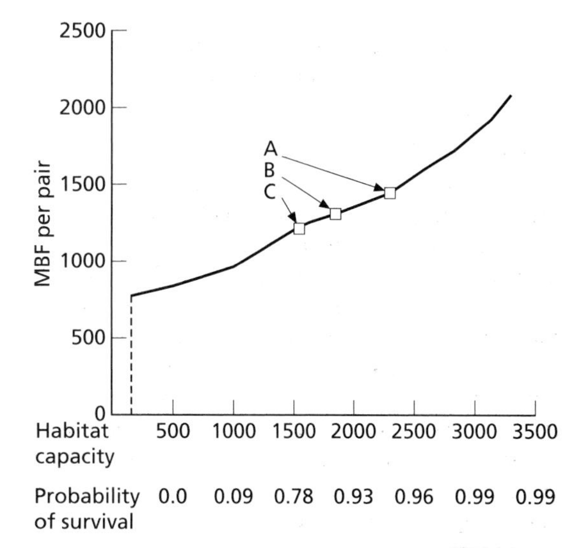

Fig. 13.4 The volume of timber harvest (million board feet) forgone to provide each habitat capacity (measured in terms of number of breeding pairs supported). (Source: Montgomery *et al.* 1994, with permission.)

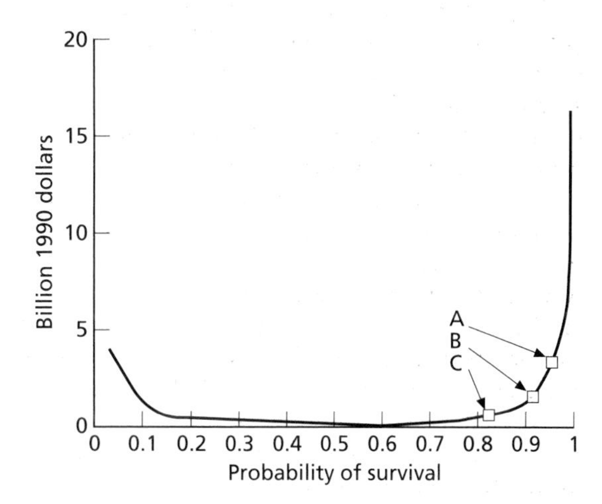

Fig. 13.5 The marginal cost incurred, in terms of forgone revenue, per unit increase in the probability of species survival. (Source: Montgomery *et al.* 1994, with permission.)

of probability. The results of their analysis for three different schemes are shown in Table 13.11.

SUMMARY OF MARGINAL ANALYSIS

What can be made of this result? Perhaps its most useful function is simply to highlight that it is marginal, rather than absolute, costs and benefits that may be a more

Proposal	A*	B†	C‡
Probability of survival, S	0.95	0.91	0.82
Habitat capacity, C (owl pairs)	2 400	1 900	1 600
Public timber harvest reduction, Q (million board feet)	2 300	1 800	1 400
Welfare loss, $W(S)$ (billion $, 1990)	46	33	21
Marginal cost, $W(S)$ (billion $, 1990), (per unit of probability)	3.8	1.4	0.6

* ISC conservation strategy on public land and restriction of timber harvest in critical habitat units.
† ISC conservation strategy on public land.
‡ USDI recovery plan. The spatial configuration of habitat areas proposed in the recovery plan was chosen to reduce impact to high-risk communities.

Table 13.11 Comparison of three proposals to conserve Northern Spotted Owl habitat. (Source: Montgomery *et al.* 1994, with permission.)

appropriate focus for attention. Without a political debate, the marginal cost curve does not indicate any one plan as being superior to any other. However, there are several further issues that arise from this problem. The distribution of the costs and benefits of any conservation plan is highly significant, as it is the local people who will bear the brunt of the economic costs in terms of lost jobs and the secondary effects of depopulation in rural areas. Meanwhile, benefits of preserving the species accrue to society as a whole, both now and in the future. This is an equity issue, and also a political one; if resource use is determined by local government in response to local voter preferences, economic arguments will often dominate. Large global existence values may exist for the species, but these do not themselves provide jobs and local incomes.

13.3 Ecological economics and protected areas

The section above concentrated on the preservation of particular species from a number of different perspectives. In this section, we consider some of the issues surrounding conservation areas rather than individual species, although these issues are evidently related, as the Northern Spotted Owl example shows most clearly. In essence, protecting areas is simply the most direct and effective means for protecting species in most cases. However, the practicality of conserving specific areas does throw up some different problems to pure species conservation.

Here we review three sets of issues. First, we review an economic valuation approach through two studies that calculated direct monetary costs and benefits associated with designated protected areas in Africa. Though limited in scope, they reveal some of the key factors involved in these calculations. Secondly, we examine some limitations of the SSSI conservation system in the UK, from both an economic and an ethical perspective. Finally, we follow through an analysis comparing the cost-effectiveness of conservation achieved through SSSI and ESA policy in the UK.

13.3.1 Costs and benefits of protected areas

As for species, the total economic value of a protected area can be estimated and compared with the costs of maintaining the area as protected. In many ways, conducting such an analysis for a piece of land or water is easier, and more meaningful, than for a single species, as many of the benefits and costs are easier to estimate, and several studies suggest that benefits of protected areas may be large and may exceed the costs. For example, de Groot (1994) suggests that the Galapagos National Park provides a monetary return of $US120 per hectare made up from direct, indirect and option values. Posner *et al.* (1981) suggested a benefit : cost ratio of 11.1 : 1 for the Virgin Islands National Parks, and for Cahuita National Park in Costa Rica a benefit : cost ratio of 9.54 : 1 (Marcondes 1981). We show two examples of these kinds of analysis below.

KENYAN PROTECTED AREAS

In an analysis of the economic costs and benefits of protected areas Norton-Griffiths and Southey (1995) tried to estimate the opportunity costs of biodiversity in Kenya. They tried to estimate a partial TEV for reserves in Kenya and assumed that the net benefits

Table 13.12 Gross revenues and net returns from agricultural and livestock production within land potential zones, 1989 year (dollars per hectare per year) ($US1 = 20.6 Kenyan shillings, 1989). (Source: Norton-Griffiths & Southey 1995.)

Land potential zone	Gross revenues	Net revenues
Per humid	118.4	38.3
High potential	411.7	150.7
Medium potential	232.0	90.7
Arable	149.4	54.2
Ranching	21.2	5.3
Pastoral	1.6	0.6

(NB) of conservation was made up as below, where OC = opportunity costs.

NB conservation = NB direct use + NB indirect use + NB non-use − OC conservation.

However, they only felt able to comment on the direct use part of the equation and aimed to compare this value with the opportunity cost of the protected areas. They assumed that direct use values were made up of the net benefits (gross revenues − costs) of tourism and forestry, and defined the opportunity costs of biodiversity conservation as being equivalent to the net benefits from the forgone potential agricultural and livestock production. The opportunity costs were estimated for each of Kenya's six land potential zones as shown in Table 13.12. By allocating each park and reserve forest into a land potential class it was possible to estimate the net returns that could potentially be achieved if the protected areas were converted to agricultural and livestock production. This figure was estimated to be $203 million for 1989.

Compared to these opportunity costs, tourism in Kenya generated about $419 million in 1989. Not all of this is related to protected areas, however, as on average tourists spend only 1.1 nights of a 14-night stay in a game park (6.1 nights in a coastal hotel, 1.9 in a Nairobi hotel, 4.9 elsewhere). Although tourism in Kenya is thus in many ways a composite good, the authors assumed that game viewing drove the tourism and they estimate that 50% of the tourism revenues should be directly attributable to wildlife parks and reserves. After considering factors such as foreign exchange retention and capital charges, it was estimated that the net returns to wildlife tourism were about $27 million. A similar analysis suggested that the net returns to forestry were in the order of $14.8 million.

From these estimates we can see that the net benefits from direct use are:

NB direct use = NB tourism + NB forestry
= $27 + 15 million
= $42 million.

We can also see that this figure is substantially lower than the net opportunity cost of $203 million. Thus, on these estimates and without considering any indirect or non-use values, the protected areas of Kenya are seen to cost the country around $161 million in forgone benefits.

UGANDAN NATIONAL PARKS AND RESERVES

In an analogous analysis Howard (1995) sought to estimate the costs and benefits associated with all protected areas in Uganda. In addition to possessing a wealth of biodiversity, Uganda is also heavily populated with a total population of 16.7 million, giving a density of 85 people per square kilometre. This places Uganda in fourth place behind Rwanda, Burundi and Nigeria in terms of population density in Africa. Currently, 32 440 km^2 of Uganda is designated as National Park, Game and Forest Reserve, equivalent to approximately 13.7% of the total area of the country, or 16.7% of its land area. The bulk of these protected areas were established during the first decades of the twentieth century, at a time when the human population was about one-fifth of its present size.

Howard analysed data from 1993–94 and concluded that on the basis of the financial flows to the Ugandan government the protected areas had a cost : benefit ratio of less than 1, and thereby their maintenance could be justified (Table 13.13). It is worth noting, though, that almost 90% of the benefits flowing to the government and its agencies were in the form of donor contributions. In the second of Howard's analyses he considered the likely costs and benefits of the protected areas which would accrue to global society over the next 25 years (Table 13.14).

Howard assumes that the donor contributions (international grants to maintain protected areas) represent the non-use values of the citizens of the donating countries. This may underestimate real non-use values. The validity of this estimation may be important, as even though the direct use value of the protected areas increases significantly, and the protected areas provide significant global benefits in terms of carbon sequestration and watershed protection, under this analysis

Table 13.13 Financial return to the Ugandan government and its agencies from protected areas in millions of $US in 1993–94 (UNP = Ugandan National Parks). (Source: Howard 1995.)

Item	Annual flow (million $)	NPV (million $)
Benefits		
Total revenues	1.007	14.193
Donor contributions	10.677	150.482
Subtotal: financial benefits	11.684	164.675
Costs		
Government capital development expenditure	−1.108	−15.616
Government recurrent expenditure	−1.169	−16.486
UNP revenue reinvested	−0.851	−11.994
Subtotal: financial costs	−3.128	−44.086
Net financial benefit (to Uganda)	8.556	120.589

Table 13.14 Projected net present value (NPV) of Uganda's protected areas internationally in millions of $US assuming a 5% discount rate and a 25-year time frame (based on 1993–94 data). (Source: Howard 1995.)

Item	Annual flow (million $)	NPV (million $)	NPV per ha ($)
Social benefits			
Timber values*	40	563.8	173.9
Tourism revenues†	16.3	229.7	70.8
Potential game utilization	0.7	9.6	2.9
Community use	33	465	143.3
Watershed benefit to fisheries	13.8	194.5	59.9
Carbon sequestration	17.4	245	75.6
Biodiversity option value	2.3	32.2	9.9
Subtotal	123.5	1 739.80	536.3
Social costs			
Financial costs to Uganda	−3.1	−44.1	−13.5
Donor contributions	−10.7	−150.5	−46.4
Crop and stock losses	−75.5	−1 064	−328.1
Opportunity cost of land	−110.6	−1 558.80	−480.7
Subtotal	−199.9	−2 817.40	−868.7
Total	−76.4	−1 077.60	−332.4

* Based on application of a shadow price for quality hardwoods of $US 200 per m^3 and 200 000 m^3 total production.
† Based on figures of foreign exchange surplus retained in Uganda (UNDP/WTO 1993).

the costs of maintaining the protected areas exceed benefits.

Some limitations of the assumptions of this analysis are important to note. By far the greatest costs in both the Kenyan and the Ugandan examples are the opportunity costs of land. This cost is not without controversy. There are problems associated with estimating the potential income that would arise from protected areas. Although analyses normally try to compare like with like—land in protected areas with land of similar type outside the reserve—generally, protected areas occupy land that is to some extent marginal, e.g. steep, high-altitude mountain slopes. It is therefore debatable what proportion of protected areas are potentially cultivatable and what incomes could be derived from them. Pearce and Moran (1994) suggest the NPV of cropping activities in developing countries vary substantially, and they estimate an NPV of $US150 per hectare in South America, $190 per hectare in India/Pakistan and $520 per hectare in Indonesia. Cattle

Table 13.15 Estimates of willingness to pay for rainforests from 'debt for nature' swaps, 1987–89. (Source: Ruitenbeek 1992.)

Project location	Protected area (km²)	Protection scenario (years)	NPV of actual transfers (× 10³ ECU)	PV area protected (× 10³ km²/yr)	Revealed willingness to pay (ECU/ km²/yr)
Beni Reserve, Bolivia (1987)	15 420	1	234	15.42	15.18
Amazonian Parks, Ecuador (1987)	10 000	1	332	10.00	33.24
St Paul Park, Philippines (1988)	57.5	9	353	0.39	910.34
Santa Rosa Park, Costa Rica (1988)	160	5	265	0.69	383.84
Monte Verde Forest, Costa Rica (1988)*	36	5	245	0.16	1 575.76
Oban park, Nigeria (1989)†	2 500	1	1 029	2.50	411.76

* This was not a debt-for-nature swap in the strict sense, the money was through WWF Canada in order to purchase land in rainforest from local farmers.
† This was not a debt-for-nature swap in the strict sense either; here a fund raising campaign was initiated in the UK through the sale of Oban Park Founder's Bonds.

ranching, on the other hand, generally provides much lower returns, e.g. $Z3.6 per hectare in Zimbabwe (Child 1984). However, even these seem high for many areas of Africa, where soils are generally poor and much farming is at subsistence level.

Secondly, estimation methods tend to assume that land is a limiting factor in the economy and that, if the protected area land were made available for settlement, it could be developed immediately to levels of productivity observed elsewhere in each district. This is probably not the case. Thus, although much of the land designated as protected area undoubtedly has agropastoral development potential, this is unlikely to be realized for some time because of constraints other than land shortages: lack of investment capital, lack of people to work the land and lack of a market for goods. Given these problems, estimates of the opportunity cost of land in analyses need to be treated with some caution.

CONCLUSIONS ON NET PRESENT VALUES OF PROTECTED AREAS

On the basis of such CBAs of protected sites, it has been pointed out by some economists that as many of the benefits of protected areas in developing countries actually accrue mainly to the developed world (for example, carbon sequestration and existence values), then the developed world should transfer funds to developing countries in order to cover the difference between the national costs and benefits of maintaining the protected areas and their global value. This idea of transferring funds to developing nations to support protected areas has been termed the *supply price*—the price at which countries would be willing to supply the world with biodiversity conservation. Some work has been done by Ruitenbeek (1992) estimating the rainforest supply price.

In this analysis Ruitenbeek considers Korup National Park in the Cameroon, which contains the oldest rainforest in Africa. As discussed above, he estimated the costs and benefits associated with a conservation project to protect the national park from increased land pressure. He concludes that it is not in Cameroon's financial interests to do so unless 5.4 million ECU is transferred to the Cameroon. This is equivalent to rainforest supply price of 1060 ECU per km² per year, and is in line with historical 'debt for nature' swaps, in which debt relief of between 15 and 1575 ECU per km² per year has been paid to governments in return for rainforest protection (Table 13.15).

13.3.2 Policy instruments for nature conservation

This section considers in more detail some effects of two particular UK policy instruments, Sites of Special Scientific Interest (SSSIs) and Environmentally Sensitive Areas (ESAs). The approaches for determining the target of spending on conservation are developed in Chapter 7. Here we are concerned with some of the issues entailed in the effective design of instruments for delivering conservation benefits.

SITES OF SPECIAL SCIENTIFIC INTEREST (SSSI) AND NEGOTIATED MANAGEMENT

The UK SSSI designation dates back to 1949, and is carried out by three statutory bodies covering England, Scotland and Wales, according to a set of ecological criteria established to protect the rarest and most important natural habitat sites in the UK. Overall, designated sites cover approximately 1.8 million hectares of land. The vast majority of this land is privately owned, mostly on farm holdings. The designation process is mandatory, in that all sites passing the set criteria must be designated SSSI status.

Many SSSIs lie in inaccessible areas where there is little threat of damage or destruction as a result of human development. However, a large number also lie close to, or on land that has a direct economic value to the owner if put to some productive use. Under UK legislation, if an SSSI site is threatened by a potential development—technically called a potentially damaging operation (PDO)—such as drainage of a boggy habitat or felling of an ancient copse, the authorities can apply for a review of the request. If the operation is deemed to cause unacceptable damage, the authority must establish a management agreement with the owner.

The management agreement has two parts: one covering the practices that are unacceptable; the other a sum for financial compensation for the owner in light of his lost opportunity cost. This sum will vary from agreement to agreement, depending on the opportunity cost estimated by the farmer, and the budgetary constraint of the authority. In the case of total disagreement, the authority can push for compulsory purchase of land on which they cannot get agreement. This has never happened in practice.

The situation facing the two parties involved in the agreement can be represented by Fig. 13.6 (Spash & Simpson 1994). The authority will have a marginal willingness to pay for a certain area of land, assuming that there is some minimum area that will make a minimum level of conservation worthwhile, and that more land improves the conservation value of the land, though at diminishing marginal rate. That is, the conservation authority will be less and less inclined to include more land into the SSSI as its total area gets bigger.

In their analysis of the dilemma facing the authority, Spash and Simpson (1994) indicate that the marginal cost of land to be taken into an agreement is the opportunity cost of the land to the landowner (MC_1). The authority itself has a willingness to pay for particular

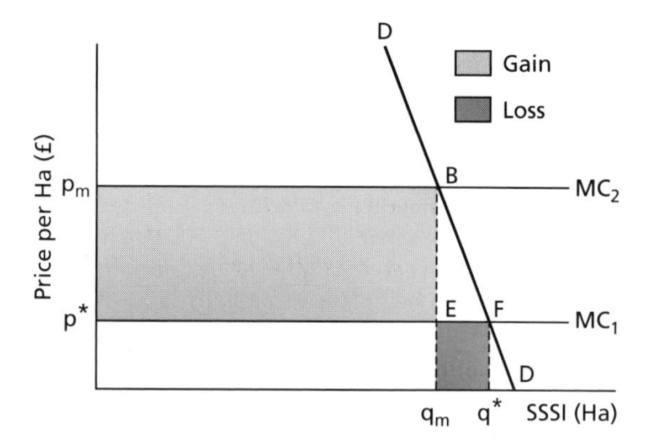

Fig. 13.6 A demand curve for designating areas of SSSIs showing gains and losses incurred following a shift in the marginal costs of land protection. (Source: Spash & Simpson 1994.)

areas of land, indicated by its demand curve for conservation (DD). If each of the stakeholders had perfect information, then the authority and farmer would agree on a payment per hectare of p*; the amount that fully compensates the farmer and matches the acceptable area of land that the authority can protect at this price.

It is clear, however, that this method of reaching agreement has several drawbacks. First, farmers need only threaten to develop land on an SSSI to force the authority into considering an agreement. Thus, there is a strategic incentive to develop land improvement schemes because they bring compensation even where no development was really intended. Second, once the bargaining process is under way, landowners have a further incentive to exaggerate their opportunity costs, because this obliges the authority to pay more per hectare. Note that this depends on an inelastic demand for the SSSI designated area; that is, that the authority is very keen to protect at least a minimum core area. This means that the demand curve is very steeply sloped; consequently, if the landowner can misrepresent his opportunity cost as MC_2, and secure this as the compensation price under the agreement, he stands to gain a total profit equal to area p^*p_mBE, while losing a sum equal only to area q_mEFq^* (the reduction in the total area that would have been protected at the real opportunity cost multiplied by the agreed payment to compensate for lost opportunity cost).

Two issues are of interest within this analysis. First, the assumption of the legislative procedures is that the landowner holds strong property rights: he or she must be offered compensation for a change in

Table 13.16 Number and area of agreements: environmentally sensitive areas (ESAs) in England, 1987–91. (Source: Ministry of Agriculture, Fisheries and Food 1990, 1991.)

	1987	1988	1989	1990	1991	Total
Number of new agreements	1 330	1 090	350	186	105	3 660
New area under agreement (hectares)	31 230	68 700	8 970	2 650	1 560	113 110
Cumulative area (hectares)	31 230	99 930	108 900	111 550	113 110	113 110
Average area of new agreements (hectares)	22.7	55.4	25.6	14.2	14.9	
Total ESA cost (£m)	4.9	9.3	9.5	10.5	11.1	
Mean cost (£ per hectare)	156.9	93.1	87.2	94.1	98.1	

proposed land use. At the same time, the authority does reserve the right in extreme cases to compulsory purchase, therefore landowners' rights are circumscribed absolutely.

Secondly, this system of selecting SSSIs is incompatible with the objectives of the authorities as laid out in their own charters, namely to protect a minimum habitat set for the benefit of the nation. As authorities are obliged to offer compensation on SSSI sites, they must make judgements about investment in new sites, what size of site to designate, and so on. The bargaining process may arrive, theoretically, at the constrained optimum set of sites under a budget constraint, but it does not recognize the requirement for the absolute size of designated sites in order to protect the nation's unique natural heritage, which is the stated purpose of the legislation. The mission of the conservation bodies is in fact a Kantian absolutist one—that these unique sites should be protected regardless of cost, because of their intrinsic merit. But the actual operation of the system has become a utilitarian one—trading off costs against benefits to maximize utility, subject to a budget constraint.

From a landowning point of view, compensation is only fair in the case where society as a whole gains from designation, but the landowner bears the full cost of lost opportunities. In the extreme case of a whole farm becoming designated, for example, the farmer might become bereft of the major part of his or her livelihood and, furthermore, with land assets worth nothing because no one can develop them once they are designated. The free market offers one way to resolve these issues, as non-governmental organizations, such as the Royal Society for Protection of Birds (RSPB) in the UK and the Audubon Society in the USA, purchase land on behalf of their members. Society members therefore pay collectively through their organizations for conservation undertaken by the organization on their behalf.

COST-EFFECTIVENESS OF DIFFERENT POLICY INSTRUMENTS: COMPARISON OF ESAs AND SSSIs

The case of SSSIs as a system designed to balance agricultural development and conservation can be compared with a more recent system adopted in the UK, the ESAs scheme. Under ESAs, farmers are offered a fixed payment for adopting certain environmental practices, which they are free to accept or reject. All farmers inside the ESA boundaries are eligible to receive payments, which are set on three levels or tiers, the value of each tier depending on the extent of the new practices that have to be adopted. These range from simple agreements to avoid the use of certain fertilizers, up to more substantial proactive conservation management. Data on the ESA scheme are given in Table 13.16.

Whitby and Saunders (1996) undertook a study of the costs associated with both the ESA and the SSSI systems, to investigate which offered a more cost-effective method of conservation. Clearly, as they acknowledge, this requires some assumption about the sorts of environmental benefit delivered by each scheme. Environmentally sensitive areas are a much broader designation, limited to 19 areas of the UK with generally valued characteristics, whereas SSSI are a much more tightly targeted system aimed at providing a minimum ecological heritage. Nevertheless, their analysis pointed out some interesting features.

In a theoretical analysis of the two systems, the associated marginal costs per protected hectare can be represented by Fig. 13.7. The basic tier 1 payment per hectare under ESAs is represented by OA, and as this is a flat rate it is paid on all land entered voluntarily under the scheme, up to area C. The total payments made are therefore OABC, and the land protected is OC. Under SSSI agreements, each agreement is negotiated separately and so marginal costs are shown rising from zero up to a maximum Z, the greatest payment the designating

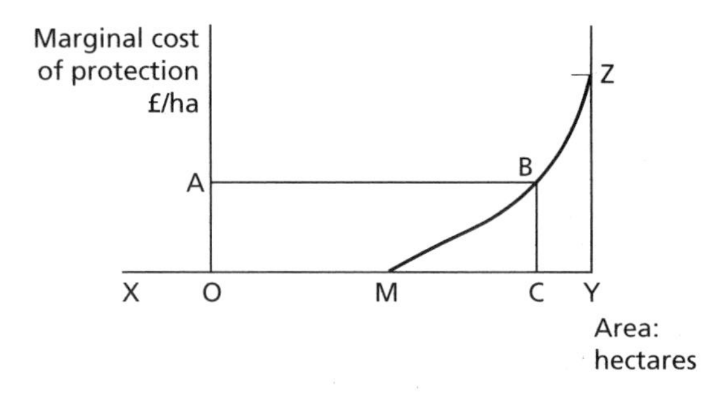

Fig. 13.7 The associated marginal costs per protected hectare for sites of special scientific interest (SSSIs) and environmentally sensitive areas (ESAs). (Source: Whitby & Saunders 1996, with permission.)

authority is prepared to make. Note that each point on this curve theoretically represents one farm, because the marginal cost per hectare is fixed on a per farm basis, not per hectare.

One controversial point emphasized by Whitby and Saunders is that the starting point for SSSI protected areas is M and not O. This is because, they argue, the SSSI system ensures by default that the inaccessible SSSIs are protected but not paid for—in other words, payments only have to be made on SSSIs that are threatened—whereas ESAs give payments regardless of whether damage was threatened or not. It is therefore fair to claim that OM is a 'natural' level of protection that does not have to be compensated because it has no opportunity cost. The blanket ESA scheme therefore appears to overcompensate certain farmers.

Note also that the marginal SSSI cost curve must cross the marginal ESA cost curve at B. This indicates that land to the right of B would not be entered into the voluntary ESA scheme because the payments offered are too low; that is, lower than the opportunity cost of developing the land from the farmer's perspective, remembering that each individual SSSI agreement must identify the farmer's individual opportunity cost for the land. Obviously we are assuming here, for theoretical purposes, that the SSSI and the ESA payments refer to the same kind of land, which therefore have the same opportunity costs associated with it. Almost certainly, this will not be true in many cases.

Figure 13.7 gives us a helpful way to consider the pros and cons of the two schemes theoretically. However, many complications should be noted. Most importantly, the issue of transaction costs are not addressed. In the case of both schemes there are start-up costs, and in the case of SSSIs each agreement will have a transaction

Table 13.17 Average compensation costs of management agreements for sites of special scientific interest (SSSIs) and environmentally sensitive areas (ESAs): southwest and southeast England (£ per hectare). (Source: Whitby & Saunders 1996, with permission.)

	Southwest	Southeast
SSSI		
Agreement area	173	228
Whole area	26	11
ESA	104	129

cost associated with it. Theoretically again, we can imagine this to be factored into the per farm payment as an additional cost, which would not change the underlying rationale. However, in empirical terms Whitby and Saunders had to accept that they could not make an acceptable estimate of transaction costs, entailing administrative time, monitoring costs, and so on.

Working within these limitations, Whitby and Saunders estimated average payments for SSSIs and ESAs (Table 13.17), and they concluded that on a per hectare basis, SSSIs were cheaper than ESAs in the level of environmental protection they afforded. Two estimates for the direct payment costs of SSSI agreements, the real data for the MZ curve in Fig. 13.7, excluding transaction costs, were also made and shown in Fig. 13.8(a) and (b).

Four final points are worth emphasizing:

1 The assumptions regarding property rights that underlie these two conservation regimes are critical. The ESA confers an additional right of ownership on farmers, namely the right to claim an additional payment from government in exchange for certain behaviour. The SSSI recognizes only the right to draw benefits from property ownership and the right to compensation by the state if this is denied in the interests of wider national conservation goals. These property rights systems are essentially politically determined. Environmentalists have often argued that ownership of an SSSI should confer responsibilities without compensation, in the way that owners of listed buildings in the UK may be barred from making adaptations to them on the grounds that they are a part of the national heritage that must be preserved. Grants may be provided only for repair or renovation, but not to compensate for the physical discomfort of leaky roofs and/or draughty windows.

2 The objectives of these conservation policies have here been assessed against extremely narrow criteria—namely, hectares protected. Conservation policies are

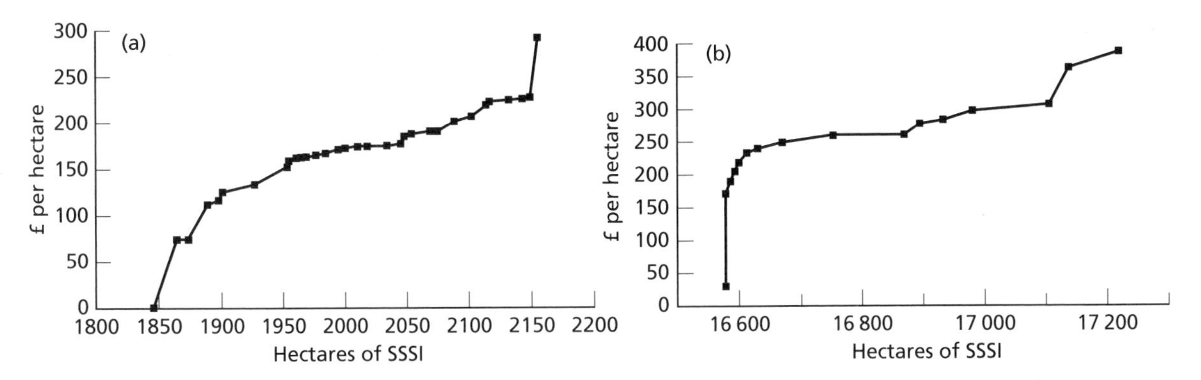

Fig. 13.8(a) and (b) Two estimates for the direct payment costs of SSSI agreements in different parts of the UK. (Source: Whitby & Saunders 1996, with permission.)

increasingly seen as part of a wider rural development policy which may have other objectives, such as income support for farming communities, employment opportunities for other rural services, and so on. Higher average payments for ESA schemes may therefore achieve these wider objectives at lower overall cost than specific SSSI agreements linked with other schemes. This is again an executive political decision. As the demand curves for environmental goods are very largely unknown, the allocation of these budgets is largely a matter of political judgement.

3 Very significant market distortions exist in the opportunity costs facing farmers, and the effectiveness and cost of both ESAs and management agreements are thereby significantly affected. Increasingly, European agricultural policy is attempting to come to terms with these distortions by decoupling income support measures from production output levels and linking them instead to environmental targets.

4 Market-established levels of environmental protection may approximate optimal budget allocations between those engaged in bargaining, but these represent constrained maximization subject to the property rights, transaction costs and budget constraints of those involved. However, as is clear in the case of SSSIs, there is no guarantee that these will produce either an ecological minimum or even, in the longer term, a sustainable allocation of resource use.

Further Reading

Mecke, G.K. & Ronald Carroll, C. (1997) *Principles of Conservation Biology*. Sinauer Associates, Massachusetts.

Milner-Gullard, E.J. & Mace, R. (1998) *Conservation of Biological Resources*. Blackwell Science, Oxford.

Munasinghe, M. & McNeely, J. (1994) *Protected Area Economics and Policy-Linking Conservation and Sustainable Development*. World Band and World Conservation Union (IUCN).

14 Pollution and Waste

'Economists argue that all the world lacks is
A suitable system of effluent taxes
They forget that if people pollute with impunity
This must be a symptom of lack of community' [Kenneth Boulding]

14.1 Introduction

The basic economic processes are production and consumption; that is, the conversion of natural resources by firms into the commodities demanded by consumers. In physical terms, this conversion is never perfectly efficient: by-products (residuals) are produced. When a residual has no economic value then it can be thought of as waste. Waste arises at all points in the economic process, from raw material extraction through to the final disposal of the commodity. It is important to note that by-products of economic processes can exist as both matter (e.g. packaging) and energy (e.g. heat generated by light bulbs in factories).

The impact of a pollutant on the natural environment is sometimes geographically distant from the point of production (the 'source'). Pollutants can be categorized in terms of their source, their transport medium (e.g. air and water) and their targets. The Swedish government sued the UK government for the emission by UK coal-burning power stations of sulphur dioxide residuals which allegedly caused acid rain in Sweden, a distant but down-wind target. Similarly, radioactivity released from the nuclear meltdown in Chernobyl in the Ukraine caused damage to agricultural and ecological systems in mountainous areas of the UK. The packaging of goods produced in China may be disposed of in US landfill sites. Some of the key factors in this chain for a pollutant are summarized in Fig. 14.1.

The subject of this chapter is pollution and waste management. We start by analysing economic theories on pollution in Section 14.2, defining the level of pollution-generation which is 'best' for society. Section 14.3 considers the various economic instruments that might be used to achieve this pollution level, and Section 14.4 the optimal choice between the alternatives. Many of the principles set out are similar to those for harvest control in Chapter 12. This theoretical section is followed in Section 14.5 by an evaluation of a real-world application of market instruments to air quality management in Los Angeles: the RECLAIM programme.

Section 14.6 introduces the second part of the chapter. The preceding sections evaluated theory and practice given a predefined pollution source, medium and target. In the case of RECLAIM, these would be a polluting firm, air, and the South Coast district, respectively. The second part of the chapter considers the various options that are available in waste management, not just for the production process but also for managing the demand for waste services. To explain the distinction between the two parts of the chapter, consider the following scenario; an aluminium manufacturer produces emissions of heat and airborne pollutants. Part 1 defines what level of airborne pollution is optimal for society and how this should be achieved, whereas part 2 questions whether incentives for increased aluminium recycling should be instigated, thereby reducing the demand for new aluminium and the associated pollutant-emission in the first place.

In Section 14.6 an overview of the waste management options available for municipal solid waste (MSW) and the environmental impacts associated with each option is presented. Section 14.7 considers the economics of waste management, and puts this in the context of the waste hierarchy. One methodological tool that is increasingly being used to test the environmental impact of various process and product options is life cycle assessment (LCA), which is the subject of Section 14.8. Finally, an application of a partial LCA to test MSW management options is presented in Section 14.9.

14.2 Pollution and social optimality

A fundamental question that needs investigation is: should society allow pollution? As described above, the economic system transforms natural resources, through the production process, into commodities for households to consume. This transformation increases

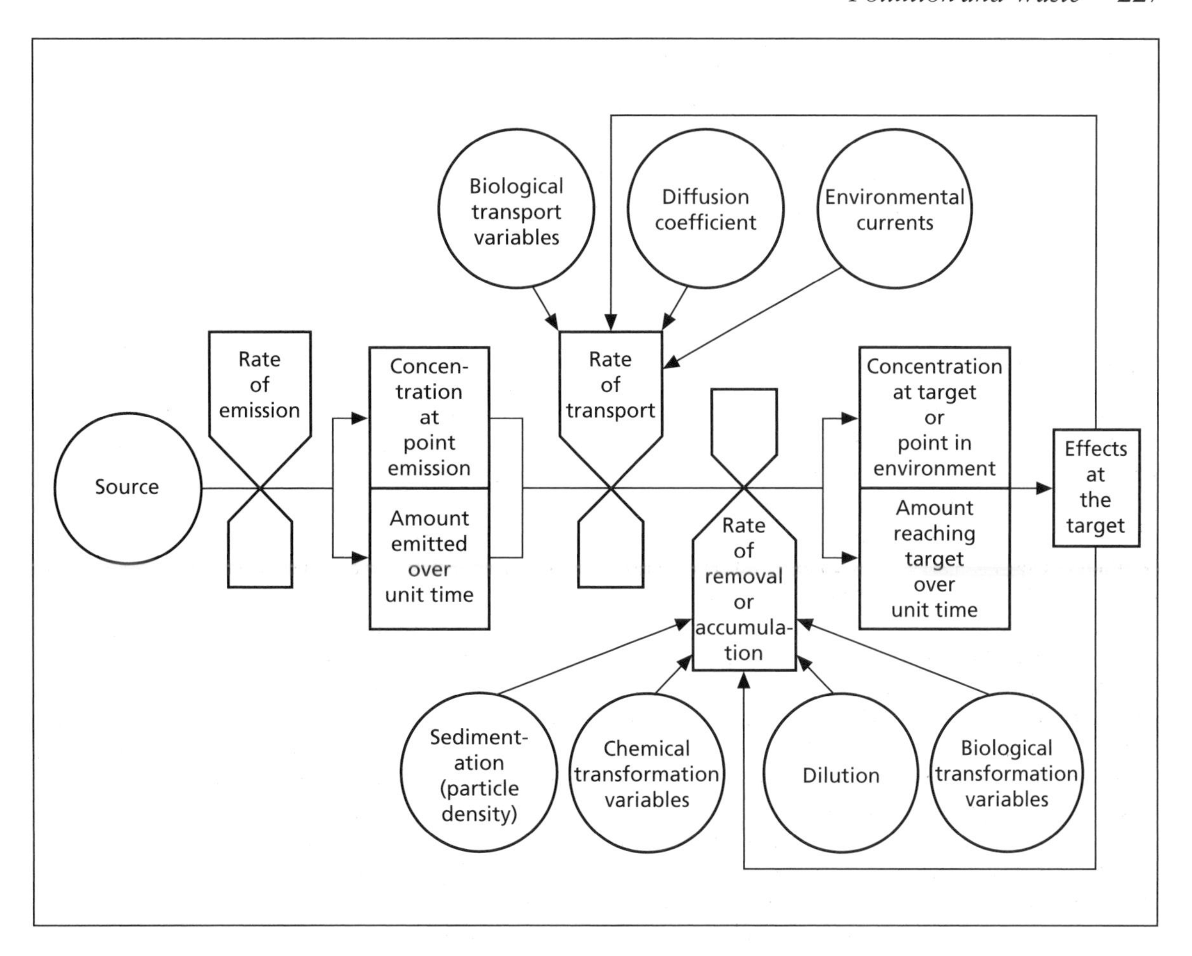

Fig. 14.1 A systems diagram of pollution processes stressing source–pathway–target linkages. (Source: Holdgate 1979.)

entropy and creates waste which may lead to pollution. An intuitive response to the problems caused by pollution may be to aim for zero pollution. However, since the production of pollution is an inevitable output of the economic process, then zero pollution would require that we had zero economic activity—an implausible situation.

Such a simple analysis does not take account of either the fact that not all waste is polluting, nor that the environment has some capacity to deal with some level of pollution without sustaining damage. This is called the assimilative capacity of the environment. Thus, an alternative solution may be for society to permit economic activity that either produces waste that is non-polluting, or pollutants that can be assimilated by the environment. This outcome would certainly be sustainable, but it would also require a radical shift in consumption patterns: a decrease in overall consumption and the stopping of the production of non-biodegradable commodities (e.g. plastics) altogether. Individuals are assumed to gain utility from consumption, and so such a shift implies reduced societal welfare, at least in the short term. Is this, then, the 'right' outcome for society?

The economic approach does not seek to ban or even minimize the level of pollution; rather it seeks to attain the optimal level—that level of economic activity, and hence pollution, that maximizes the total benefits to society. To understand this consider Fig. 14.2. This shows the costs and benefits associated with increasing levels of economic activity, and hence pollution. Two curves are drawn: the marginal abatement costs (MAC) and the marginal social costs (MSC). Economists use these curves in order to attempt to optimize the level of pollution.

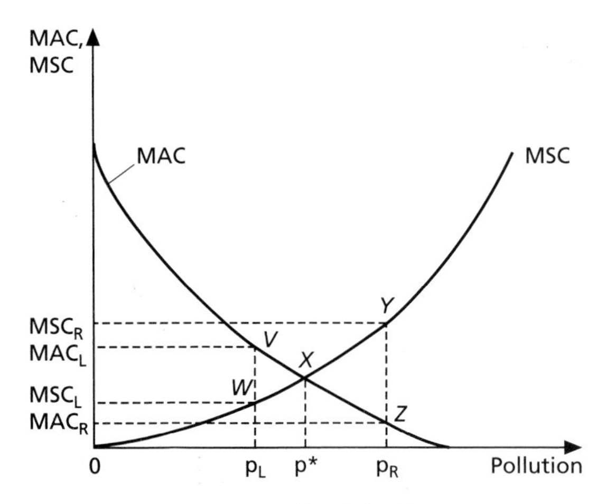

Fig. 14.2 The socially optimal level of pollution.

The MAC curve shows the *private* costs to the firm itself of cutting pollution by one extra unit for different pollution levels, net of private benefits. The private costs might be the installation of end-of-pipe clean-up technologies, or the introduction of a pollution management scheme; the private benefits might arise from 'green' consumers choosing environmentally friendly products and thereby increasing the firm's profits.

The MAC curve is downward-sloping. When pollution levels are relatively high, say at p_R, the cost to the polluter of abating one extra unit is relatively low (MAC_R), whereas at p_L the extra cost is relatively high (MAC_L). This is because the profit-maximizing firm chooses to adopt the cheapest 'fixes' first to abate pollution; for example, a waste management scheme might reduce pollution by one unit at p_R and have a relatively low cost. If forced to abate further it may be necessary to install expensive clean-up technologies, so that the extra (marginal) abatement costs at p_L are relatively high.

The MSC curve depicts, for different pollution levels, the extra cost to society—in this case everyone except the polluting firm—of the firm producing one extra unit of pollution. The MSC is upward-sloping. To justify this consider the effects on society of air pollution emitted from two identical firms, one located in a heavily urbanized and polluted conurbation, and the other in a relatively unpolluted rural setting. This corresponds with air pollution levels p_R and p_L, respectively. When ambient air pollution levels are low (p_L), the effect of an extra unit of air pollution on welfare is relatively low (MSC_L), whereas the last thing that those spluttering in the urban smog (p_R) want is one extra unit of pollution (MSC_R). Their higher disutility is expressed as a higher MSC at p_R compared with p_L.

Although the MAC curve and the MSC curve will be downward- and upward-sloping respectively, in most cases, their exact derivation depends on various factors. For MAC, these include: the state of technology; incentives for 'green' reform; the penalties associated with failing to comply with legislation; and the probability of being caught.[1] For MSC, there are various ways of valuing the damage caused by pollution. These are discussed in Chapter 6.

From these curves we can identify the optimum societal pollution level, which lies at p*. Why is this? Perhaps the best justification is to consider any other pollution level. Bear in mind that the firm is itself part of society, because it will generally be owned by households. Any abatement costs are then ultimately borne by society. Therefore, any increase or decrease in the level of abatement costs to the firm is valued exactly the same as any given increase or decrease in social costs. This assumption simplifies our analysis: if MSC is greater than or less than MAC for some level of pollution then that level cannot be socially optimal.

Consider p_R, which is to the right of p*. If society were to choose this level of pollution p_R then the associated MSC_R is higher than MAC_R. For the pollution unit p_R the net loss in societal welfare is the vertical distance *YZ*. For all the units between p* and p_R, MSC exceeds MAC. The total net loss to society of choosing p_R pollution is then given by area *XYZ*. How do we determine this? Consider Fig. 14.3 which shows the area *XYZ* on a larger scale.

p_1 is one pollution unit less than p_R, p_2 is one unit less than p_1, and p_3 one unit less than p_2, etc. The loss to society from choosing p_1 is given by Y_1Z_1, the loss from p_2 is Y_2Z_2, and the loss from p_3 is Y_3Z_3, etc. The societal loss associated with pollution units closer to p* is less than those further away, but there is a loss, nevertheless. From inspection, if we add up these losses and make each pollution unit infinitesimally small, then *XYZ* represents the total loss to society from choosing p_R instead of the socially optimal p*. The same argument can be applied for a choice of pollution level somewhere

[1] In our analysis, we assume that there is perfect monitoring and enforcement of regulatory policies. This is obviously a simplifying assumption which might not apply for two distinct reasons. First, the firm might knowingly breach regulations; and, secondly, the firm might not accurately monitor its pollution outputs.

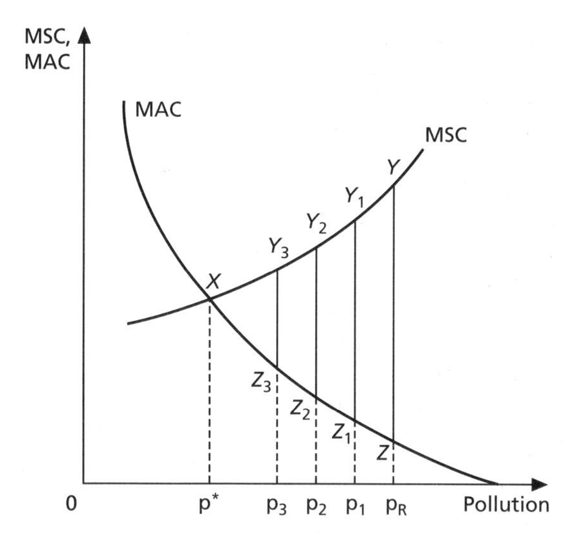

Fig. 14.3 The societal loss from choosing too high a pollution level.

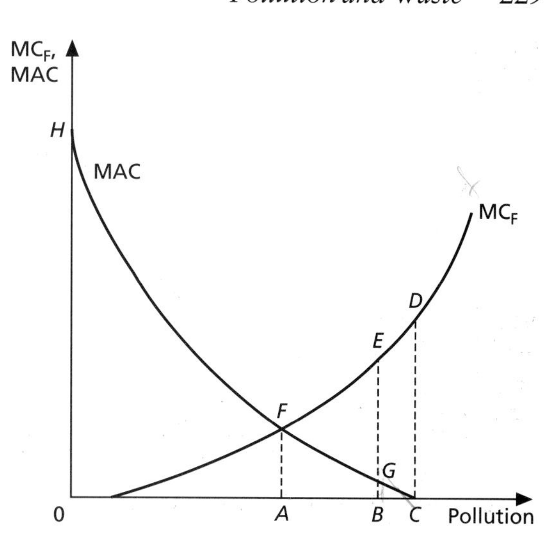

Fig. 14.4 The Coase theorem.

to the left of p*, say at p_L. The loss to society of choosing p_L is *VWX* in Fig. 14.2.

14.3 Instruments for achieving the social optimum level of pollution

So far we have assumed that society can simply choose some level of pollution. There are different ways of achieving a target level, each with its associated pros and cons. Broadly speaking, these instruments can be split into two categories: *market-based*, and *command-and-control*. Market-based instruments use some form of pricing to determine the amount of pollution produced, whereas command-and-control methods simply set regulations. Pigouvian taxes and tradable pollution permits are the two main market-based instruments, whereas quality or emissions standards are the main command-and-control instruments. What all these instruments share in common is the need for state intervention to achieve the 'right' outcome. But is such intervention always necessary? Before considering the three instruments in depth, we first consider a situation in which there is no need for such intervention.

14.3.1 The Coase theorem

The Coase theorem (named after the economist, Ronald Coase) proposes that, under conditions of competition, if the property rights to a resource affected by externalities are designated to any one party and free negotiation between the parties involved is possible, then the resultant resource allocation will be Pareto efficient, i.e. no one can be made better off without making someone else worse off. We go through the reasoning behind the theorem below.

Consider Fig. 14.4. Assume that the polluter is a profit-maximizing steel firm which could release its waste into a watercourse. Downstream, there is a fish farmer whose production decreases as the steel firm's pollution increases. The marginal cost of pollution to the fish farmer is given by MC_F. This represents the relationship between loss of fish and pollution; this is assumed to be a steady continuous relationship.[2]

First consider the case where the steel firm is free to pollute as much as it wants to: the steel firm is said to have the *property rights* to the environment in this case. It can use the environment just as it pleases. If it chooses to pollute at level C, the cost of the last incremental unit of pollution to the fish farmer is high (*CD*), whereas the incremental benefit to the steel firm of disposing of this unit to the watercourse, as opposed to abating it, is virtually zero. If the two parties could negotiate, then it would be worth the fish farmer's while to actually

[2] The assumption of a 'continuous' damage function, in which fish losses increase smoothly with increasing concentration of pollutant, may often be unrealistic. If the damage function shows discontinuities (i.e. sudden large increases in fish losses when a critical pollution concentration threshold is passed) the marginal trade-off model is inappropriate.

pay the steel firm to abate this unit. For all pollution units from *C* to *A*, there are potential gains from such negotiations, because the extra income from the fish saved and harvested, given by MC_F, is greater than the costs of abatement, given by MAC. The gains become smaller as the pollution level approaches *A*.

How much compensation is appropriate? The steel firm would not abate unit *B* unless it received at least *BG* in compensation, and the fish farmer would not be willing to compensate more than *BE*; the compensation for abating unit *B* must satisfy these two conditions in order for an agreement to be profitable to both parties. If the bargain lies somewhere between *G* and *E* then both parties are better off: a 'win–win' situation. The total potential gain from abating *CA* pollution and polluting to the level 0*A* is *CDF*, to be split between the two parties.

Interestingly, we can follow a similar argument and reach the same result, 0*A* pollution, by giving the fish farmer the property rights instead of the steel firm. In this case the potential gains from negotiation are 0*HF*. The difference in this case is in the distribution of wealth between steel firm and fish farmer, but the allocation of the scarce resource (the environment) between polluting and non-polluting activities is the same: the 'optimal' pollution level, at *A*.

Giving the property rights to the polluter might appear unethical; the fish farmer is then compensating the polluter for abating pollution. The Organization for Economic Co-operation and Development (OECD) apply the 'polluter pays' principle, which would imply giving the fish farmer the property rights. However, the direction of transfer of compensation is less significant than the outcome—a socially optimal pollution level being achieved through private negotiation.

Although this result is known as the Coase theorem, Coase himself was at pains to point out that whether this optimal outcome is achieved in practice depends primarily on the *transaction costs* of the negotiation. Transaction costs are simply the costs of carrying out market transactions—things such as discovering and checking prices, bargaining, checking quality of goods before buying, and so on. The situation analysed above has only two parties: the steel firm and the fish farmer. This makes the bargaining process relatively straightforward.

If, however, there are many affected parties then the transaction costs might be so high that bargaining is no longer feasible. If the steel firm's pollution of the watercourse caused hazardous sludge to be washed up onto a beach used by local townsfolk, then negotiating with and then compensating each affected inhabitant would not be a realistic proposition. Further, this sludge might contaminate the beach to the extent that future generations could not use it. It is impossible to negotiate with affected parties who are not yet alive. The possibility of optimal resource allocation is limited by these factors, so the Coase theorem is certainly not applicable to all environmental problems.

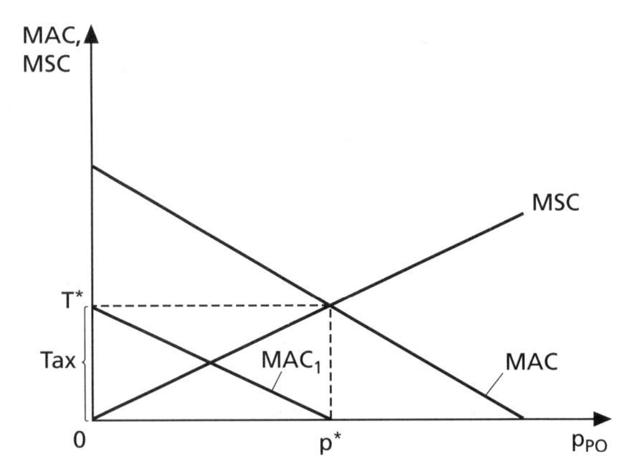

Fig. 14.5 Standard, Pigouvian tax and tradable pollution permits.

14.3.2 State intervention to achieve the social optimum

This section considers some of the tools that a state regulator can use to attempt to obtain a socially optimal pollution level. The principal tools are standards (S), Pigouvian taxes (PT) and tradable pollution permits (TPP). In what follows we refer to Fig. 14.5.

For both S and TPP, the state regulator simply dictates that the maximum pollution level that is allowed is p^*. The difference between the two instruments is that TPP is a market-based instrument, whereas S is command-and-control. Under a TPP scheme, a firm which is releasing pollution to an environmental sink, say a particular watercourse, must gain permission to do so in the form of a pollution permit. The aggregate number of permits corresponds with the socially optimal pollution level (p^*).

There are two alternative mechanisms to distribute these permits. The permits might be 'grandfathered'; that is, allocated for free, based on some criterion, such as previous pollution levels. Alternatively, they could be auctioned at the beginning of the scheme. A polluter might then choose to buy or sell TPPs—they are *tradable*—thereby giving the firm some flexibility in the amount it pollutes. The individual firm should buy permits if the MAC for the last unit of pollution emitted is higher than the market permit price, and sell permits if the MAC is lower. The aggregate pollution to the watercourse remains at p*, but the *distribution of pollution outputs* across firms might change as a result of this trade.

The same outcome (p*) can be achieved by simply setting a standard for all the firms in the industry to achieve. This standard is set so that aggregate pollution is p*. Assume that there are three firms affecting the watercourse. A standard might be set such that each firm's pollution level must not exceed one-third of p*. There is no flexibility for the polluter under S. The benefits of this flexibility are demonstrated in Section 14.4.

A third instrument is a Pigouvian tax which, like TPP, is a market-based instrument. Consider Fig. 14.5 again and the scenario with the steel firm. If the firm had to pay a tax per unit of pollution output at a rate of 0T*, how would its behaviour change? The answer is that the MAC curve faced by the firm would shift from MAC to MAC_1, i.e. the vertical distance 0T*. If the firm abates one given unit of pollution then it does not have to pay the tax on that unit. Before the application of the tax, the firm would pollute (unregulated) $0p_{PO}$ units; with the tax, it is cheaper for the firm to abate all units from $0p_{PO}$ to p* rather than pay the tax for each of these pollution units. Without further state intervention, the steel firm chooses to pollute 0p* units as this is the point at which MAC_1 crosses the axis. As the Pigouvian tax is a market-based tool, it allows the firm some flexibility; if the firm's MAC changes, say because of innovation in abatement technology, then the firm can respond by choosing to abate more and consequently pay less tax.

14.4 The economically optimal choice of instrument

There are two main elements in selecting the best pollution control instrument in economic terms. The first is that market-based instruments are generally preferable to command-and-control. The second is that, under conditions of uncertainty, the choice of TPP vs. PT depends upon the relative susceptibility of the polluter and society to misspecifications (incorrect estimation) in the MAC and MSC curves. We treat each element in turn.

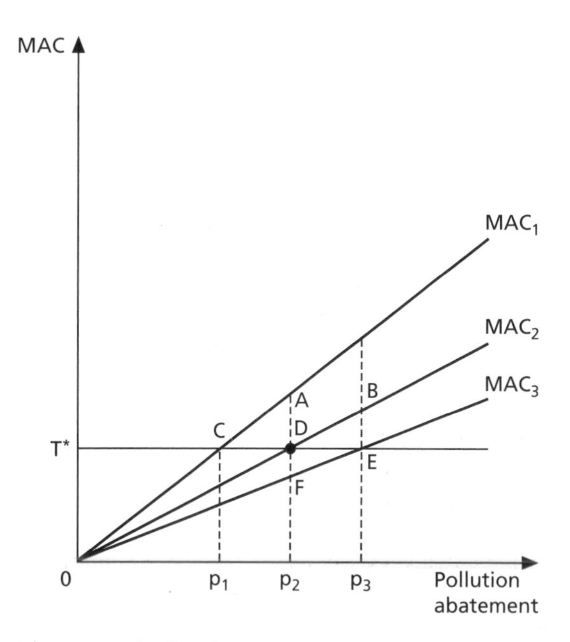

Fig. 14.6 Market-based instruments vs. command-and-control.

14.4.1 Market instruments vs. command-and-control

In order to see why the inherent flexibility of market-based instruments is preferable to command-and-control, we consider a scenario with three steel firms polluting the watercourse. The important consideration is that each firm has a different level of abatement efficiency (MAC_1, MAC_2 and MAC_3 shown in Fig. 14.6), and thus each curve has a different slope. As we have pollution *abatement* on the horizontal axis, the three MAC curves are upward sloping. Firm 1 is the least efficient abater of pollution, perhaps an older firm with less efficient machinery, and firm 3 the most efficient. The MAC curves have been drawn so that the distance p_1 to p_2 is the same as the distance p_2 to p_3. This has been done merely to simplify the explanation; the basic result does not depend on it.

Assume that the state regulator wants to achieve a socially optimal level of pollution, and that this would require the industry to abate pollution. The level of this total abatement has been identified as three times the quantity $0p_2$. This abatement could be achieved by setting a standard: each of the three firms must abate pollution $0p_2$. An alternative market-based tool to achieve the same outcome is a Pigouvian tax applied at level 0T* per unit of pollution. With this tax rate, firm 1 abates $0p_1$, firm 2 abates $0p_2$ and firm 3 abates $0p_3$. As $p_1p_2 = p_2p_3$, total abatement is three times $0p_2$.

Although the total pollution abatement is the same for the two instruments, the *costs* of achieving this abatement are not. The area under a marginal cost curve shows the total costs, because the marginal cost is the cost per unit, and adding together each marginal cost gives the total cost. For the standard, the total costs to firm 1 are $0Ap_2$, to firm 2 are $0Dp_2$ and to firm 3 are $0Fp_2$. For the tax, the total costs are $0Cp_1$, $0Dp_2$ and $0Ep_3$, respectively. How do the total aggregate abatement costs compare? The total costs to firm 2 are the same under both tools—so in comparing costs we can consider only firms 1 and 3. If the tax is cheaper then $(0Cp_1 + 0Ep_3)$ must be less than $(0Ap_2 + 0Fp_2)$. For firm 1, the extra abatement costs for a standard, over and above the costs for the tax, are given by p_1p_2AC. For firm 3, the extra abatement costs of the tax, over and above those for the standard, are p_2p_3EF. As p_1p_2AC is bigger than p_2p_3EF, the total abatement costs for the industry are higher for the standard compared with the tax.

This scenario is somewhat contrived, but the explanation holds in general; in order to achieve a given level of pollution abatement, it is cheaper to use a market-based tool than to use a command-and-control tool. So far we have discussed two alternative market-based tools: TPPs and Pigouvian taxes. We now consider which of these two instruments is better.

Under conditions of certainty, where the MAC and MSC curves are well-defined and their respective positions known, if the market for TPPs is functioning perfectly, and there are no transactions costs, then the price of the TPP equals 0T*, i.e. the Pigouvian tax rate. However, under conditions of uncertainty—the real world—one instrument might be preferred over the other. This is the subject of Section 14.4.2.

14.4.2 Optimal instrument selection under uncertainty

If there is uncertainty over MAC and/or MSC, the TPP instrument is similar to a standard in that, given perfect monitoring and enforcement, it ensures a given level of pollution; a tax does not. However, this does not necessarily imply that TPPs are always preferable. The optimal instrument choice depends on the sensitivity of the two affected parties—society and the polluting firm—to misspecifications in the curves. Intuitively, the output

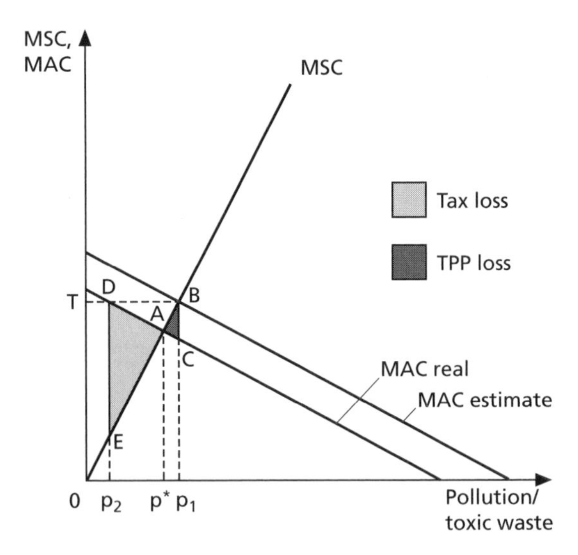

Fig. 14.7 Optimal choice under uncertainty: relatively steep marginal social costs.

level of a hazardous pollutant, such as toxic waste, should not be at the polluter's discretion. The corollary of this is that, for non-hazardous pollution, the flexibility implied by a Pigouvian tax can be optimal. These two scenarios imply, respectively, a steep MSC curve and a shallow MSC curve.[3]

In order to justify the intuition consider Fig. 14.7, which shows the case for toxic waste. There are two MAC curves: one estimated, one real. The state regulator applies either a TPP system or a Pigouvian tax system based on the *estimated* MAC. Thus, the tax level is set at 0T per unit pollution, and $0p_1$ TPPs are distributed. However, the real socially optimal pollution level is 0p*. As the estimate of MAC does not coincide with real MAC, there is some loss in social welfare regardless of which instrument is selected. For TPPs, the loss is ABC; for all the pollution units from p_1 to p*, MSC is greater than real MAC, and the vertical distance represents societal loss. A Pigouvian tax based on estimated MAC is set at 0T. The polluter responds to this tax rate by emitting $0p_2$ pollution, i.e. less than the optimal 0p*. The societal loss for the tax system is ADE: for all pollution units from p_2 to p*, MAC real

3 'Steep' and 'shallow' depend on the scale used on the horizontal axis. But for a given scale—say tonnes of pollutant—toxic waste has a steeper MSC than non-toxic residential garbage.

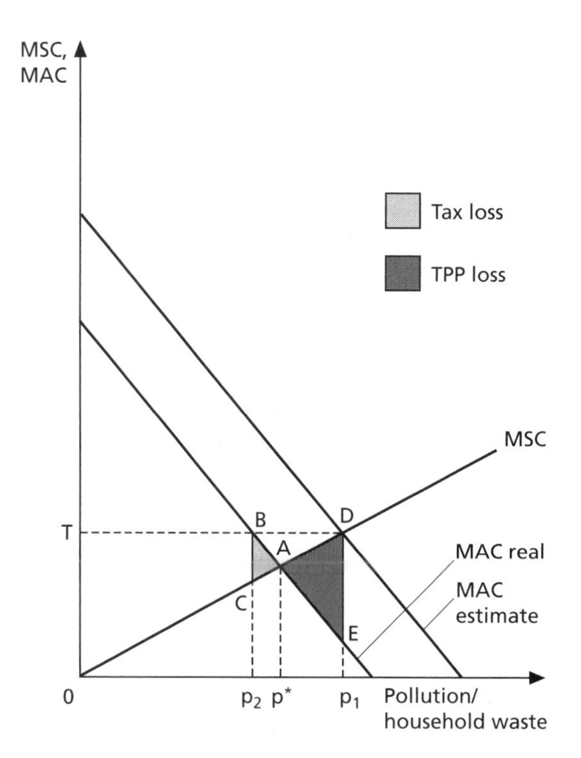

Fig. 14.8 Optimal choice under uncertainty: relatively shallow marginal social costs.

exceeds MSC. The area ADE exceeds ABC, and thus the losses to society from choosing a tax system in this case exceed those from using a TPP system.

The result is the same regardless of whether MAC is overestimated or underestimated: it is the relative slopes of MSC and MAC which determine optimal selection. If MSC is steep relative to MAC then TPPs are better, and if relatively shallow then taxes are preferred. Figure 14.8 shows this latter scenario. The loss associated with TPP (ADE) in this case exceeds the loss to the tax (ABC).

The final element of our analysis is to consider misspecifications in MSC. As the polluter's behaviour is not determined in any form by MSC, the losses associated with a tax system are in fact exactly the same as those for TPPs. In Fig. 14.9, the tax level is set at 0T and the number of permits distributed is $0p_1$, based on the estimated MSC. Even though estimated MSC differs from real MSC, the firm still abates up to the point where the tax 0T equals MAC, at $0p_1$. Thus, the societal losses are the same for both instruments.

The following section considers some of the advantages and disadvantages of the two instruments overall.

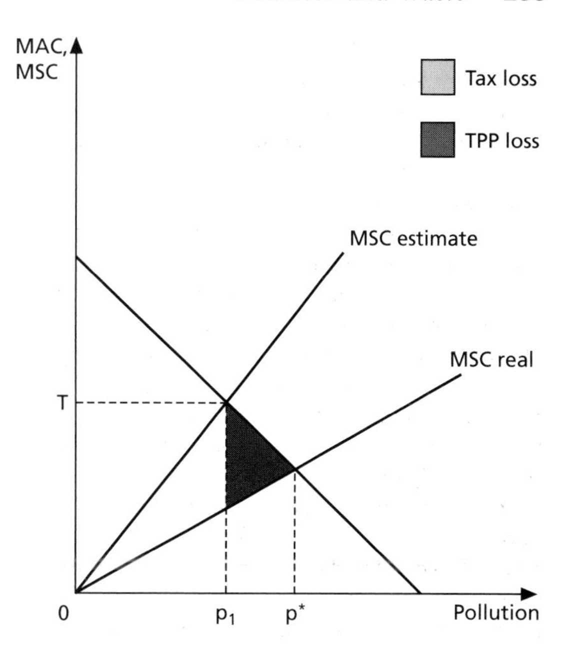

Fig. 14.9 Misspecification in marginal social costs.

14.4.3 Taxes vs. tradable pollution permits: pros and cons

Both instruments have their advantages and disadvantages. The benefits of a TPP system over a tax system are as follows.

1 *Accommodation to industry change.* The state regulator need not intervene if the number of polluters changes, as each new entrant can (theoretically) buy permits from the market and sell if exiting.

2 *Accommodation to inflation.* A fixed tax per unit pollution becomes less significant over time if there is an increase in price levels in general. If this occurs, and no inflation adjustment is made, then over time the aggregate pollution output for the tax system increases progressively from the socially optimal initial level. Under the TPP system, the price of permits in the market increases without regulatory intervention.

3 *Acceptability to industry.* Historically, firms are accustomed to the command-and-control system, and having a maximum pollution output level dictated to them. If TPPs are 'grandfathered', then there is no cost to the polluter and it has a preset 'standard' which it can modify through buying and selling permits.

On the other side, the benefits of a tax system over the TPP alternative are as follows.

1 *Public revenue.* Most taxes cause distortions in the economy, and yet it is necessary to raise public revenue.

It is perhaps more appropriate to raise this revenue through internalizing the pollution externality. The 'polluter pays' principle argues that those using environmental sinks should pay.

2 *Impediments to trade in TPPs.* There may be significant search costs for firms in obtaining TPPs.

3 *Hoarding.* If holding a permit is necessary for pollution output it becomes a necessary condition of production. Hoarding permits might then be a viable strategic mechanism to artificially restrict competition, as permits are then a barrier to entry to the industry.

Now that the economic theories of pollution have been presented, we turn to examine a real-world application.

14.5 Pollution regulation in practice: the RECLAIM scheme

Air quality has been on the political agenda for over 50 years in Los Angeles because of photochemical smog. However, despite much activity, command-and-control has not been sufficiently effective for Los Angeles to meet federal targets set in the Clean Air Act of the 1960s. This is one reason why an alternative market-based approach to pollution regulation was pioneered —the Regional Clean Air Management (RECLAIM) programme—which was adopted in October 1993.

The concept of RECLAIM was simple; it aimed to achieve a given air emissions limit from a particular facility while dictating neither which pieces of equipment would be regulated, nor how, nor to what degree. Further, this emissions limit would decline over time in line with the region's overall emissions reductions requirements.

RECLAIM introduced flexibility for polluters in two ways: first, they could determine the optimum way to achieve a given emissions level; and, secondly, they could choose to emit more than the limit set by the regulator if they bought the requisite extra emission permits. This latter option depends on trades in the permits market. If the polluter decides that lower abatement than that permitted by the emission limit is optimal, then they can buy emission reduction permits; these are sold by those facilities which pollute less than their emission limit. If the facility operatives are better informed than the regulator about their emissions and abatement technologies, then this type of permit system should lead to efficient pollution control. This depends upon the important caveat that the market functions well and that transactions costs are relatively low.

With a command-and-control system, there is little incentive for polluting firms to inform the regulator about better abatement technologies than those in place, because the regulator is then likely to force the firm to adopt them. If a firm does exceed the standards there is no credit given. Indeed, if this pollution prevention is at the expense of profits, then the firm is likely to be lowering its competitiveness. The firm would be shooting itself in the foot by revealing potential improvements.

Thus, under command-and-control, only the employees of the regulatory body and the manufacturers of abatement technologies are looking for potential environmental improvements. Under RECLAIM, all facility operators of polluting firms would be seeking out these improvements. There is said to be *incentive compatibility*: there is an incentive for the firm to behave in a way that the regulator would want it to behave.

Five basic principles for the RECLAIM programme were agreed as a result of meetings between California Air Resources Board, the US Environmental Protection Agency, both large and small firms, environmental pressure groups, trade unions, academia (including economists), and representatives from the commodity markets and stock markets.

1 Three independent markets would be designated: nitrogen oxides (NO_x); sulphur oxides (SO_x); and volatile organic compounds (VOCs).

2 The initial emissions allocation would be set at the 1994 emissions target level in the Air Quality Management Plan; a clean-up plan that had already been adopted.

3 Mandatory federal targets would be met for NO_x and SO_x by 2003, and for VOCs by 2010. These were to be achieved by reducing the emissions permitted by the markets on an annual basis.

4 The amount of ozone reduction achieved by the emissions trading programme must be at least equal to that expected under command-and-control.

5 The reporting and enforcement of emissions levels must be at least on a par with command-and-control.

When these principles were implemented the permit market scheme covered about 70% of total emissions for both the NO_x and SO_x markets. For NO_x, this implied 370 facilities out of a total of about 32 000 in the South Coast Air Quality Management District (SCAQMD). The sources for which RECLAIM was mandatory included petrochemical producers and refiners, aerospace and military installations, and construction materials manufacturers. Most other smaller stationary sources could voluntarily enter the scheme.

Table 14.1 Variation in emission reductions between facilities. (Source: Lents & Leyen 1996.)

Annual reduction rates by number of facilities (1994–2000)	NO_x	SO_x
2%	93	10
2 to < 6%	70	16
6 to < 10%	89	3
10 to < 16%	118	11
Total	370	40

14.5.1 Reducing emission levels

One of the principal implementation problems for the NO_x and SO_x markets was the setting of a 'fair' starting point, and it was agreed to apply the highest year of reported emissions between 1989 and 1991, reduced by any percentage required through legislation applied up to 1993. Subsequent to establishing these starting conditions, the SCAQMD provided targets for aggregate annual emissions reductions from the initial 1994 levels: a 75% reduction in NO_x and a 60% reduction in SO_x by 2003. However, simply legislating that each facility should have its emissions reduced by this fixed percentage would not be economically optimal as the initial levels of technology and scope for emissions reduction were different. The marginal abatement costs were not identical across facilities, and therefore enforcing the same annual emissions reductions would penalize facilities that were relatively efficient in 1994.

Allowing for this, the procedure was to assign a percentage reduction for each equipment type, and then calculate a facility level emissions entitlement. This generally implied a three-stage procedure of information provision and clarification between the regulator and facility operators. The net result was that the variation in emissions reductions were significant between facilities, as demonstrated in Table 14.1.

14.5.2 The trading system

The easier the conditions for trading, the more the trades and the lower the transactions costs of the programme. The unit of exchange (RECLAIM Trading Credit or RTC) was simply one tonne of pollutant. Every facility had to hold sufficient RTCs to cover their emissions or they would be fined. As each facility had its own unique rate of reduction of emission cap, it was not possible to have multiyear credits; plants were assigned a certain (decreasing) number of RTCs for each year from 1994 until 2003. After 2003, the balance of RTCs remain constant. Because these RTCs were assigned in a batch in advance, the facilities were free to trade, say, 1999 RTCs in 1995. This dual cycle ameliorates the problem of permit hoarding.

14.5.3 Operating RECLAIM

RECLAIM was adopted in October 1993. The first cycle of permits was issued by January 1994; there was a 10% appeal rate although most complaints were resolved through clarifications and minor amendments. The second cycle went out by July 1994 and included a dozen smaller facilities which had chosen to enter the permits market voluntarily.

There was a clearly defined procedure for facilities in the interim phase between command-and-control and the installation of continuous emission monitoring, and hence the fully fledged market system. In this interim period of one year, the larger facilities provided the SCAQMD with monthly emissions reports which were compared with the command-and-control concentration limits. During this period, assistance was provided to industry in the form of workshops and 'how to' manuals, and conferences provided a meeting point for buyers and sellers of emissions-monitoring technologies.

One outcome which the SCAQMD wanted to achieve was transparency and public accessibility. This was achieved through an annual programme audit which was in the public domain[4] and provided dozens of performance features. A second desired outcome was that industry endorsed the programme; despite the large reduction targets in SO_x and NO_x, two-thirds of firms have given their endorsement to RECLAIM. Thirdly, Lents and Leyden (1996) claim that the performance outcomes have been achieved at half the cost of the command-and-control alternative. To date, RECLAIM has been a good example of how economic theories, when adapted for local conditions and opportunities, can improve social welfare.

14.6 Municipal solid waste

Much of the debate about waste and recycling is

[4] The RECLAIM programme has a web-site which provides performance updates: http://www.aqmd.gov

dominated by municipal solid waste (MSW): solid waste refuse collected by municipal authorities (or subcontractors) from households and small-scale commercial enterprises. In fact, in the UK, MSW constitutes only about 6% of the aggregate waste produced in the late 1980s (Fig. 14.10).

The UK defines controlled waste as the sum of MSW, demolition and construction waste, industrial waste, and waste from energy production. Of the aggregate waste of 330 million tonnes in the late 1980s, 120 million tonnes was controlled waste, and therefore of particular concern in terms of waste management. Although the definition of 'controlled waste' varies from country to country, Table 14.2 gives some indication of the pattern of controlled waste generation across some developed countries, and demonstrates a strong trend towards increasing output. In the context of the total world output of waste, this trend is likely to persist for two reasons: first, population levels are growing; and, secondly, average per capita income levels in general are also increasing. As Table 14.3 shows, an increase in income positively correlates with an increase in the output of MSW. The other types of waste are controlled through independent legislation; for example, mining waste through planning regulation with consent being obtained only if the mining operator provides satisfactory plans for disposal.

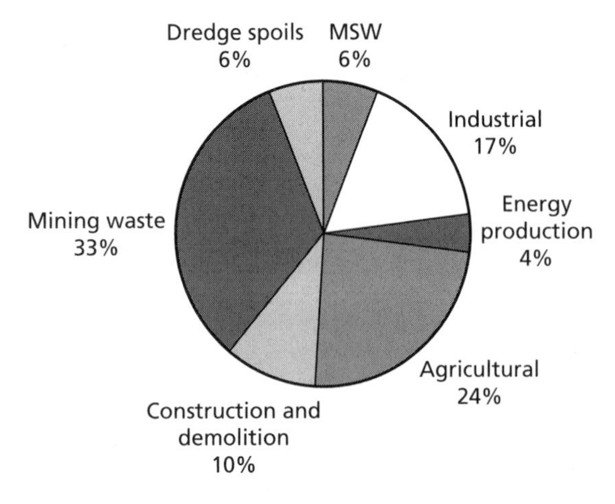

Fig. 14.10 Wastes arising in England and Wales in the late 1980s. (Modified from Brisson 1997.)

The possible means of treatment of MSW is the subject of the following sections. Each treatment option has its associated environmental impacts. They are considered in turn. Table 14.4 gives a breakdown of the current methods of disposal of MSW across different countries.

Table 14.2 Changes in controlled waste arising across certain developed nations in millions of tonnes per year. (Source: OECD 1989b, 1993.)

Country	Millions of tonnes per year 1985	1990	Percentage change 1985–90
Canada	91	37.4*	–59
USA	976	8 385	760
Austria	16.1†	56	248
Belgium	11.1‡	32.1§	190
Finland	19.7	21.2	8
France	19.7	21.2	8
Germany	98.4¶	109.9	12
Greece	7.7	15.0	95
Italy	50	89.1**	78
Luxembourg	0.3	6.7	214
The Netherlands	19.0	29.0	53
Norway	4.2	6.0	43
Spain	33.5††	48.3	44
Switzerland	2.5	6	140
UK	81.7¶	121	48

* Does not include industrial waste.
† 1983.
‡ 1980.
§ 1988.
¶ 1984.
** 1991.
†† 1986.

Table 14.3 Controlled waste arising. (Source: Pearce & Turner 1994.)

	Low income ($350/capita/year)	Middle income ($1950/capita/year)	High income ($17 500/capita/year)
MSW arisings per capita	200 kg/capita/yr	300 kg/capita/yr	600 kg/capita/yr
MSW/$ income	0.51 kg/$	0.15 kg/$	0.03 kg/$

MSW, municiple solid waste.

Table 14.4 Disposal of municipal solid waste by method across various countries. (Percentage of total.) (Source: World Resources Foundation 1995.)

Country	Landfill	Incineration	Recycling	Composting
Austria	65	11	6	18
Belgium	43	54	3	0
Canada	80	8	10	2
Denmark	29	48	19	4
Finland	83	2	15	0
France	45	42	3	10
Germany	46	36	16	2
Greece	100	0	0	0
Ireland	97	0	3	0
Italy	74	16	3	7
Japan*	20	75	2	1
Luxembourg	22	75	2	1
The Netherlands	45	35	16	5
Norway	67	22	7	5
Portugal	85	0	0	15
Spain	65	6	13	17
Sweden	34	47	16	3
Switzerland	12	59	22	7
UK	90	8	2	0
USA	67	16	15	2

* Municiple solid waste levels in Japan are calculated after the removal of recyclables.

14.6.1 Municipal solid waste disposal options

LANDFILL/DUMPING

The most common method for dealing with waste is dumping. In the nineteenth century, unregulated dumps emerged outside cities, stimulating disease and vermin. The current practice in the developed world is to site landfills away from geological strata of high permeability to avoid leaching. This can be taken a stage further in 'sanitary' landfill which are lined with impermeable materials to prevent leakage into ground water. The solid waste that arrives at a landfill site is spread and compacted in layers, within 'cells', and each layer is covered with soil at frequent occasions, at least once per day, to restrict vermin. The compacting process is sometimes augmented by shredding and bailing, but the expense of this procedure means that most waste is not treated in this manner.

There are two alternative landfill types: wet and dry. In a wet landfill, the site is set up so that the leachate can circulate to aid biodegradation which reduces the volume of waste. In the case of the dry landfill, the leachate is pumped out to prevent degradation and toxin development. The application of the wet option is not as wide-spread as that of the dry option, although the UK's national waste strategy states a preference for the wet option.

The environmental costs of landfill are primarily linked with the biodegradation process. This process releases methane (CH_4) and carbon dioxide (CO_2) which both contribute to global warming. Methane is potentially explosive and thus its release can be dangerous. If the leachate arising from biodegradation is not controlled then it can enter surface waters, being especially toxic when it combines with heavy metals and pesticide residues. Even the leachate from household waste can be dangerous: ethyl carbonate is a stabilizer for cosmetics; triethyl phosphate is a plasticizer for resins, plastics and solvents; dicyclohexyl phthalate is a plasticizer for rubber. Each chemical is hazardous for human health, as well as for other species. Apart from these direct reactions and immediate heath hazards, there is a longer term consequence of landfilling and dumping; the land is then not readily available for alternative uses, such as agricultural production.

INCINERATION

When waste is dumped at an unregulated site it often self-ignites. This was the case in developed countries before the twentieth century and is still the case today in the developing world. The first incinerators were developed in 1874 in Nottingham in the UK, but early versions mainly failed owing to the high moisture content of waste. Further, they released unpleasant odours and gritty smoke. These problems led to many incinerators being closed down. It was not until the 1970s that incineration regained some popularity as a waste management option as a result of waste-to-energy facilities being constructed. These plants can capture some of the energy from incineration and thereby generate electricity and heat. Some of the environmental problems have been addressed by using acid-gas scrubbing and particulate-removal equipment.

Two types of incinerator currently exist: mass-burners and those using refuse-derived-fuel (RDF). Mass burners do not require waste pretreatment, apart from the removal of bulky items. However, this type is less energy-efficient than the RDF type. The RDF type requires the following processing of the input waste stream:

1 Waste is shredded to reduce bulk and increase consistency. Ferrous metals can be recovered using magnets.
2 The shredded output is split between inorganic waste, which is sent to landfill, and organic.
3 The organic residue is secondary shredded.
4 It is then either densified into pellets which can be co-fired with wood and coal to generate power, or pulverized to a powder which can be slurried with oil and burned.

Another option for the RDF type is biological gasification. In this case, the organic waste is subject to anaerobic digestion by bacteria. This produces biogas consisting of methane and other gases which, through the removal of carbon dioxide, hydrogen sulphide and water, can be used as a substitute for natural gas. The residue can be oxidized, leaving an ash that can be composted or landfilled.

The separation processes required for RDF have other benefits apart from improving energy efficiency as compared with mass burning; less air pollution is emitted with less trace toxic metals. However, carbon dioxide, sulphur oxide and nitrous oxides are still released, as well as the occasional release of harmful dioxins. On the credit side of the environmental equation, incineration leads to a large reduction in the volume of waste (10% of the original) and weight of waste (20–30% of original). This, coupled with heat recovery, can make incineration an appealing option environmentally.

RECOVERY AND COMPOSTING

Composting is the degradation of plant biomass without energy recovery. The recovered product may be used as a soil fertilizer. The quality of product is dependent on the presorting procedure. In some communities in the Netherlands, Sweden and Denmark, households are required to separate their organic waste from the remainder to facilitate the sorting process.

RECYCLING

The reason for recycling is to save on the use of virgin raw materials, therein reducing both resource use and emissions outputs from the production process. In the developing world, waste sites are still populated by 'scavengers' who sort through waste to find reusable and saleable waste. This used to be the occupation of 'rag-and-bone' men in developed countries, but this is not generally the case today. The potential savings from recycling are large for certain categories of waste output, as can be seen from Table 14.5.

Table 14.5 Potential percentage savings from recycling. (Source: Bartone 1990.)

	Aluminium	Steel	Paper	Glass
Energy use	90–97	47–74	23–74	4–32
Air pollution	95	85	74	20
Water pollution	97	76	35	—
Mining wastes	—	97	—	80
Water use	—	40	58	50

The case for recycling aluminium and steel is particularly strong. However, these figures do not necessarily imply that recycling is always a good environmental option for aluminium and steel; this depends on the transport costs and reprocessing costs.

14.7 Waste hierarchy and the economics of municipal solid waste management

Recycling is a preferred disposal option in terms of the waste hierarchy, the subject of this section. The various waste management options have been arranged in a theoretical ladder which orders them in terms of sustainability; this ladder is termed the waste hierarchy. The hierarchy is as follows, ranked in terms of decreasing sustainability:

1 waste minimization;
2 reuse;
3 recovery (e.g. materials recycling, composting);
4 incineration with energy recovery;
5 treated disposal (e.g. incineration without energy recovery); and
6 untreated disposal (e.g. sea dumping).

Section 14.9 applies a modified version of the LCA methodology (discussed in Section 14.8) to rank the options in terms of environmental impact, and thereby tests the hierarchy. However, it is noteworthy that this hierarchy has already received official governmental endorsement and legislation has been passed to promote those options higher up the ranking.

In the UK, the waste hierarchy was recommended by the Royal Commission on Environmental Pollution in 1993, calling for a waste management strategy. The consultation draft of the UK government's Waste Strategy for England and Wales suggests targets for the year 2000 that are aligned with the waste hierarchy: stabilize household waste production to 1995 levels (waste minimization); increase recycling levels to 25% of the domestic waste stream (recovery); and reduce the proportion of waste going to landfill by 10% over 10 years (untreated disposal). One of the UK government's objectives is to promote financial incentives for firms to move up the waste hierarchy. We consider two market instruments: a landfill tax and recycling subsidies.

14.7.1 Landfill tax

The intention of introducing a landfill tax in the UK was announced in the Chancellor of the Exchequer's Budget speech in November 1994. After various amendments, the landfill tax was formally introduced in the UK in October 1996, and defined in the UK Finance Act 1996. The tax is a charge on the disposal of material waste at a registered landfill site.

In order to test the economic efficiency of the landfill tax, it is important to identify the polluter for MSW. Generally, households and small firms are the polluters, but one or two intermediary agents operate to collect the MSW and then to dispose of it to landfill. In this situation there is *incentive incompatibility*: a goal is defined but the economic incentives do not lead to realizing the goal. If households are charged a standard flat rate for waste disposal which is not dependent on the quantity of refuse each individual household produces, there is no incentive for individuals to throw away less. Under this scenario, there is a free-rider problem: everyone would be better off if total MSW quantities were reduced, but the actions of any one household in isolation is insignificant in terms of the total.

If waste were only to be collected in official disposal bags which households had to buy from the collection authority, then households would have an incentive to reduce their individual volume of waste by composting, recycling and reducing waste-intensive consumption. The unfortunate corollary of this is that incentives for fly tipping (illegal disposal), burning of combustibles and putting waste in public litter facilities are also increased.

14.7.2 Recycling subsidies

The principal benefits from recycling are realized not at a local level but at national and global levels. Recycling an aluminium can displaces production and the pollution associated with this production, but it is unlikely that this displacement improves the local environment at the place of disposal. However, recycling is generally carried out at the local level and is generally currently a loss-making activity as compared with the financial costs of landfill disposal. A subsidy which is proportional to the weight of waste recycled and which is financed from national government might be a suitable instrument to stimulate local collection authorities to favour more recycling than currently takes place. However, used in isolation this market instrument does not correct the incentive incompatibility mentioned above.

The economic justification for state intervention in terms of the waste hierarchy is given in Fig. 14.11. The socially optimal level of recycling occurs where the marginal benefits of last unit quantity recycled are equal

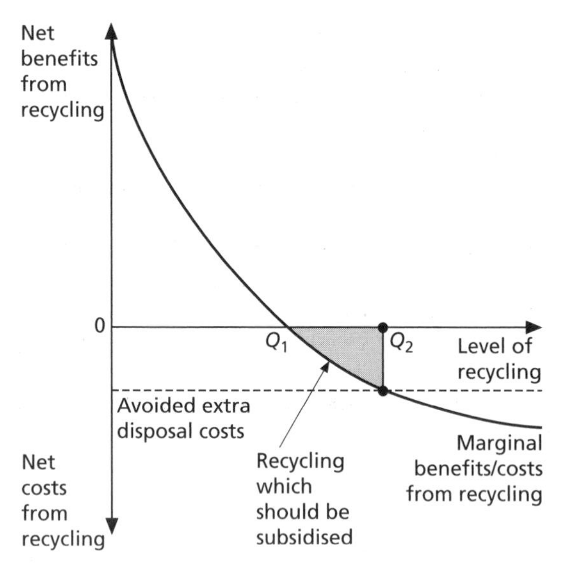

Fig. 14.11 The optimal level of recycling.

to the marginal costs. The benefits of recycling are the sum of the economic value of the displaced production and the avoided economic cost, including environmental cost, of an alternative disposal option. The costs of recycling are the costs of separating the recyclable waste, and the total costs of the recycling process.

In Fig. 14.11, the free market equilibrium level of recycling is given at Q_1. The socially optimal level of recycling is given at Q_2, which accounts for the total avoided costs of disposal. From this analysis, it is apparent that waste management is subject to market failure, and therefore some intervention is socially optimal. However, this does not imply that waste minimization and recycling should be maximized and untreated disposal option minimized, as policies set up to achieve these dual goals in all circumstances might be inefficient.

For example, consider the scenario where a household lives in a remote area with low population density. The demand for land is likely to be low relative to an urban conurbation, and so some of the costs of landfill are likely to be lower. If the nearest recycling site is far away then the landfill option might be cheaper in terms of total economic costs, which include environmental costs. It is just this kind of scenario which lends itself to the application of LCA in order to rank alternative products or processes in terms of their respective environmental impacts. The LCA methodology is set out in the next section, with a case study application to waste management options analysed in Section 14.9.

14.8 Life cycle assessment

Life cycle assessment[5] is an analytical tool used to capture, in quantitative terms, the environmental impact of a product or process from raw material extraction to final waste disposal and decomposition: a 'cradle-to-grave' analysis. It is intended as an objective and transparent tool to choose between alternative products or policies, based on their environmental impacts. There are many types of environmental impacts associated with the choice of a particular product or policy; for example, effects on global warming, fossil fuel depletion, renewable resource use, etc. These impacts are often not easily comparable. In terms of environmental impact, is the emission of a tonne of sulphur dioxide to the air better or worse than the emission of a tonne of phosphates to a water course? Life cycle assessment attempts to make such diverse impacts commensurable in a standardized methodology. Further, the LCA study should report the assumptions made.

The life cycle approach was first developed in the late 1960s and early 1970s (VITO 1995). Life cycle assessment is a development of the energy analysis methodology that was used for assessing the behaviour of extended industrial systems (Boustead 1993). This energy modelling, also known as resource and environmental profile analysis (REPA) requires the construction of balanced flow charts to describe the industrial process, and requires a calculation of the consumption of raw materials and the generation of waste. The number of applications of the REPA methodology increased as a result of the oil crises of the 1970s. After a long period of relative inactivity, the application of life cycle techniques was revived in the 1980s in response to environmental concerns.

The demand for LCA applications today comes from various sources. 'Green' consumers might want to express a preference in the market for environmentally friendly commodities. However, there have been instances where false or misleading claims have been made by firms—termed 'greenwashing'. For instance, a washing powder manufacturer might market its product as 'chlorine-free' to win some of the 'green' consumer vote, even though chlorine was never used in either its production process or that of competitors. Faced with confusing information, consumers are inclined towards seal-of-approval environmental schemes that

[5] We treat life cycle analysis as synonymous with life cycle assessment, following VITO (1995).

Fig. 14.12 The EC Ecolabel.

are externally regulated. These schemes are underpinned by LCA and exist worldwide: for example, the European Community's Ecolabelling Scheme (Fig. 14.12); Germany's 'Blaue Engel', first developed in 1978; Canada's Environmental Choice Programme; and Japan's EcoMark. Such schemes are significant in that they offer commercial viability to LCA applications.

14.8.1 The life cycle assessment methodology

In 1993, the LCA working group of the Society of Environmental Toxicology and Chemistry (SETAC) published a code of practice (SETAC 1993) based on the recommendations of more than 50 LCA experts from 13 countries. This was an attempt to harmonize LCA methodologies. The complete SETAC LCA methodology is split into four interrelated phases. These stages are as follows:

1 *Goal definition and scoping*. The purpose of the study is defined in a clear and unambiguous manner as well as the mechanics of reaching agreements on the final contents.

2 *Inventory*. The data collection stage: flows of energy, material inputs and emissions are defined for the entire product life cycle.

3 *Impact assessment*. The data gathered from the inventory stage are translated into corresponding environmental impacts (classification and characterization). These loadings are then weighted to allow comparison under a single measure (valuation).

4 *Improvement assessment*. Options for reducing aggregate loading are identified and evaluated.

Each stage is discussed in more detail in the sections below.

GOAL DEFINITION AND SCOPING

The structure, scope and content of the LCA is influenced both by the anticipated target audience and the function of the research. Life cycle assessment applications are data intensive, and so establishing the boundaries for inquiry is critical for efficiency. For instance, in analysing the transport costs of a product or process, how far back should the LCA go: emissions from lorries; or the production and disposal of lorries? If lorry production is included, then should the environmental impacts of the construction of the plant also be in the LCA boundary? There is a need to define a geographical boundary too: are effects to be assessed in terms of impact on a local, regional, national or international level? If any misspecification of this boundary omits significant data then the results may be systematically skewed; conversely, collecting and manipulating data on insignificant environmental loadings is a waste of research energies.

If a direct comparison between alternatives is the established aim of the study, then the LCA might be streamlined by contracting its scope. Most studies to date have been of this kind; for example, disposable vs. reusable nappies. Processes that are common to both products might be usefully omitted from analysis without jeopardizing the validity of the final results.

At this stage, the definition of an appropriate functional unit is required. For instance, in an LCA of milk containers (glass vs. cardboard carton) it would not be methodologically correct to have the functional unit as 'one tonne of glass/cardboard' as one tonne of glass will not hold as much milk as the equivalent weight of cardboard cartons. A more appropriate functional unit in this case would be the amount of glass/cardboard

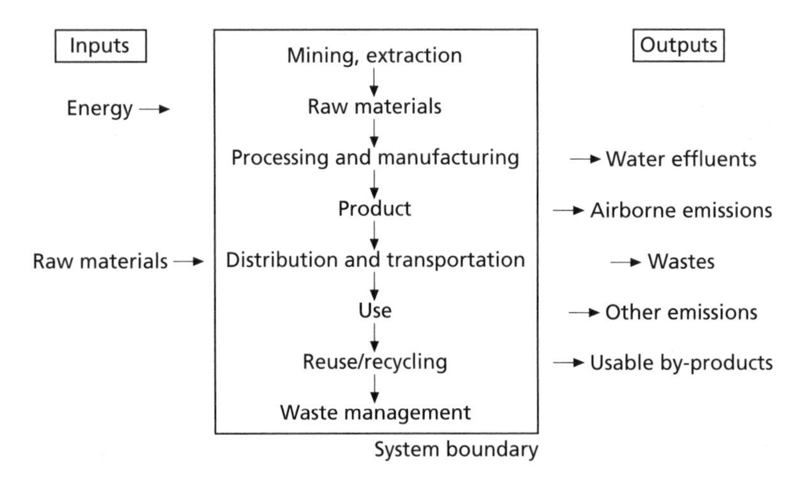

Fig. 14.13 Schematic view of the life cycle of a product. (Source: VITO 1995, with permission.)

needed to hold a given quantity of milk. Appropriate designation of the functional unit gives a 'level playing field' for a comparison of environmental loadings between alternatives.

INVENTORY ANALYSIS

This is the principal data collection stage of an LCA study. It entails identifying and then gathering quantifiable data relating to the material and energy inputs into a product across its entire life cycle and any associated emissions, discharges and wastes. This procedure has three stages, outlined below.

1 *Defining the life cycle.* This implies 'dissecting' the product into its component parts, which in turn are composed of several raw materials. All intermediate transport steps between the various stages should be accounted for in the analysis. A schematic view of the life cycle of a product is presented in Fig. 14.13.

2 *Process tree evaluation.* For each of the materials identified in the definition stage, a complete process tree is drawn, using the whole life cycle of the material from raw material extraction to ultimate elimination.

3 *Inventory profiling.* Identical inputs and outputs (e.g. carbon dioxide emissions, iron ore) are aggregated across the entire life cycle. This should generate a quantified measure of impacts in terms of a vector of inputs–outputs for one functional unit of the product being studied.

IMPACT ASSESSMENT

At the impact assessment stage, the raw data on resource use and emissions are translated into respective potential environmental impacts (e.g. global warming, acidification, etc.). This procedure tends to reduce the number of categories of data, and so allows an estimate of the relative environmental friendliness of the product or process to be made. SETAC splits the impact assessment phase into four distinct steps: classification; characterization; normalization; and valuation.

1 *Classification.* This is a process of data aggregation into several impact categories. Any input or emission which contributes to a designated impact is listed in that category. Thus, if an input of iron ore to production effects both global warming and human toxicity then it appears in both impact categories. SETACs defined impact categories are given in Table 14.6.

2 *Characterization.* After qualifying which inputs and emissions affect each impact category, the next task is to quantify these impacts, a process which SETAC terms characterization. Thus, given data limitations, the potential contributions of emissions and inputs to the different impact categories are estimated. One estimation method is the use of equivalence factors. In this method, no direct impact is calculated, only the potential impact relative to a reference substance. One benefit of such a methodology is the availability of equivalence factor databases[6] which can aid the har-

[6] Established databases include: Global Warming Potential (CO_2 equivalent); Ozone Depletion Potential (CFC-11 equivalent); Acidification Potential (SO_2 equivalent); Eutrophication Potential (phosphate equivalent); Photochemical Ozone Creation (ethylene equivalent) (VITO 1995).

Table 14.6 The SETAC standardized classification table. (Source: SETAC 1993.)

	General areas for protection		
Specific impact categories	Resources	Human health	Ecological health
Resource depletion			
Depletion of abiotic resources	plus		
Depletion of biotic resources	plus		
Pollution			
Global warming		(plus)	plus
Ozone depletion		(plus)	(plus)
Human toxicity		plus	
Ecotoxicity		(plus)	plus
Photochemical oxidant formation		plus	plus
Acidification		(plus)	plus
Eutrophication			plus
Degradation of ecosystems			
Land use			plus

Plus, direct potential impact; (plus), indirect potential impact.

monization of results across LCA studies. Further, the calculation is simple: the sum of input–emission quantities multiplied by their respective equivalence factors. Such databases and calculations are beginning to become available in computer-generated spreadsheet form. Calculation of the aggregate impacts for each category (the effects score) results in the *environmental profile* of the product or system. This profile can then be presented as a summary of various environmental impacts across the life cycle of the product or process.

3 *Normalization.* As the various equivalence factors used in calculating the environmental profile are different in terms of units and orders of magnitude, the overall environmental friendliness of the product is difficult to grasp. Normalization is a procedure that attempts to make the various effects score more meaningful by relating them to the magnitude of the environmental problem in a given time frame.

4 *Valuation.* At the valuation stage, the various impact categories are weighted. Multiplying each normalized effects score by its respective weighting results in one number—the *environmental index.* Examples of standardized weighting schemes include the Swiss eco-point system, the Dutch environmental index method and the Swedish EPS-system.

IMPROVEMENT ASSESSMENT

An analysis of an LCA study should provide options for reducing the environmental burdens or impacts of the system under study. These options might be used as a research and development tool to reduce environmental impact at the product design stage.

14.8.2 Problems with life cycle assessment

Some of the problems associated with the LCA methodology are: parameter designation; assignable cause; data use and comparison; and inappropriateness of one 'correct' answer.

PARAMETER DESIGNATION

At the preliminary stage, it is necessary to define the scope, boundaries and functional units of the study. These decisions have a subjective element. Having said this, Boustead (1993) demonstrates that, under normal circumstances, the (stock) impact associated with equipment production is usually negligible in comparison to the (flow) impact from its use. Thus, this tertiary level can generally be excluded without significantly biasing the results.

ASSIGNABLE CAUSE

Many production processes produce more than one useful output; this requires that a qualitative valuation be made as to the proportion of aggregate impact that should be attached to the various products and by-products. There are various alternatives.

- *Natural causality.* If there is a causal relationship linking specific emissions or resource consumptions with specific loadings then an objective assignment should result. An example is sulphur content in inputs leading to SO_x production.
- *Economic value.* The monetary value of the products or by-products might be used, as such assignment takes into account the reasons for producing the environmental impact in the first place. Such assignment is contentious in that monetary values might vary, resulting in a change in assignments without a physical change in the production process.
- *Arbitrary.* For example, a 50/50 split.

There is no hard-and-fast rule as to which method is most appropriate for an LCA application, therefore implying some subjectivity.

DATA USE AND COMPARISON

The data used in the LCA are likely to be of variable quality and quantity which can introduce problems.

INAPPROPRIATENESS OF ONE 'CORRECT' ANSWER

The fact that the LCA procedure ends up with one number for environmental impact has pros and cons. In comparing products (e.g. milk containers) it is tempting for a decision-maker to assume that one option has 'won' over the other. A fuller answer is that one option is better given the assumptions applied and documented in the LCA. These assumptions are likely to be contentious. Transparency ensures that the assumptions and limitations are stated in the study but these are often not considered by decision-makers.

14.9 Life cycle assessment application to municipal solid waste management

The study outlined in this section comes from Brisson (1997). She describes her study as a 'partial economic life cycle assessment'; it differs from a conventional LCA in that pollution damage estimates are measured in monetary terms. In this sense it parallels the methodological approach of cost–benefit analysis (Chapter 8). The life cycle inventory is assessed for waste management options, and monetary valuations of damage estimates are entered if available. This valuation procedure thus partially substitutes for the third impact assessment stage of the LCA. The study considers 12 European countries,[7] but we concentrate on the UK for the sake of brevity.

14.9.1 Goal definition and scoping

As the LCA had a large scope—testing the waste hierarchy—the boundary limits are necessarily narrow, therein omitting a number of environmental impacts. Further, appraisal of two of the options (reuse and waste reduction) is omitted owing to the complexities of estimating private costs. The four alternatives to be ranked are: landfill; incineration; composting; and recycling. For the first three options, the environmental impact associated with the production and consumption of the commodities which comprise the MSW are outside the boundary, as it is assumed that they are likely to be unaffected by the waste management option chosen. These impacts are included in the boundary conditions for recycling as they are avoided through the choice of this option.

Common to all four options is the appraisal of environmental impacts arising from collection, separation, processing and waste disposal. Transport is a critical impact. The study sets the boundary for analysis as the environmental impact of the energy consumed in transport, and includes air pollution and road accidents. Thus, secondary/tertiary impacts—for example, the production of lorries and the construction of the lorry manufacturing plant—are excluded. Current practice is assumed, as opposed to estimates of future conditions. The functional unit is 1 tonne of MSW, with environmental impacts measured as kilogrammes per tonne of MSW.

The study appraised the waste options in the context of the breakdown of waste for each study nation. The figures for the UK are given below in Table 14.7, along with those for Germany for comparison.

14.9.2 Life cycle inventory

In formulating the life cycle inventories for the four waste management options, Brisson (1997) states the following caveats and assumptions.

- Following the influential work of White *et al.* (1995) each household consists of 2.5 persons with an aggregate waste generation of 870 kg per household per year.

[7] These countries are Belgium, Denmark, France, Germany, Greece, Ireland, Italy, Luxembourg, the Netherlands, Portugal, Spain and the UK.

Table 14.7 Estimated composition of municipal solid waste. Percentage of total by type of material/component. (Source: Coopers & Lybrand 1996, in Brisson 1997.)

		Material						
Country	Year	Organic	Board	Paper/Glass	Metals	Plastic	Textiles	Other
UK	1991	20	33	9	8	6	4	20
Germany	1992	31.6	24	8	5.6	8.8*		22

* In Germany the 8.8% figure refers to textiles and plastic combined.

- For 'bring' systems of recycling, as opposed to collection, consumers drive to recycling banks. The average car journey is 1.8 km.
- The average fuel consumption per tonne for waste collection and transport to bulking depots is 10 L.
- Bulky and garden waste is treated separately, with 3.4 trips per household per annum of 2.56 km each way made to deposit such waste in centralized civic amenity landfill sites.

RECYCLING

- For recycling with kerbside collection, collection comes from standard wheelie bins and blue boxes weighing 15 and 1.6 kg, respectively, and with an average lifespan of 10 and 7.5 years, respectively. Bins containing biowaste are washed out six times a year, whereas those containing solid waste are not washed.
- The electric energy used per tonne of waste input is around 19 kWh.
- Diesel input is 1 L per tonne of waste input.
- Losses during sorting are 5% for blue boxes and 10% for wheelie bins, and the lost proportion is landfilled.
- The reprocessed and recycled material is a suitable replacement for the virgin product.

COMPOSTING

- For the UK, there is no composting.
- For other nations, it is assumed that sorting is carried out at the source of waste generation, and that 50% of the collected waste produces marketable compost; the rest is landfilled.

INCINERATION WITH ENERGY RECOVERY

- Energy efficiencies are 20% for electric.
- There is a 90% recovery rate of ferrous metals from incinerator ash.
- The remaining residues are landfilled 60 km away.

LANDFILLING

Three scenarios are considered: landfill without recovery of gas; landfill with gas collection but no energy recovery (gas is flared); energy generation from landfill gas. The first option is the one that we report.

DISPLACED ENERGY

There are two alternative scenarios used in determining the value of the energy displaced; that is, replacing other energy production by using energy from incineration. The first scenario is the marginal power plant case wherein the energy source being displaced is the least efficient one on the national grid, usually old coal-fired power stations. The second scenario inputs values for the average European Union power plant in terms of environmental efficiency.

TRANSPORT AND ACCIDENTS

A value is inputted into the calculations for the costs associated with road accidents. These figures are based on statistics for the average number of casualties per kilometre and statistical designation of the value of a human life. For the UK, the value of a mortality is given as 2 541 099 ECU, and for a serious injury 106 875 ECU.

14.9.3 Some study results

The outcome of the study for three distinct scenarios are summarized in Table 14.8. The aggregate ranking for the waste management options for all 12 nations are:

Table 14.8 Net external costs associated with present municipal solid waste (MSW) management practices in the European Union (income elasticity = 0.3). (Source: Brisson 1997.)

	Belgium	Denmark	France	Germany	Greece	Ireland	Italy	Luxembourg	Netherlands	Portugal	Spain	UK
						(ECU/tonne MSW)						
Present: mixed refuse collection, bring system for recyclable and organic materials												
Landfill												
No gas recovery	8	6	15	6	20	9	6	6	4	20	16	4
Gas flared	7	5	15	5	19	8	5	6	4	19	16	4
Energy generation (displacing old coal)	6	4	14	5	19	7	5	5	3	18	15	3
Energy generation (displacing average EU electricity)	7	5	15	5	19	8	5	6	3	19	15	4
No transfer	6	5	13	4	18	8	5	5	3	17	14	3
Incineration												
Electricity generation (displacing old coal)	−20	−18	−20	−25	—	—	−14	−12	−23	—	−13	−18
Electricity generation (displacing average EU electricity)	17	12	23	16	—	—	12	19	12	—	22	11
Recycling	−245	−157	−282	−190	−44	−204	−136	−193	−189	−17	−111	−170
Composting	47	22	25	22	—	—	—	27	16	127	49	—
Present: co-collection of mixed refuse and recyclable and organic materials (blue box)												
Landfill	6	4	12	4	17	8	5	4	3	16	13	3
Incineration												
Electricity generation (displacing old coal)	−27	−18	−21	−26	—	—	−15	−11	−25	—	−11	−18
Electricity generation (displacing average EU electricity)	11	12	22	16	—	—	12	20	11	—	26	11
Recycling	−263	−161	−278	−195	−89	−149	−147	−201	−193	−79	−108	−176
Composting	7	13	19	12	—	—	—	16	3	127	13	—
Present: separate collection of mixed refuse and recyclable and organic materials (wheelie bins)												
Landfill	6	5	13	5	n.m.	n.m.	n.m.	5	3	n.m.	14	3
Incineration												
Electricity generation (displacing old coal)	−27	−18	−20	−25	n.m.	n.m.	n.m.	−11	−25	n.m.	−14	−18
Electricity generation (displacing average EU electricity)	11	12	23	16	n.m.	n.m.	n.m.	21	12	n.m.	22	11
Recycling	−230	−145	−153	−181	n.m.	n.m.	n.m.	−184	−183	n.m.	−41	−170
Composting	42	17	79	15	n.m.	n.m.	n.m.	20	6	n.m.	100	—

n.m., this option has not been modelled for this country.
—, this option is not presently employed in this country.

1 recycling;
2 incineration, if energy recovered displaces old coal-fired generation;
3 landfill;
4 incineration, if energy recovered displaces average European electricity generation; and
5 composting.[8]

It is interesting to note that composting fares worst once the total environmental costs, accounted for in the study, are considered. Further, incineration with energy recovery is only better than landfill if the marginal (worst) generator is displaced. Recycling is not only the best solution but also has positive net environmental benefits associated with it. A caveat that needs to be considered is that the analysis is based on current recycling. The recycling that takes place now is typically the easiest and most worthwhile, and diminishing returns to increased investment in recycling infrastructure are likely to apply, even though economies of scale might exist with respect to collection.

Further reading

Allenby, B.R. & Richard, D.J. (eds) (1994) *The Greening of Industrial Ecosystems*. National Academy Press, Washington.
Nurnberger, K. (1999) *Prospects, Poverty and Pollution. Managing the Approaching Crisis*. Zed Books, London.
Rose, C. (1990) *The Dirty Man of Europe. The Great British Pollution Scandal*. Simon & Schuster, London.
Welford, R. (1995) *Environmental Strategy and Sustainable Development. The Corporate Challenge for the 21st Century*. Routledge, London.

[8] There was no large-scale civic composting facility in the UK at the time of the study.

References

Abelson, P.W. & Markandya, A. (1985) The interpretation of capitalized Hedonic Prices in a dynamic environment. *Journal of Environmental Economics and Management* **12**, 195–206.

Ackoff, R. (1977) Optimisation + objectivity = opt out. *European Journal of Operational Research* **1**, 1–7.

Akcakaya, H.R. (1994) RAMAS/METAPOP: viability analysis for stage-structured metapopulations. *Applied Biomathematics*, Setauket, New York.

Alexander, S. (1968) *Beauty and Other Forms of Value*. Thomas Y. Crowell, New York.

Amacher, G., Hyde, W.F. & Kanel, K. (1991) *Household Fuelwood Demand and Supply in Nepal's Hills*. Economic Research Service (draft), World Bank, Washington DC.

Anderson, D. (1987) *The Economics of Afforestation: a Case Study in Africa*. A World Bank Occasional Publication, Johns Hopkins University Press, Baltimore, MD.

Anderson, L.G. (1995) Privatising open access fisheries: individual transferable quotas. In: *Handbook of Environmental Economics* (ed. D. Bromley). Blackwell Science, Oxford.

Aristotle (1975) *The Nicomachean Ethics* (trans. H.G. Apostle). Reidel, Dordrecht, Holland.

Armstrong, S.J. & Botzler, R.G. (1993) *Environmental Ethics: Divergence and Convergence*. McGraw-Hill, New York.

Arrow, K., Solow, R., Portney, P.R. *et al.* (1993) *Report of the NOAA Panel on Contingent Valuation*. Report to the General Counsel of the US National Oceanic and Atmospheric Administration, Resources for the Future, Washington DC.

Bailey, A., McGregor, M.J., Davies, B.B. *et al.* (1997) Development and validation of a framework for evaluating the trade-offs between the benefits and disbenefits of agriculture. Report to the Ministry of Agriculture, Fisheries and Food, Scottish Agricultural College, Edinburgh, UK.

Banford, N., Knetsch, J. & Mauser, G. (1977) *Compensating and Equivalent Measures of Consumers Surplus: Further Survey Results*. Department of Economics, Simon Fraser University, Vancouver.

Barrow, P., Hinsley, A.P. & Price, C. (1986) The effect of afforestation on hydroelectricity generation: a quantitative assessment. *Land Use Policy* **3**, 141–151.

Bartone, C. (1990) Economic and policy issues in resource recovery from municipal solid wastes. *Resources, Conservation and Recycling* **4**, 7–23.

Bateman, I.J. (1993) *Research methods for valuing environmental benefits*. Paper presented at the UE Workshop, Identification and Valuation of Public Benefits from Farming and Countryside Stewardship, Brussels, Belgium.

Bateman, I.J. & Turner, R.K. (1992) *Evaluation of the environment: the contingent valuation method*. CSERGE GEC Working Paper 92–18, CSERGE Norwich.

Becker, L.C. (1977) *Property Rights: Philosophical Foundations*. Routledge and Keegan Paul, London.

Beckerman, W. (1974) *In Defence of Economic Growth*. Jonathon Cape, London.

Beckerman, W. (1992) *Small Is Stupid*. Gerald Duckworth, London.

Begg, D., Fischer, S. & Dornbusch, R. (1998) *Economics*, 5th edn. McGraw-Hill, Maidenhead.

Begon, M., Harper, J.L. & Townsend, C.R. (1990) *Ecology: Individuals, Populations and Communities*. Blackwell Scientific Publications, Oxford.

Bentham, J. (1970/1789) *Introduction to the Principles of Morals and Legislation*. Methuen, London.

Bergland, O., Magnussen, K. & Navrud, S. (1995) *Benefits Transfer: Testing for Accuracy and Reliability*. Discussion Paper 95–03, Department of Economics, Agricultural University of Norway.

Berk, R.A. & Berk, S.F. (1979) *Labour and Leisure at Home: Content and Organisation of the Household Day*. Sage, Beverly Hills, CA.

Bishop, R.C. (1978) Endangered species and uncertainty: The economics of a safe minimum standard. *American Journal of Agricultural Economics* **60**, 10–18.

Bishop, R.C. & Heberlein, T.A. (1979) Measuring values of extra market goods: are indirect measures biased? *American Journal of Agricultural Economics* **61**, 926–930.

Black, M. (1970) *Margins of Precision. Essays in Logic and Language*. Cornell University Press, Ithaca, NY.

Blaug, M. (1985) *Economic Theory in Retrospect*. Cambridge University Press, Cambridge.

Blaug, M. (1992) *The Methodology of Economics*, 2nd edn. Cambridge University Press, Cambridge.

Bockstael, N.E., McConnell, K.E. & Strand, I.E. Jr (1991) Recreation. In: *Measuring the Demand for Environmental Quality* (eds J. Braden & C.D. Kolstad). North Holland, Amsterdam.

Bookchin, M. (1980) *Towards an Ecological Society*. Black Rose Books, Montreal.

Bookchin, M. & Foreman, D. (1991) *Defending the Earth: a Dialogue Between Murray Bookchin and Dave Foreman.* Black Rose Books, Montreal.

Boon, P.J., Holmes, N.T.H., Maitland, P.S., Rowell, T.A. & Davies, J. (1997) A system for evaluating rivers for conservation (SERCON): development, structure and function. In: *Freshwater Quality: Defining the Indefinable* (eds P.J. Boon & D.L. Howell). HMSO, Edinburgh.

Boulding, K.E. (1966) The economics of the coming spaceship Earth. In: *Environmental Quality in a Growing Economy* (ed. H. Jarrett), pp. 3–14. Resources for the Future/Johns Hopkins University Press, Baltimore.

Boulding, K.E. (1972) New goals for society. In: *Energy, Economic Growth, and Society* (ed. S.H. Schurr), pp. 139–151. Johns Hopkins University Press, Baltimore.

Boustead, I. (1993) *Keynote address: the significance of life cycle analysis.* Life Cycle Analysis and Eco-Assessment in the Oil Industry. Institute of Petroleum, London.

Brisson, I.E. (1997) *Externalities in Solid Waste Management: Values, Instruments and Control.* Samfund, Økonomi & Miljo (SØM) publikation nr. 20.

Brody, M.S. & Kealy, M.J. (1995) Foreword to Special Issue 'Issues in ecosystem valuation improving information for decision making'. *Ecological Economics* **14**, 67–70.

Bromley, D. (1989) *Economic Interests and Institutions.* Basil Blackwell, Oxford.

Bromley, D. (1991) *Environment and Economy: Property Rights and Public Policy.* Blackwell Scientific Publications, Oxford.

Bromley, D. & Chapagain, D.P. (1984) The village against the centre: resource depletion in South Asia. *American Journal of Agricultural Economics* **66**, 868–873.

Brooks, R. (1988) *The net economic value of deer hunting in Montana.* Montana Department of Fish, Wildlife and Parks, Helena, Montana.

Brookshire, D.S., D'Arge, R.C., Schultze, W.D. & Thayer, M.A. (1982) Valuing public goods: a comparison of survey and hedonic approaches. *American Economic Review* **72**, 165–177.

Brookshire, D.S., Randall, A. & Stoll, J.R. (1980) Valuing increments and decrements in natural resource service flows. *American Journal of Agricultural Economics* **62**, 478–488.

Bryant, C. & Cook, P. (1992) Environmental issues and the national accounts. *Economic Trends* **469**, 99–122.

Buck, S.J. (1989) Multi-jurisdictional resources: testing a typology for problem-structuring. In: *Common Property Resources: Ecology and Community-Based Sustainable Development* (ed. F. Berkes), pp. 127–147. Belhaven Press, London.

Burningham, K. (1995) Attitudes, accounts and impact assessment. *Sociological Review* **43**, 100–122.

Cameron, R. (1993) *A Concise Economic History of the World.* Oxford University Press, Oxford.

Caughley, G. & Gunn, A. (1996) *Conservation Biology in Theory and Practice.* Blackwell Science, Oxford.

Cernea, M. (ed.) (1991) *Putting People First: Sociological Variables in Rural Development*, 2nd edn. Oxford University Press for the World Bank, Oxford.

Chambers, R. (1994a) The origins and practice of participatory rural appraisal. *World Development* **22**, 953–969.

Chambers, R. (1994b) Participatory rural appraisal (PRA): analysis of experience. *World Development* **22**, 1253–1268.

Charnes, A., Cooper, M.W. & Ferguson, R. (1955) Optimal estimation of executive compensation by linear programming. *Management Science* **1**, 138–151.

Child, G. (1984) Managing wildlife for people in Zimbabwe. In: *National Parks, Conservation and Development* (eds J. McNeely & K. Miller). Smithsonian Institute, Washington DC.

Ciriacy-Wantrup, S.V. (1968) *Resource Conservation: Economics and Policies.* 3rd edn, University of California Press, Berkeley, CA.

Cistuli, V. (1996) Economic valuation of environmental goods: basic concepts. In: *Environment and Sustainability Issues in Agricultural Policy Analysis*, pp. 41–74. FAO/UNEP, Cairo.

Clawson, M. (1959) *Methods of Measuring the Demand and Value of Outdoors Recreation.* Resources for the Future, Washington DC.

Clawson, M. (1977) *Decision Making in Timber Production, Harvest and Marketing.* Research Paper R-4, Resources for the Future, Washington, DC.

Clayton, A.M.H. & Radclife, N.J. (1996) *Sustainability: A Systems Approach.* Earthscan, London.

Coase, R. (1960) The problem of social cost. *Journal of Law and Economics* **3**, 1–44.

Cobb, C. & Cobb, J. (1995) *The Green National Product.* University of Americas Press, Lanham, MD.

Cole, H.S. & Curnow, R.C. (1973) An evaluation of the world models. In: *Thinking About the Future: a Critique of the Limits to Growth* (eds H.S.D. Cole, C. Freeman, M. Johado & K.L.R. Pavit). Chatto & Windus, London.

Commission of the European Communities (CEC) (1992) Towards sustainability: a European Community Programme of policy and action in relation to the environment and sustainable development, Vol. II. Commission of the European Communities, Brussels.

Commons, J. (1968/1926) *The Legal Foundations of Capitalism.* Macmillan, New York.

Conrad, J.M. (1995) Bioeconomic models of the fishery. In: *The Handbook of Environmental Economics* (ed. D. Bromley), pp. 405–432. Blackwell Science, Oxford.

Cooke, W.O. (1985) *Quantitative Methods for Management Decisions.* McGraw-Hill, New York.

Coopers, & Lybrand, Centre for Social and Economic Research on the Global Enviroment (CSERGE) and Economics for the Environment Consultancy (EFTEC) (1996) *Cost-Benefit Analysis of the Different Municipal Solid Waste Management Systems: Objectives and Instruments for the Year 2000.* Report to the European Commission DGXI, Brussels.

Costanza, R. (1991) *Ecological Economics; the Science and Practice of Sustainability.* Columbia Press, New York.

Costanza, R., Segura, O. & Martinez-Alier, J. (eds) (1996) *Getting Down to Earth. Practical Applications of Ecological Economics.* Island Press, Washington DC.

Costanza, R., d'Arge, R., de Groot, R. *et al.* (1997) The value of the world's ecosystem services and natural capital. *Nature* **387**, 253–260.

Costanza, R., d'Arge, R., de Groot, R. *et al.* (1998) The value of the world's ecosystem services: putting the issues in perspective. *Ecological Economics* **25**, 67–72.

Coursey, D.L., Schulze, D.W. & Hovis, J. (1983) *A comparison of alternative valuation mechanisms for non-market commodities,* University of Wyoming, WY.

Crowards, T.M. (1998) Safe minimum standards: costs and opportunities. *Ecological Economics* **25**, 303–314.

Daly, H.E. (1973) The Economics of the Steady State. In: *Toward a Steady State Economy* (ed. H.E. Daly). W.H. Freeman, San Francisco, CA.

Daly, H.E. (1991) Elements of environmental macroeconomics. In: *Ecological Economics, the Science and Practice of Sustainability* (ed. R. Constanza) pp. 32–46. Columbia Press, New York.

Daly, H.E. (1992) *Steady-State Economics,* 2nd edn. Earthscan, London.

Daly, H.E. (ed.) (1973) *Towards a Steady State Economy.* W.H. Freeman, San Francisco, CA.

Daly, H.E. & Cobb, J.B. Jr (1989) *For the Common Good. Redirecting the Economy Towards Community, the Environment and a Sustainable Future.* Green Print, London.

Daubert, J.T. & Young, R.A. (1981) Recreational demands for maintaining instream flows: a contingent valuation approach. *American Journal of Agricultural Economics* **63**, 666–676.

Dee, N., Baker, J.K., Drobny, N.L., Duke, K.M., Whitman, I. & Fahringer, D.C. (1973) Environmental evaluation system for water resource planning. *Water Resources Research* **9**, 523–535.

Department of the Environment (1989) *Environmental Assessment: a Guide to the Procedures.* HMSO, London.

Descartes, R. (1996/1641) *The Discourse on the Method and Meditation on First Philosophy* (ed. D. Weissman). Yale University Press, New Haven, CT.

Desvousges, W.H., Johnson, F.R., Dunford, R.W., Boyle, K.J., Hudson, S.P. & Wilson, K.N. (1992) *Measuring Natural Resource Damages with Contingent Valuation: Tests of Validity and Reliability.* Cambridge Economics, Cambridge, MA.

Dietz, T. (1984) Social impact assessment as a tool for rangeland management. In: *Developing Strategies for Rangeland Management* (ed. National Research Council/National Academy of Sciences). Westview Press, Boulder, CA.

Duffield, J. (1988) *The net economic value of elk hunting in Montana.* Montana Department of Fish, Wildlife and Parks, Helena, MT.

Duffield, J., Loomis, J. & Brooks, R. (1987) *The net economic value of fishing in Montana.* Montana Department of Fish, Wildlife and Parks, Helena, MT.

Easterlin, R. (1997) *Growth Triumphant: The Twenty-First Century in Historical Perspective.* University of Michigan Press, Ann Arbor, MN.

Edwards-Jones, G. & Ibrahim A-Asiz, I. (1997) Computer aided training on the environmental impacts of agricultural development projects: an Egyptian case-study. In: *Application of Systems Approaches at the Farm and Regional Levels* (eds P.S. Teng, M.J. Kropf, H.F.W. ten Berge, J.B. Dent, F.P. Lansigan, H. van Laar), pp. 413–426. Kluwer Academic Publishers, Dordrecht.

Edwards-Jones, G., Carlyle, E.E. & Howells, O. (1996) The use of knowledge-based systems for evaluating the conservation and amenity value of woodlands: a case-study from southern Scotland. *Arboricultural Journal* **20**, 299–312.

Ehrlich, P.R. (1988) *The Machinery of Nature.* Paladin, London.

Eiswerth, M.E. & Haney, J.C. (1992) Allocating conservation expenditures: accounting for inter-species genetic distinctiveness. *Ecological Economics* **5**, 235–249.

Etzioni, A. (1988) *The Moral Dimension.* Free Press, New York.

Farnsworth, N.R. & Soejarto, D.D. (1985) Potential consequence of plant extinction in the United States on the current and future availability of prescription drugs. *Economic Botany* **39**, 231–240.

Fiedler, P.L. & Jain, S.K. (1992) *Conservation Biology: the Theory and Practice of Nature Conservation, Preservation and Management.* Chapman & Hall, London.

Finsterbusch, K. (1980) *Understanding Social Impacts: Assessing the Effects of Public Projects.* Sage, London.

FoE (Friends of the Earth) (1995) Towards a Sustainable Scotland: Conference Proceedings. *Towards a Sustainable Scotland,* 7 November 1995, Dunfermline, Scotland.

Foster, J. (ed.) (1997) *Valuing Nature: Economics, Ethics and Environment.* Routledge, London.

Freire, P. (1968) *Pedagogy of the Oppressed.* Seabury Press, New York.

Freudenberg, W.R. & Keating, K.M. (1985) Applying sociology to policy: social science and the environmental impact statement. *Rural Sociology* **50**, 578–605.

Garrod, D.J. & Jones, B.W. (1974) Stock and recruitment relationship in the NE Atlantic cod stock and the implications for management of the stock. *Journal Conseil International pour l'Exploration de la Mer* **173**, 128–144.

Garrod, G.D. & Willis, K.G. (1990) *Contingent Valuation Techniques: A Review of their Unbiasedness, Efficiency and Consistency.* Countryside Change Initiative: Working Paper No. 10. Countryside Change Unit, University of Newcastle upon Tyne.

Gaus, G.F. (1990) *Value and Justification: the Foundations of Liberal Theory.* Cambridge University Press, Cambridge.

George, H. (1979/1879) *Progress and Poverty.* Robert Schalkenbach Foundation, New York.

George, S. & Sabelli, F. (1994) *Faith and Credit. The World Bank's Secular Empire*. Penguin, London.

Georgescu-Roegen, N. (1971) *The Entropy Law and the Economic Process*. Harvard University Press, Cambridge, MA.

Geraghty, P.J. (1993) Environmental assessment and the application of expert systems: an overview. *Journal of Environmental Management* **39**, 27–38.

Gilbert, A.J. & Janssen, R. (1998) Use of environmental functions to communicate the values of a mangrove ecosystem under different management regimes. *Ecological Economics* **25**, 323–346.

Gill, G.J. (1991) *But how does it compare with the real data?* RRA Notes, Sustainable Agriculture Programme, International Institute for Environment and Development, London **14**, 5–14.

Goldsmith, F.B. (1991) The selection of protected areas. In: *Management of Temperate Communities for Conservation* (eds I.F. Spellerberg, F.B. Goldsmith & M. Morris). Blackwell Scientific Publications, Oxford.

Goodpaster, K. (1978) On being morally considerable. *Journal of Philosophy* **75**, 308–325.

Gordon, H.S. (1956) The economic theory of a common property resource: the fishery. *Journal of Political Economy* **62**, 124–142.

Gramlich, E.M. (1990) *A Guide to Benefit–Cost Analysis*. Englewood Cliffs, NJ.

Grandstaff, S. & Dixon, J.A. (1986) Evaluation of Lumpinee Park in Bangkok, Thailand. In: *Economic Valuation Techniques for the Environment: a Case Study Workbook* (eds M.M. Hufshmidt & J.A. Dixon). Johns Hopkins University Press, Baltimore, DC.

Griffin, J. (1986) *Well-Being*. Clarendon Press, Oxford.

Griliches, Z., ed. (1971) *Price Indexes and Quality Changes*. Harvard University Press, Cambridge, MA.

de Groot, R. (1994) Functions and values of protected areas: A comprehensive framework for assessing the benefits of protected areas to human society. In: *Protected Area Economics and Policy. Linking Conservation and Sustainable Development* (eds M. Munasinghe & J. McNeely). World Bank and World Conservation Union (IUCN), Washington DC.

Grotius, H. (1972/1604) *Freedom of the Seas* (trans. R. Magoffin). Arno Press, New York.

Guariso, G. & Page, B. (eds) (1994) *Computer Support for Environmental Impact Assessment*. North-Holland, Amsterdam.

Haimes, Y.Y., Das, P. & Sung, K. (1979) *Multiobjective analysis in the Maumee River Basin: a case study*. Report submitted to National Science Foundation, Water Systems Engineering Program, Case Western Reserve University, Cleveland, OH.

Hammack, J. & Brown, G.M. Jr (1974) *Water Fowl and Wetlands: Toward Bioeconomic Analysis*. Resources for the Future/Johns Hopkins University Press, Baltimore.

Hanley, N., MacMillan, D., Wright, R.E. *et al.* (1998) Contingent valuation versus choice experiments: estimating the benefits of Environmentally Sensitive Areas in Scotland. *Journal of Agricultural Economics* **49**, 1–15.

Hanley, N. & Spash, C. (1993) *Cost–Benefit Analysis and the Environment*. Edward Elgar, Aldershot.

Hardin, G. (1968) The tragedy of the commons. *Science* **162**, 1243–1248.

Hargrove, E.C. (1989) *Foundations of Environmental Ethics*. Prentice-Hall, Englewood Cliffs, NJ.

Harrod, R. (1948) *Towards a Dynamic Economy*. St Martins Press, London.

Hartwick, J.M. & Olewiler, N.D. (1986) *The Economics of Natural Resource Use*. HarperCollins, New York.

Helliwell, D.R. (1967) The amenity value of trees and woodlands. *Arboricultural Journal* **1**, 128–131.

Helliwell, D.R. (1985) *Planning for Nature Conservation*. Packard Publishing, Chichester.

Hirsch, F. (1976) *The Social Limits to Growth*. Harvard University Press, Cambridge, MA.

HMSO (1990) *This Common Inheritance: Britain's Environmental Strategy*. HMSO, London.

HMSO (1991) *Economic Appraisal in Central Government: A Technical Guide for Government Departments*. HMSO, London.

Hodgson, G. & Dixon, J.A. (1988) *Logging versus fisheries and tourism in Palawan: an environmental and economic analysis*. EAPI Occasional Paper No. 7. East–West Centre, Honolulu.

Hodgson, G. & Dixon, J.A. (1992) Sedimentation damage to marine resources: environmental and economic analysis, resources and environment. In: *Asia's Marine Sector* (ed. J.B. March). Taylor & Francis, New York.

Hoevenagel, R. (1990) *The validity of the contingent valuation method: some aspects on the basis of three Dutch studies*. Paper Presented at the First Annual Meeting of the European Association of Environmental and Resource Economists (EAERE), Venice, April 17–20, 1990.

Holdgate, M.W. (1979) *A Perspective of Environmental Pollution*. Cambridge University Press, Cambridge.

Holling, C.S. (ed.) (1978) *Adaptive Environmental Assessment and Management*. John Wiley and Sons, New York.

Hotelling, H. (1931) The economics of exhaustible resources. *Journal of Political Economy* **30**, 137–175.

Howard, P.C. (1995) *The economics of protected areas in Uganda: costs, benefits and policy issues*. MSc thesis, University of Edinburgh.

Howells, O. & Edwards-Jones, G. (1997) A feasibility study of reintroducing wild boar (*Sus scrofa*) to Scotland: would existing woodlands support minimum viable populations? *Biological Conservation* **81**, 77–89.

Hume, D. (1978/1777) *A Treatise of Human Nature* (with notes by P.H. Nidditch), 2nd edn. Oxford University Press, Oxford.

Huppert, D.H. (1990) *Managing Alaska's groundfish fisheries: history and prospects*. University of Washington Institute for Marine Resources Working Paper, May, University of Washington, Washington DC.

Hurlbert, S.H. (1971) The non-concept of species diversity: a critique and alternative parameters. *Ecology* **53**, 577–586.

IUCN (1994) *IUCN Red List Categories*. IUCN, Gland, Switzerland.

Jackson, T., Marks, N., Ralls, J. & Stymne, S. (1997) *Sustainable economic welfare in the UK 1960–96*. Centre for Environmental Strategy, University of Surrey.

Jacobs, M. (1991) *The Green Economy*. Pluto, London.

Jacobs, M. (1996) Sustainability and 'the market': a typology of environmental economics. In: *Markets, the State and the Environment* (ed. R. Eckersley). Macmillan, London.

Jevons, W.S. (1871) *The Theory of Political Economy*. Macmillan, London.

Kahn, H., Brown, W. & Martel, L. (1976) *The Next 200 Years*. Morrow, New York.

Kahneman, D. & Knetsch, J.L. (1992) Valuing public goods: the purchase of moral satisfaction. *Journal of Environmental Economics and Management* **22**, 57–70.

Kamenetzky (1992) The economics of the satisfaction of needs. In: *Real Life Economics* (eds P. Ekins & M. Max-Neef), pp. 181–196. Routledge, London.

Kant, I. (1949/1785) *Fundamental Principles of the Metaphysics of Morals* (trans. L. White Beck). University of Chicago Press, Chicago.

Keeney, R.L. (1982) Decision analysis: an overview. *Operations Research* **30**, 803–838.

Keeney, R.L. & Raiffa, H. (1976) *Decisions with Multiple Objectives: Preferences and Value Trade-Offs*. John Wiley and Sons, New York.

Kenny, A. (1973) Mental health in Plato's Republic. In: *The Anatomy of the Soul* (ed. A. Kenny), pp. 1–27. Oxford University Press, Oxford.

Kim, S.H. & Dixon, J.A. (1986) Economic evaluation of environmental quality aspects of upland agricultural projects in Korea. In: *Economic Valuation Techniques for the Environment: a Case Study Workbook* (eds J.A. Dixon & M.M. Hufschmidt). Johns Hopkins University Press, Baltimore.

Knetsch, J.L. & Sinden, J. (1984) Willingness to pay and compensation demanded: Experimental evidence of an unexpected disparity in measures of value. *Quarterly Journal of Economics* **99**, 507–521.

Krajewski, C. (1989) Phylogenetic relationships among cranes (Gruiformes: Gruidae) based on DNA hybridization. *Auk* **106**, 603–618.

Krebs, C.J. (1978) *Ecology. The Experimental Analysis of Distribution and Abundance*. Harper & Row, New York.

Krutilla, J.V. (1967) Conservation reconsidered. *American Economic Review* **57**, 777–786.

Krutilla, J.V. (1969) *On the economics of preservation or development of the lower portion of Hell's Canyon*. Draft report to the Federal Power Commission, Washington DC.

Krutilla, J.V. & Fisher, A.C. (1985) *The Economics of Natural Environments* (revised edn). Johns Hopkins University Press for Resources for the Future, Baltimore.

Kula, E. (1998) *History of Environmental Economic Thought*. Routledge, London.

Kuuluvainen, J. & Tahvonen, O. (1995) The economics of natural resource utilization. In: *Principles of Environmental and Resource Economics* (eds H. Folmer, H.L. Gabel & H. Opschoor). Edward Elgar, Aldershot.

Lacy, R.C. (1993) VORTEX: a computer simulation model for population viability analysis. *Wildlife Research* **20**, 45–65.

Lacy, R.C. & Clark, T.W. (1990) Population viability assessment of eastern barred bandicoot. In: *Management and Conservation of Small Populations* (eds T.W. Clark & J.H. Seebeck). Chicago Zoological Society, Brookfield, IL.

Lacy, R.C., Flesness, N.R. & Seal, U.S. (1989) *Puerto Rican parrot population viability analysis*. Report to the US Fish and Wildlife Service. Apple Valley, MN. Captive Breeding Specialist Group, Species Survival Commission, IUCN.

Lancaster, K.J. (1971) *Consumer Demand: a New Approach*. Columbia University Press, New York.

Layland, R. & Glaister, S. (eds) (1994) *Cost–Benefit Analysis*. Cambridge University Press, Cambridge.

Lee, N. & Dancey, R. (1993) The quality of environmental impact statements in Ireland and the UK: a comparative analysis. *Project Appraisal*, **8**, 31–36.

Lents, J.M. & Leyen, P. (1996) RECLAIM: Los Angeles' new market-based smog cleanup program. *Journal of Air and Waste Management* **46**, 195–206.

Leonard, H. & Zeckhauser, P. (1986) Cost–benefit analysis applied to risks. In: *Values at Risk* (ed. D. McLean). Rowman and Littlefield, Totowa, NJ.

Leopold, A. (1949) *A Sand County Almanac*. Oxford University Press, Oxford.

Leopold, L.B., Clark, F.E., Hanshaw, B.B. & Balsley, J.R. (1971) *A procedure for evaluating environmental impact*. US Geological Survey Circular 645, Department of Interior, Washington DC.

Lindenmayer, D.B., Lacy, R.C., Thomas, V.C. & Clark, T.W. (1993) Predictions of the impacts of changes in population size and environmental variability on Leadbeater's possum, *Gymnobelideus leadbeateri* McCoy (Marsupialia: Petauridae) using population viability analysis: an application of the computer program VORTEX. *Wildlife Research* **20**, 67–86.

Lipsey, R.G. (1995) *An Introduction to Positive Economics*, 8th edn. ELBS/Oxford University Press, Oxford.

Locke, J. (1967/1690) *Two Treatises of Government*, 2nd edn. (ed. P. Laslett). Cambridge University Press, Cambridge.

Loomis, J. & Cooper, J. (1988) *The net economic value of antelope hunting in Montana*. Montana Department of Fish, Wildlife and Parks, Helena, MT.

Loomis, J.B. & White, D.S. (1996) Economic benefits of rare and endangered species: summary and meta-analysis. *Ecological Economics* **18**, 197–206.

Luken, R., Johnson, F. & Kibler, V. (1992) Benefits and costs of pulp and paper effluent controls under the Clean Water Act. *Water Resources Research* **28**, 665–674.

MAFF (1990) *UK ESA Statistics*. Ministry of Agriculture, Fisheries and Food, London.

MAFF (1991) *Ministerial Information in MAFF (MINIM) 1990*. Ministry of Agriculture, Fisheries and Food, London.

Mainome, M. (1985) Using multi-criteria evaluation in developing solid waste plans. In: *Evaluation of Complex Policy Problems* (eds A. Faludi & H. Voogd). Delftse Uitgevers Maatschcappij, Delft.

Mallmann, C.A. (1973) *On the satisfaction of human aspirations as the development objective*. Paper presented at the Symposium on Science, Technology and Human Values, Mexico City, 2–3 July.

Malthus, T.R. (1989/1807) *An essay on the principle of population, or, A view of its past and present effects on human happiness: with an inquiry into our prospects respecting the future removal or mitigation of the evils which it occasions* (ed. P. James). Cambridge University Press, Cambridge.

Marcondes, M.A.P. (1981) *Adaptcion de Una Metodologia de Evaluacion Economics Aplicada Al Parque Nacional Cahuita, Costa Rica*. Centro Agronomico Tropical de Investigation y Ensennanza (CATIE), Serie Tecnia no. 9.

Marshall, A. (1890) *Principles of Economics: an Introductory Volume*. Macmillan, London.

Marshall, K. & Edwards-Jones, G. (1998) Reintroducing capercaillie (*Tetrao urogallus*) into southern Scotland: identification of minimum viable populations at potential release sites. *Biodiversity and Conservation* **3**, 275–296.

Masera, M. (1991) *Contents and Phases of an EIA study*. Proceedings of the Eurocourse Environmental Impact Assessment, JRC, Ispra, 30 September–4 October 1991.

Maslow, A.H. (1954) *Motivation and Personality*. Harper & Row, New York.

Max-Neef, M. (1992) Development and human needs. In: *Real Life Economics* (eds P. Ekins & M. Max-Neef), pp. 197–213. Routledge, London.

Meadows, D.H., Meadows, D.L. & Randers, J. (1992) *Beyond the Limits: Confronting Global Collapse, Envisioning a Sustainable Future*. Chelsea Green, Post Mills, VT.

Meadows, D.H., Meadows, D.L., Randers, J. & Behrens, W.W. (1972) *The Limits to Growth: a Report for the Club of Rome's Project on the Predicament of Mankind*. Universe Books, New York.

Merchant, C. (1980) *The Death of Nature: Women, Ecology and the Scientific Revolution*. Harper & Row, San Francisco, CA.

Mill, J.S. (1848) *Principles of Political Economy: with some of their applications to social philosophy*. J.W. Parker, London.

Mill, J.S. (1962/1861) *Utilitarianism* (ed. M. Warnock), Collins, London.

Misomali, E.D. (1987) *Economic cost–benefit analysis in relation to fuelwood plantations in Malawi*, BSc dissertation, Department of Forestry and Wood Sciences, Bangor.

Mitchell, R.C. & Carson, R.T. (1989) *Using surveys to value public goods: the contingent value method*. Resources for the Future, Washington DC.

Montgomery, C.A., Brown, G.M. & Adams, D.M. (1994) The marginal cost of species preservation: the Northern Spotted Owl. *Journal of Environmental Economics and Management* **26**, 111–128.

Moore, G.E. (1971/1903) *Principia Ethica*. Cambridge University Press, Cambridge.

Muir, J. (1898) The wild parks and forest reservations of the West. *Atlantic Monthly*, **LXXXI**, 483.

Munasinghe, M.C. (1993) *Environmental Economics and Sustainable Development*. World Bank Environment Paper No. 3. The World Bank, Washington DC.

Myrdal, G. (1958) Values in social theory. In: *Values in Social Theory: a Selection of Essays on Methodology* (ed. P. Streeten). Harper, New York.

Naess, A. (1973) The shallow and the deep, long-range ecology movement: a summary. *Inquiry* **16**, 95–100.

Naughton-Treves, L. & Sanderson, S. (1995) Property, politics and wildlife conservation. *World Development* **23**, 1265–1275.

NCC (1989) *Guidelines for Selection of Biological SSSIs*. Nature Conservancy Council, Peterborough.

Neher, P.A. (1990) *Natural Resource Economics: Conservation and Exploitation*. Cambridge University Press, Cambridge.

Newcomb, K. (1984) *An economic justification for rural afforestation: the case of Ethiopia*. Energy Department Paper 16, World Bank, Washington.

Nilsson, S.G. & Nilsson, I.N. (1976) Hur skal naturomraden varderas? Exempel fran fagallivet I sydsvenska sjoar. *Fauna Och Flora* **4**, 136–144.

Norberg Hodge, H. (1992) *Ancient Futures: Learning from Ladakh*. Rider, London.

Norgaard, R.B., Bode, C. & Values Reading Group (1998) The value of ecosystem services: whose values? *Ecological Economics* **25**, 37–39.

Norton, B. (1991) Thoreau's insect analogies: or, why environmentalists hate mainstream economists. *Environmental Ethics* **13**, 235–251.

Norton-Griffiths, M. & Southey, C. (1995) The opportunity costs of biodiversity conservation in Kenya. *Ecological Economics* **12**, 125–139.

Nozick, R. (1974) *Anarchy, State and Utopia*, Basic Books, New York.

OECD (1989a) *Environmental Policy Benefits: Monetary Evaluation*. OECD, Paris.

OECD (1989b) *OECD Environmental Data, Compendium 1989*. OECD, Paris.

OECD (1993) *OECD Environmental Data, Compendium 1993*. OECD, Paris.

Parton, J. (1864) *Life and Times of Benjamin Franklin*, Vol. 1. Mason Brothers, New York.

Passmore, J. (1974) *Man's Responsibility for Nature: Ecological Problems and Western Traditions*. Duckworth, London.

Paton, H.J. (1948) *The Moral Law*. Hutchison University Library, London.

Pearce, D. (1991) *Assessing the Returns to the Economy and to Society from Investments in Forestry*. Forestry Expansion—A study of technical, economic and ecological factors, Paper No. 14, Forestry Commission, Edinburgh.

Pearce, D. & Moran, D. (1994) *The Economic Value of Biodiversity*. Earthscan, London.

Pearce, D.W. (1983) *Cost–Benefit Analysis*. Macmillans, London.

Pearce, D.W. (1989) *Environmental Policy Benefits: Monetary Valuation*. OECD, Paris.

Pearce, D.W. & Turner, R.K. (1990) *Economics of Natural Resources and the Environment*. Harvester Wheatsheaf, Hemel Hempstead.

Pearce, D.W. & Turner, R.K. (1994) *Market-based instruments and developing economies*. Paper read at ESRC sponsored seminar: Whose Environment? New Directions in Solid Waste Management, Birmingham, UK, 25–27 May.

Pepper, D. (1996) *Modern Environmentalism: An Introduction*. Routledge, London.

Pimental, D. (1998) Economic benefits of natural biota. *Ecological Economics* 25, 45–47.

Pinchot, G. (1910) *The Fight for Conservation*. Seattle University of Washington Press, Seattle, WA.

Plato (1987) *The Republic* (trans. D. Lee), 2nd edn. Penguin, London.

Plumwood, V. (1993) *Feminism and the Mastery of Nature*. Routledge, London.

Polyani, K. (1965) *The Great Transformation*, Beacon Press, Boston, MA.

Posner, B., Cuthbertson, E.C., Towle, E. & Reeder, C. (1981) *Economic Impact Analysis for the Virgin Islands National Park*. St Thomas, US Islands Resources Foundation, Virgin Islands.

President of the United States (1993) Message from the President of the United States transmitting the Convention on Biological Diversity, with annexes, given at Rio de Janeiro, June 5, 1992, and signed by the United States in New York on June 4, 1993, US Government Printing Office, Washington DC.

Pretty, J.N. (1994) Alternative systems of inquiry for sustainable agriculture. *Institute of Development Studies Bulletin* 25, 37–48.

Pretty, J.N. (1995a) Participatory learning for sustainable agriculture. *World Development* 23, 1247–1263.

Pretty, J.N. (1995b) *Regenerating Agriculture: Policies and Practice for Sustainability and Self-Reliance*. Earthscan, London.

Price, C. (1989) *The Theory and Application of Forest Economics*. Basil Blackwell, Oxford.

Ramsey, F. (1928) A mathematical theory of saving. *Economic Journal* 38, 543–559.

Ranwell, D.S. (1969) pers. comm. cited in: Ratcliffe, D.A. (1986) Selection of important areas for wildlife conservation in Great Britain: the Nature Conservancy Council's approach. In: *Wildlife Conservation Evaluation* (ed. M.B. Usher). Chapman & Hall, London.

Ratcliffe, D.A. (1977) *A Nature Conservation Review*. Cambridge University Press, Cambridge.

Reid, D. (1995) *Sustainable Development: an Introduction*. Earthscan, London.

Ricker, W.E. (1975) *Computation and Interpretation of Biological Statistics of Fish Populations*. Bulletin 191, Fisheries and Marine Service, Environment Canada, Ottawa.

Rivers, M.J. & Buchan, D. (1995) Social assessment and consultation: New Zealand cases. *Project Appraisal* 10, 181–188.

Robbins, L. (1935) *An Essay on the Nature and Significance of Economic Science*. Macmillan, London.

Rolston, H. (1982) Are values in nature subjective or objective? *Environmental Ethics* 4, 125–151.

Romero, C. & Rehman, T. (1985) Goal programming and multi-criteria decision-making in farm planning: some extensions. *Journal of Agricultural Economics* 36, 171–185.

Rosen, S. (1974) Hedonic prices and implicit markets: product differentiation in pure competition. *Journal of Political Economy* 82, 34–55.

Rosenau, P.M. (1992) *Post-Modernism and the Social Sciences: Insights, Inroads and Intrusions*. Princeton University Press, Princeton, NJ.

Rousseau, J.J. (1973/1768) *The Social Contract and Discourses*. Everyman Classics, London.

Rowe, R.D., d'Arge, R.C. & Brookshire, D.S. (1980) An experiment on the economic value of visibility. *Journal of Environmental Economics and Management* 7, 1–19.

Ruitenbeek, H.J. (1992) The rainforest supply price: a tool for evaluating rainforest conservation expenditures. *Ecological Economics* 6, 57–78.

Russell, B. (1935) *Religion and Science*. Oxford University Press, Oxford.

Saaty, T.L. (1987) How to make a decision: the analytical hierarchy process. *European Journal of Operational Research* 48, 9–26.

Sagoff, M. (1988) *The Economy of the Earth*. Cambridge University Press, Cambridge, MA.

Sahlins, M. (1972) *Stone Age Economics*. Aldine Atherton, Chicago, IL.

Samples, K.C., Dixon, J.A. & Gowen, M.M. (1986) Information disclosure and endangered species valuation. *Land Economics* 62, 306–312.

Schulze, W.D., d'Arge, R.C. & Brookshire, D.S. (1981) Valuing environmental commodities: some recent experiments. *Land Economics* 47, 151–169.

Schumacher, E.F. (1973) *Small Is Beautiful*. Abacus, London.

Schumacher, E.F. (1979) *Good Work*. Abacus, London.

Sen, A. (1987) *On Ethics and Economics*. Basil Blackwell, Oxford.

SETAC (1993) *Guidelines for Life Cycle Assessment: a Code of Practice*, 1st edn. SETAC, Brussels.

Shafer, E.L., Carline, R., Guldin, R.W. & Cordell, H.K. (1993) Economic amenity values of wildlife: six case studies in Pennsylvania. *Environmental Management* 17, 669–682.

Simon, J.L. & Kahn, H. (1984) *The Resourceful Earth: A Response to Global 2000*. Basil Blackwell, Oxford.
Singer, P. (1975) *Animal Liberation*. Avon, New York.
Singer, P. (ed.) (1985) *In Defence of Animals*. Basil Blackwell, Oxford.
Slee, B. & Snowdon, P. (1995) An economic appraisal of rural development forestry in Scotland. *Scottish Agricultural Economics Review* 9, 9–19.
Smith, A. (1966/1776) *The Wealth of Nations*. Everyman, London.
Smith, A. (1982/1760) *The Theory of Moral Sentiments* (eds D.D. Raphael & A.L. Macfie). Liberty Press, Indianapolis, IA.
Smith, V.K. (1989) Taking stock of progress with travel cost recreation demand methods: theory and implementation. *Marine Resource Economics* 6, 279–310.
Smith, V.K. (1993) Nonmarket valuation of environmental resources: an interpretative appraisal. *Land Economics* 69, 1–26.
Smith, M.A. (1995) Community impact agreements, mechanisms for change management: the Niagra experience. *Project Appraisal* 10, 189–196.
Sobral, M.M., Hipel, K.W. & Farquhar, G.J. (1981) A multi-criteria model for solid-waste management. *Journal of Environmental Management* 12, 97–110.
Soulé, M.E., ed. (1986) *Conservation Biology. The Science of Scarcity and Diversity*. Sinauer Association, Sunderland, MA.
Soulé, M.E., ed. (1987) *Viable Populations for Conservation*. Cambridge University Press, Cambridge.
Spash, C. & Simpson, I. (1994) Utilitarian and rights-based alternatives for protecting Sites of Special Scientific Interest. *Journal of Agricultural Economics* 45, 15–26.
Spellerberg, I.F. (1992) *Evaluation and Assessment for Conservation*. Chapman & Hall, London.
Starkie, D.N.M. & and Johnson, D.M. (1975) *The Economic Value of Peace and Quiet*. Heath, Lexington, MA.
Stigler, G.J. (1946) *Production and Distribution Theories: the Formative Period*. Macmillan, London.
Stroud, D.A., Mudge, G.P. & Pienkowski, M.W. (1990) *Protecting Internationally Important Bird Sites*. Nature Conservancy Council, Peterborough.
Taylor, P. (1986) *Respect for Nature: a Theory of Environmental Ethics*. Princeton University Press, Princeton.
Taylor, C.N., Hobson, C.B. & Goodrich, C. (1995) *Social Assessment: Theory, Process and Techniques*. Centre for Resource Management, Lincoln University, Canterbury, New Zealand.
Teckle, A. (1992) Selecting a multi-criterion decision-making technique for watershed resources management. *Water Resources Bulletin* 28, 129–139.
Thayer, M. (1981) Contingent valuation techniques for assessing environmental impacts: further evidence. *Journal of Environmental Economics and Management* 8, 27–44.
Tisdell, C. (1990) Economics and the debate about the preservation of species, crop varieties and genetic diversity. *Ecological Economics* 2, 77–90.
Traill, B. (1988) The rural environment: what role for Europe. *Land Use and the European Environment* (eds M.C. Whitby & J. Ollerenshaw), pp. 78–86. Belhaven, London.
Trice, A.H. & Wood, S.E. (1958) Measurement of Recreation Benefits. *Land Economics* 66, 589–597.
UNDP/WTO (1993) *Integrated Tourism Master Plan. Project No. UNDP/WTO UGA 91/010 Report*. 3 vols. United Nations Development Programme and World Tourism Organisation, Madrid.
United Nations (1993) *Integrated Environmental and Economic Accounting, Interim Version*. Handbook of National Accounting, Series F, No. 61, Deparment of Economic and Social Development, Statistical Division, New York.
Uphoff, N. (1992) *Learning from Gal Oya: possibilities for participatory development and post-newtonian social science*. Cornell University Press, Ithaca.
Usher, M.B. (1986) *Wildlife Conservation Evaluation*. Chapman & Hall, London.
Usher, M.B. (1989) Nature conservation in the UK. *Journal of Applied Ecology* 26, 813–824.
Vaze, P. (1996) Environmental accounts: valuing the depletion of oil and gas reserves. *Economic Trends* 510, 36–45.
Vaze, P. & Balchin, S. (1998) The pilot United Kingdom environmental accounts. In: *Office for National Statistics UK Environmental Accounts 1998*, pp. 7–27. Office for National Statistics, TSO, London.
Viaammse Instelling voor Technologisch Onderzoek (VITO) (1995) *Life Cycle Assessment*. Stanley Thornes, Cheltenham.
Vickery, W.S. (1953) An exchange of questions between economics and philosophy. In: *Economic Justice* (ed. E.S. Phelps), pp. 35–62. Penguin, Harmondsworth.
Wackernagel, M. & Rees, W.E. (1997) Perceptual and structural barriers to investing in natural capital: economics from an ecological footprint perspective. *Ecological Economics* 20, 3–24.
Ward, W.A. & Loomis, J.B. (1986) The travel cost demand model as an environmental policy assessment tool: a review of the literature. *Western Journal of Agricultural Economics* 11, 164–178.
Weisbrod, B. (1964) Collective consumption services of individual consumption goods. *Quarterly Journal of Economics* 78, 471–477.
Whitby, M. & Saunders, C. (1996) Estimating the supply of conservation goods in Britain: a comparison of the financial efficiency of two policy instruments. *Land Economics* 72, 313–325.
White, L. (1967) The historical roots of our environmental crisis. *Science* 155, 1203–1207.
White, P.R., Franke, M. & Hindle, P. (1995) *Integrated Solid Waste Management: a Lifecycle Inventory*. Blackie, London.

Whitehead, A.N. (1929) *The Aims of Education and Other Essays*. Macmillan, London.

Wilen, J.E. (1976) *Common property resources and dynamics of over-exploitation: the case of the North Pacific fur seal*. University of British Columbia Resources Paper No. 3, University of British Columbia.

Willig, R.D. (1976) Consumer's surplus without apology. *American Economic Review* **66**, 587–597.

Wilson, E.O. (1980) Article in *Harvard Magazine*, Jan–Feb edn.

Wilson, E.O. (1988) *Biodiversity*. National Academy Press, Washington DC.

Wilson, J.O. (1992) Socio-economic justice. In: *Real Life Economics* (eds P. Ekins & M. Max-Neef), pp. 54–59. Routledge, London.

Winpenny, J.T. (1991) *Values for the Environment*. HMSO, London.

World Commission on Environment and Development (WCED) (1987) *Our Common Future*. Oxford University Press, Oxford.

World Resources Foundation (1995) Information sheet. *Warmer Bulletin* **44**, Tonbridge, UK.

Zeleny, M. (1976) *Multiple Decision Decision-Making*. Springer-Verlag, Berlin.

Zhou, Z. & Pan, W. (1997) Analysis of the viability of a giant panda population. *Journal of Applied Ecology* **34**, 363–374.

Index

Note: page numbers in *italics* refer to figures; those in **bold** refer to tables